LEXICON ON GEOGRAPHY OF DEVELOPMENT

About the Author

Saroj K. Pal who was a professor in physical geography at Delhi University, is long acknowledged as a leading specialist on physical and mathematical aspects of geography in India. In his natural competence on physical geography, he has lectured and researched for more than 42 years including 37 years at Delhi University as well as at Jawaharlal Nehru University.

Since geography is a bifocal subject having its domains in social aspects also in addition to physical, Prof. Pal has developed a keen interest in "human action-oriented approach" for dealing with socio-economic processes which when accompanied with enlargement of spatial functions lead to development. His publication in 1999 in two volumes on eight issues of socio-economic processes in India regarding their structures and growth in the last 50 years since 1950, provided him the impetus to work on a new strand of enquiry for analysing whether a country is on a developmental path or not, that requires the totality approach of geographical historiography.

LEXICON ON GEOGRAPHY OF DEVELOPMENT

By

SAROJ K. PAL

CONCEPT PUBLISHING COMPANY, NEW DELHI-110059

Cataloging in Publication Data—DK
Courtesy: D.K. Agencies (P) Ltd. <docinfo@dkagencies.com>

Pal, Saroj K. (Saroj Kumar), 1939-
Lexicon on geography of development / by Saroj K. Pal.
p. cm.
Includes bibliographical references (p.)
Includes index.
ISBN 8180692108 (HB)
ISBN 8180692140 (PB)

1. Economic geography. 2. Economic development. I. Title.

DDC 330.9 21

ISBN : 81-8069-210-8 (HB)

First Published 2005

Published and Printed by
Ashok Kumar Mittal
Concept Publishing Company
A/15-16, Commercial Block, Mohan Garden
NEW DELHI-110059 (India)

Phones : 25351460, 25351794
Fax : 091-11-25357103
Email : publishing@conceptpub.com

PREFACE

Economic and political systems do operate within an environmental and demographic framework which they do not themselves create : this framework is the bounded permanence at a given location (*i.e.,* place-bound fixity or, *spatial framework*) of geography into which the flux and flow of capital circulation (or, capital mobility) brings different labour processes at different time periods to configurations of a general social relation which is dynamic and powerful in the influence it wields. Within this continuum from changes over time leading to growth and to progress as reflected in spatial framework, geography studies the aspects of development in spatial terms. *Development*, to the extent that it denotes a "progressive change for the betterment that takes place in land (space) and people (society) of an area" is a *socio-economical process*—a "creator and organizer of space". And, in its turn, spatial reality at any point of time provides us with a dimension not only for evaluating the nature and efficacy of the social process, but also for influencing and directing it in a desired manner. It is this possibility of *spatial transitivity* which creates a general context of spatial differentiation and local variability that enables geography to provide new insights into the "development process" .

Development of an area is an image of reality regarding a geographical landscape of spatial relations and organizations linked in spatial division of labour and of functions. Its abstraction is not possible directly other than by the non-visible valuation of the land due to social significance put on land—the latter provides a newer pattern of territorial division of labour over time where specialization and competence overshadow the earlier object dependence on natural environment so that inadequacies of natural resources in man's immediate geographic space, if there is any, get compensated through inter-regional transfer of resources operated by *trade.* The *more developed the region becomes,* the *more pronounced becomes the manifestations of such specialized economic functions.* Again land as a *site* is an intransitive object, yet it operates as a transitive medium of socio-economic variations that has forced a growing re-evaluation of its role in the development process of the site/area. Hence, while significance can place different values on similar plots of land, land in turn as an element of geographic space confers new and differential values on various members of *society* and its spatial differences are convened (or, arranged) under the temporal sequence.

Development in the context of *coming out of underdevelopment of land, economy and societies* is a relatively new field of inquiry and it opens the way to study Countries of the South which has become an important context in the last 50 years or so since the end of colonialism in the aftermath of the World War II. The classical discourse on development studies which has its roots in the late 1940s and was institutionalized in 1950s and 1960s, is having a multitude of autonomous inter-disciplinary field containing a set of theoretical cores with political economies as a respected member of the family rather than a single, unified, monolithic methodological approach in economics. But the theories of development as a general way of viewing the world, were not articulated until after World War II when the decolonization brought a polarized world of Developed *vs.* Underdeveloped or The North *vs.* The South. The theoretical schools of development which are experimented on the undeveloped world are hotly debated—competing yet in dialogue with the capitalist economy which have been practiced in developed world since the eighteenth century—they fall into two very broad paradigms and like the different geometries in mathematics, each paradigm begins with its own set of postulates or assumptions.

On one hand, there are those who see "underdevelopment" as a primary condition out of which all countries must evolve if they are to attain development. There are numerous variations on this. In one view the Countries of Global South may be perceived as stuck to precapitalist forms of production, such as feudalism bringing a patchwork of self-sufficient domains with little trade or mercantilism in an undeveloped spatial system of market towns. Other theorists emphasize the lack of decisive characteristics in the South that have emerged in the North, such as advanced technology, investment capital, entrepreneurial proficiency, and high levels of education. Underdevelopment, then, is intrinsic to these countries themselves, and their development is believed to come out of kinds of rigidities present in them through a diffusion of Western traits for a relatively smoothly functioning economy. This point of view might be termed the "modernization paradigm". Although the history and effects of colonialism and imperialism are not denied, the emphasis is on the view of the "structuralist school" that these conditions within individual Countries of the South will prevail.

The second school of thought *on development of the underdeveloped* belongs to the *"dependencistas"* of the sixties and seventies*: their paradigm takes an opposite point departure.* From this perspective, *underdevelopment was caused by Western expansion which persists because of the unequal power relationship between the North and the South. The focus here is not on the internal structure of the individual country but the country's place in the international system; the causes of the underdevelopment of a particular country are, thus, external* and not in the context of their inherent problems. *Some variations of this perspective downplay the role of the nation-states altogether, emphasizing instead structural blocks, multinational enterprises,* MNEs *or impersonal economic forces. Countries are condemned to impoverishment not because they lack technology or capital but because of their placement within the structure of world capitalist system. It is not likely, then, that development will take place through transfers from the North; indeed, such transfers are motivated by the self-interest of industrial countries which simply reinforce dependency and powerlessness on the South.*

The nineteen eighties meanwhile witnessed the re-emergence of economic factors which acquired a "sway" and an ascendancy rivalling that of the "modernization paradigm" of the sixties. Thus in the closing decade of the twentieth century, *globalization as influenced by neo-liberals,* the *neo-statist/non-statist economic policies came to characterize the economic landscape of the "real" world* by a diversity and multiplicity of themes and approaches. While the "fast developers" of Latin American countries, alongwith South Asia and Southern Europe are still indeed *dependent* on the metropolitan countries, and the Sub-Saharan countries are having a regressive economy, but the emergence of the four East Asian countries as the Newly Industrialized Countries (NICs) point out a "new politics of dependence" through *export-oriented and free-market policies* while the *dependency theories* have tended to be long on rather theorizing and description but short on analysis and concrete prescription which can grapple with developmental problems.

Thus there is almost a Tower of Babel situation and in working on this book by an approach of *geographical historiography to development, the author aims to stretch in geography the seminal work done by the theorists on their developmental ideas of the underdevelopeds through an analytical overview.* Since there is very little contribution from geography to this debate on development, the information and theories have been drawn from *almost every division of social sciences, ranging from political economics and sociology and demographies to human ecology. Initially highly intimidated by the sheer amount of material available, the author was hardly unaware of the hubris of venturing out of his own field of geography into the lion's dens of other disciplines. But what is realized is that the world is of a piece, and the process of dividing it arbitrarily into multitudes of disciplines, seems increasingly artificial since various perspectives fit together rather nicely and constantly reinforce each other*—thus history led easily and naturally into geography and political science, which in its turn involves economics, not separately but all knit together *to which the author has called geographical historiography. Nevertheless, writing this was a humbling experience. In short, this slim volume is offered as a starting-point for further thought and discussion rather than as a set of final conclusions.*

The foregoing tells the purpose of this book. Thus in thinking about the process of development since the 1950s in the post-war period, the task is to see the way the normatively defined goals of development get rooted for it to become an important step so that economic growth brings a "sea-level of changes" at one individual as well as at societal level of growth in the world. The book starts with embryonic process of development so it charts on : Ontogeny of Spatial Development (Chapter One) and its Perspectives on the Changing World : The Global Countries of the North and the South (Chapter Two). Chapter Three looks at the Theories of Underdevelopment: Modernization and Dependency Paradigms. Chapter Four considers the Countries of Global South in International Setting of Trade. This chapter discusses that participation in foreign trade is an integral facet of the total growth phenomena of a country but this has not been crowned with conspicuous success particularly for the LDCs of the South. Chapter Five examines the World-Systems Theory of Polarized Growth. Chapter Six discusses the Economic Approach for Development through Capital Formation which is continued in by considering the Feasibility of Endogenous Economic Development Approaches adopted by most of the Countries of the South in the Post-1950s. It makes a review of development of capitalism since the days of Adams Smith and David Ricardo upto present day economy's context of Monopoly Capitalism and Neo-colonialism. Chapter Seven looks at Neo-Classical Answer to ISI Failure, it examines the New Economic Model adopted by the East Asian countries. Finally while trying to capture the changing face of development through these seven chapters, an overall conclusion is drawn on in Chapter Eight, in which it has been realized that in the name of raising issues of development, there have been situations where there has been imprecision or lacking in universal applicability for Countries of the South, so that the subject matter ends up asking questions which are not possible to answer but at the same time the core of the theories regarding development is still intact, not abominable, though they themselves need developing.

SAROJ K. PAL

CONTENTS

1

ONTOGENY OF SPATIAL DEVELOPMENT

The term "spatial development" is a concept regarding the progressive changes which takes place over time in a bounded area which may be a nation-state (or a region/a sub-nation within the nation) through the socio-economic interactions—thus an area changes for the better in terms of output, for example, by changes in land use from the present to a more economically productive activity like woodland (forestry) changing to cropland (farming) because of an *interaction* caused by "factors of production", that is, combination of the opportunities and the *stock resources* (the land and minerals) available in the environment (living as well as workspace) of the area in terms of its "inputs" the *stock resources* and *qualities and potentials of the people* inhabiting the area (through their activities) as "production process" seeking to transform itself in a normative mode as suited to the *unique geographical historiography* of the area. Thus by one simple but possible relationship between two inputs of labour and capital as well as the local land endowment, we get the so-called aggregate production function as the *landscape development function*. In addition, this interaction has to take cognizance of the "externalities" which occurs in it so that we require to set the area (constructed and experienced as material ecological artefacts and intricate networks of social relations) within a broader, more flexible and more dynamic perspective. This perspective is provided by the *political economy approach* inherent in economic development theory. Political economy approach through capital circulation is having a *definite penchant* for achieving uneven sectoral and geographical development so as to force a definite competitiveness which can "properly relate itself with the unique features that characterize the area-variability at different scales by inquiring with the economic and social relations, with the spheres ot production and consumption rather than equilibrium, and with an historical dimension rather than without one" (Fei and Ranis, 1964)[1].

Regarding economic landscape of an area, its development is an image of reality, but it surrounds the diversity of interrelationships of economic organizations implicated by demographic, political, cultural, social and technological changes contingent with the area's local and historical factors, hence the abstraction of the area's development is not possible directly, neither any single economic theory exists to explain it. Hence there is a methodological problem in identifying spatial development within a given body of themes and beliefs rather than with collection of subject matters and problems to be tackled. Fortunately or unfortunately, in discursive power of description of development of a region, geography do not share a single, unified way of thinking about, its subject matter, that is, the land and its people. Thus it can be ascribed that study on development in a spatial framework is not a "coherent" or self-contained discourse, but it is hoped not undisciplined. In that way it is no less broad than spatial organization of geography, the principles of which underlie spatial development and are generally applicable all over the world. So though spatial organization of countries are unique to themselves by production of their differences, subjectivities and social orders, but the way the space is organized, there is somewhat common which is more important. In economic system, different land uses compete for land, and those which are able to generate the highest financial return wins out, unless of course, there may be a less productive use like a "public park" [2] which is having a public realm to be protected by special laws. Hence in organizing the subject matter of study of spatial development, following Abler *et al.* (1971),[3] an approach of "circularity causal relationship" or "positive feedback" between spatial structure and spatial process (as in between production and demand context of economy) is adopted, so that spatial processes are generated by dialectic interactions between local endowments (*i.e.,* land or, natural resources, wholly tangible, being embodied in observable material form) on one hand and economic and cultural value *cum* institutional/societal rules (or, variables) to satisfy people's needs and desires, on other hand. These interactions create spatial structures which in turn over time changes with certain continuities. Understanding the structural and processional consequences of human spatial behaviour has now become an absolutely necessary condition for world's development and that spatial context is one of the essential features of development. Again since space is a "social and cultural *construct* as well as a *context*", hence geography requires to have a comprehensive "historical" understanding of the processes that lead to the dominant patterns of spatial development. Thus overall trend in geography for development studies is away from a cross-sectional, static/time-slicing presentation in terms of sectoral pattern (like agricultural, industrial etc.) towards a *geographical historiography* approach. By involving phases of the world economy as integrated with the macro-economic activities of nation-state, this emerging trend shifts geography towards a more comprehensive *historical understanding of the processes* (that lead to dominant patterns) freeing itself from its traditional approaches of cross-sectional or, "time-slice" modelling of sectoral (*i.e.*, agricultural, industrial etc.) and regional economic landscapes which fall short in explaining the variances in world (economic) landscapes to a broader, more flexible and progressive elaboration of a more dynamic context of geography of development. Thus the present aim in studying "development in geography" is devoted to exploring in detail *the geographical path dependence* of economic activities regarding how over time the patterns of economic activities, shaped by the historical contingencies of one period, meet with changed organizational and technological conditions (in an area) and leave a round of "creative" destruction that leaves an indelible imprint on the area's economic landscape (Knox and Agnew, 1994)[4].

The term "development" whether a notion or a practice, is a comprehensive proposition no aspect or niche of society is beyond its sway which rules out any cut-and-dried definition and as an idea it is applicable to any field of study related to a range of phenomena traditionally associated with *activities.* But this concept of development was rarely used before the process of decolonization resulting in emergence of independent nations. Till the early fifties of the twentieth century, the theorists interpreted economic development as denoting growth in per capita real income. Lately, development in the context of *coming out of underdevelopment of land, people and economy* as experienced in the *less-developed countries*, has become an important context in the last 50 years of the twentieth century or so since the end of colonialism in the post-World War II.[5] Understanding the experiences of the former with the analysis of economic phenomena in the *developed countries*, form a framework of study which alludes to an increasing concern for a deeper understanding of the geographical historiography of the processes for creating both the overall context of spatial differentiation between the less-developed and the developed countries as well as the regional or local variability across them.

In traditional meaning of development, the spatial character of an area is understood exclusively in economic terms of "growth"; the latter is indicated by *change* (for a country or a region) over a time-period, say a year or more and when economic growth of an area by change in *output* becomes more than that of population of the area, then economic growth in simplest sense may be said to have occurred through *labour productivity*. If the unit of analysis is the nation-state as a whole, then in calculating the total output of the economy at national level, the country's inputs are its stock resources of land, workers and capital— now if economic growth is understood as change in gross production at national level, so that we take into account the country's total amount of all *goods* and all *services* of whatever sort and by whomsoever produced in arriving at the familiar *Gross Domestic Product* (GDP); the latter when divided by the aggregate population, is known as the GDP per capita in a country. When *net income flow from abroad* is added to GDP, we have with *Gross National Product* (GNP), and this GNP per capita provides a measure of the country's *average real income per person* based on market valuations.[6]

Thus if "output" is called *gross national product* (GNP), and if the inputs to output are workers and capital resources (*i.e.*, *labour*, *L* and *capital*, *K* as *variable* inputs) then to get more *output*, *Y*, we have to increase the amount of labour/the amount of capital, or both, alongwith the local endowment, *land*, *A* which being a fixed resource, is assumed, a constant. It is a common misconception to imagine that the economies which are portrayed through the growth models assumed are necessarily operating at top efficiency. All that is assumed, in fact, is that the degrees of efficiency or inefficiency, whatever it may be, does not change during the period under consideration. Thus "growth" means more of the same type of output involving greater inputs, for example, if a country is said to be growing at 5 per cent per annum, it means that the total volume of its national output, GNP is increasing at this rate; now if population of this country is growing at 3 per cent per annum, this means that the country is producing more than its requirements so that its output per capita (or, the average standard of living) and its economy is growing at 5–3 = 2 per cent per annum. In short, in thinking about development, some degree of simultaneous economic growth is needed.

But since the term "development" as an idea is applicable to any field of study, and it is relatively common to distinguish it from "growth", though they are used interchangeably in some cases. While their common usage is occasionally acceptable, it can confuse. The two terms have different meanings, implicit on general use and explicit in what follows. In particular, if economic growth means more inputs, it may mean putting all the eggs in the same basket, and the perils of such economic growth is that it may not be beneficial to all, so that the characteristics of *physical capital goods* or natural stock of resources and *technology* as the "factors" and "the means" of producing human-made capital respectively (by raising output from natural stock of resources) differs profoundly. For example, well endowed with natural resources, Brazil, the biggest country in South America, has the ninth largest *Gross Domestic Product* (GDP) in the world.[7] Yet the country remains wrecked by poverty and inequality. UN (United Nations) figures show that 20 per cent of the population controls 80 per cent of the nation's wealth, and according to government statistics the wealthiest one per cent of the populace accounts for half of the national income. At least half of the country's workforce in this country of 167 million people earns less then the minimum wage of $ 77 per month. It is remarked : "Brazil is not a poor country, it is an unjust one". As Brazil demonstrates, wealth alone — wealth without equity — is not prosperity. The inequities of Brazilian society are particularly stark when it comes to land distribution. The country's legacy of colonialism has left three per cent of the population holding nearly two-thirds of Brazil's arable land. According to the Brazilian government, 30 per cent of Brazilian farmers own just 20 acres of land or less. In contrast, the country's largest farms, those of 2,000 acres or more, comprise only 1.6 per cent of all farms but sit on 53.2 per cent of the usable land. Another 4.8 million rural families — approximately 25 million people in a country with a total population of 167 million — have no land at all and survive as temporary labourers. Perhaps worst of all, much of Brazil's more than 1.2 billion acres of arable land lies unused. At least 40 per cent of agricultural land lies fallow or, at best, is used only for cattle grazing. Among the largest farms — those geared for the export market or held only for speculative reasons — an estimated 88 per cent of the land is permanently idle. The twin injustices of land concentration and idle farms are largely responsible for the chronic poverty that plagues Brazil.

While technology is non-rival, which means that its use in one employment does not necessarily preclude its use in another, but capital goods are rival. Again since human knowledge is not an exhaustible resource: every product can be improved hypothetically infinite number of times and hence it does not face the same limitations to accumulation as physical capital goods are—thus new technologies appear together with new firms, which crowd out old firms, and this means *creative self-destruction.* But, the analysis of economic growth is highly aggregated and is concerned mainly with measuring growth in limited number of economic variables featuring only few inputs and identifying their interrelationships such as between the national income growth rate and the speed of capital formation (Hayami, 1997).[8] But development is a much wider concept than the growth of per capita income alone: growth might be said a necessary condition as an aggregate measure of growth or per capita income but it pays no attention to how that output is "distributed" amongst its population: it also says nothing about the composition of its output (whether the goods are consumption goods, investment goods or public goods such as education and health provision), and it gives no indication of the physical, social and economic environment in which the output is produced. In short, the economic growth rates of nations can not be taken as measures of increase in the welfare of societies because the well-being of people is a much more inclusive concept that integrate variegated composition of the economic, the differences in sectoral context of the societal as well as the political considerations rather than the level of income alone.

Gross National Product (GNP) as an index measured for a set of key activities: goods and services (rather than service indices simultaneously), facilitates the state of well-being of a country. Hence GNP as a measure of national economic activity, is a heavily-loaded indicator which facilitates communication regarding the state of well-being of a country at an international level. So any increase in terms of provision of goods and services is almost a necessary condition for economic development. This is particularly so, for an industrialized country having a high GNP per capita, growing in economic terms and "thus" improving further, so that its process of production of goods and services generate the capacity as well as of sustaining an annual increase in its gross value-loaded indicator, and this country is said to be a developed country, (or, in short, DC). "Keynes study" of development of the industrialized countries in the North revealed that early growth had been accompanied by increased inequality initially but he assumed that as the "modernization" gradually encompassed these countries, the standard of living of the poor increased so that poverty does no longer remained self-perpetuating, but social balances get achieved and inequality would diminish. Keynesian principle of "balanced growth between key sectors of economy" became and "continues to be a rational for development policies based on modernization principle and on growth of GNP. Essentially the idea was: Don't worry about inequality because it would automatically self-correct as growth continues".[9]

Thus traditionally, if any increase in terms of provision of goods and services is taken almost a necessary condition for development, and applied for an industrialized country having a high GNP per capita geared to Euro-American materialism and mass consumption, sustaining in economic terms of GNP growth per capita, say, from 5 per cent to 7 per cent per year was considered development. A by-product of this growth is that share of agriculture in the GNP would decline and that of manufacturing and services would increase. One group of researchers put it, "A national economy is considered developed if it has high level of labour productivity". But according to another group, like a rich man putting his wealth to wrong purposes, this labour productivity may be, does not increase the standard of living of the masses so that no development encompasses aspects of the society (other than the economic), since according to them, development in general goes beyond this to include changes in the composition of output and in the relative sizes of the contributions of the various inputs to the productive process. Lately, developmental studies have taken an interesting and significant turn.

Growth in GNP over time facilitates the state of well-being of a country. Thus any increase in terms of provision of goods and services is almost a necessary condition for development—the absence of economic growth makes difficult to imagine a process of continuing development. When economic "growth" takes place in an area at a given time in addition to increase in its GNP, there are favourable changes in production techniques and in consumer behaviour, so that a number of interrelated social benefits are realized as *indicators* in levels of living (including nutrition and health-care, literacy and higher education, civil liberty and life expectancy, etc.), this growth is known as *progress* and when the latter takes place plus a diminution of social tensions between groups within the society (by attaining zero unemployment and the like), it is known as *development*. Thus we find a *continuum from growth to development*: while *growth* is an economic process, a set of increases in quantities produced: for example, if the process of production of goods and services generate the capacity as well as of sustaining an annual increase in its GNP at 5 per cent to 7 per cent or more, its per capita GNP may become double in less than 10 to 14 years (depending upon 7 or 5 per cent growth), implying a more well-being and may be less poverty! But even when such growth takes place, there may be an absence both of the structural changes to make complementary growth in other economic sectors of the country and lack of institutional changes to diffuse gains in other economic sectors of the country as well as gains in real incomes across all sectors of the population, so that the problem of unemployment, inequality in income and poverty may remain the same or deteriorate, then such a situation might be classed as "economic growth with negative (or without) development". For example, this latter can take place in case land reform measures adopted may get discount the subsistence agriculture for the spread of plantation agriculture, this would force off the peasant's family of its land, this compels the family to enter the wage market of labour so that the family's money income may increase, but its security and ability to feed itself may diminish immeasurably.

These sort of situations call for a "new economic order" in which redistribution in GNP involve changes in the composition of an economy's outputs and inputs. These changes in turn which form part of development's focus on the overall improvement in the human habitat condition of people in a setting of impoverished majority is known as spatial development in geography. It is just like an area having a high annual rainfall with seasonal variation, but the intensity and above all the unreliability of the rainfall doesn't make the region well-watered. For example, a situation like "growth without development" may describe a country which can have a rise in exports of primary products from large estates owned by foreign owners, but this rise in real income does neither induce any complementary growth in other economic sectors nor there is any institutional changes to diffuse the gains among all sectors of the country's poulation. Conversely, it is possible to envisage "development with negative growth" which might occur where a major restructuring in production relations through such means like *land reform* which creates conditions for future development, but at a cost of reduced GNP due to disruption of the previous production and distribution system for a time being. Hence what is to be understood that even if a purely quantitative "growth" takes place in an area at a given time within the framework of the existing structure or system for development to be accompanied as a process there has to be a "progress", so that a number of interrelated social benefits (or well-being) are realized as *indicators* in levels of living (including nutrition, health, literacy and higher education, life expectancy). Thus in the above two possible situations like "growth without development" and "development with negative growth without progress" are experienced in an area or a country may, then we find a continuum from growth to progress in the latter because of stress on land reform as an internal policy to give the entitlement to land for the peasantry. Thus while *growth* is an economic process, a set of increases in quantities produced, *development*, as mentioned earlier, goes beyond this to include changes in the composition of output and in the relative sizes of the contributions of various inputs to the productive process. Hence economic development is usually conceived as a "socio-economic process" in nature in which growth as an economic process operates in society not only by quantitative expansions but also by changes in non-quantitative factors such as a "social construct" (or social formation, made up of specific proportions of different social classes) so that if we follow this usage, economic growth is considered a "quantitative aspect of economic development".[10]

Thus, when development takes place, in addition to economic growth, there are favourable changes in production techniques and in consumer behaviour and when development becomes primarily a social phenomenon based upon economic growth, it is defined as *progress*, so that the latter is development plus a diminution of social tensions between groups within a society. In context of developing economies, since the range of economic development is very wide particularly in context of developing economies, the study of economic development must investigate the influences of institutional and cultural factors on economic growth as well as the impacts of economic growth on these factors. For example, Countries of the North with developed economies include relatively high productivity and income, a labour force engaged mostly in service and in industrial occupations, and demographic characteristics that include even-aged, nearly stable or growing population that are mostly urban. Nearly all of Western European countries, Japan, Australia, Canada and USA have long-standing status as countries of developed economies, while East Asian countries like Singapur, Taiwan, South Korea and Malaysia which have fretically growing economies through "export-oriented manufacturing zones" and have been recently (since 1976) in tabulations of economically developed countries as *Newly Industrialized Countries* (NICs) but they do have few of the *social characteristics*[11] of developed countries. Again the world's two most populous countries—China and India—support their growing economies that are becoming more diversified and productive. In these countries a demographic and economic transition features slowing population growth and urban expansion as their burgeoning service and industrial sectors participate in the rapidly global marketplace (Marsh and Grossa, 2002).[12]

Hence "the economies of the world nations lie more on a broad continuum" and within this continuum "from growth to progress", it is generally recognized that the "laws of economics" as the *growth-mediated* processes on one hand and "practical (that is socialist) politics" or *support-led* processes on other hand as the *power* are set to influence very strongly the pattern and pace of development. Yet economic and political systems operate within an environmental and demographic framework which they do not they themselves create : this framework (discussed earlier) is the *spatial framework* which is dynamic and powerful in the influences it wields. Within this continuum from changes leading to growth and to progress, geography studies the aspects of development in spatial terms. For example, by analogy, one wants to carry out differences between "growth" and "development" in spatial context of a region which is characterized by agricultural economy, and if growth involves changes in overall aggregates of agricultural economy, but spatial structure of the region remain composed of the same static phenomena, then the region will remain as *uniform region* relatively throughout, but in other context if the economic growth may bring spatial development, that is, changes in functional capacities bringing spatial interactions in the region which link places with flow of people, goods, trade and information creating *nodal regions*—the latter are delimited on the basis of spatial interaction. Thus the "nodal regions" are the result of spatial organization in contrast to the "uniform regions", latter being composed of homogeneous phenomena throughout.

Ideologies of Development

For all these enhanced capabilities, economic growth is a "means" for more of the same type of output in creating "wealth" which increases the range of human choice, for example, if economic growth provides man having greater control over environment, then the economic growth is greatly to be desired since thereby man's freedom is increased. "If (human) freedom is what development advances, then there is a major argument for concentrating on that overarching objective, rather than on some particular means, or some specially chosen list of instruments. Viewing development in terms of expanding substantive freedoms directs attention to the *ends* that make development important, rather than merely to some of the *means* that, *inter alia*, play a prominent part in the process" (Sen, 1999).[13] Growth also makes structural adjustments easier—for example, people are more likely to accept the closing of a factory if other sectors of the economy are growing. But whether its contribution leads to the dialectic of quantitative and qualitative changes known as "economic development" of an area, it has to be judged by more throughgoing changes : changes in social and technical relations of production—like in the expansion of social services, for example, education and health carc (and in many cases *social safety net*) that economic growth may make possible (through strong employment orientation)—hence any country likes to raise the level of public health and has the expectation of raising its per capita income, then it would "*pretty daft*" (daft = mad) to try to achieve this through raising its income per head, rather than going directly for these objectives through public policy and social change, as Sri Lanka has done and equally well-exemplified by China and Kerala in India, which all have had very rapid reduction in mortality rates and enhancement of living conditions, without much economic growth". Hence this recognition of economic growth, that of and by itself, it does not automatically lead to the wider, normatively defined goals of changes at societal levels for development which societies strive.

As early as in 1971, Denis Goulet argues for three components to be taken as true meaning of development in his book, *The Cruel Choice : A New Concept on the Theory of Development* [14] : he calls *life-sustenance* (in terms of such basic needs as food, housing, and clothing), *self-esteem* (through minimal education, in not being exploited by others and not having insidious and harmful policy which can impair its welfare) and *freedom* (so that people can determine their destiny by remaining free from being imprisoned on the margin of subsistence with no education and no skills). Amartya Sen in his book, *Development as Freedom* (1999) argues in a similar vein to Goulet that Economic Growth should not be viewed sensibly as an end in itself, but as the means to the achievement of a much wider set of objectives so that the lives what we lead get enhanced and we enjoy more freedom. Expanding the freedoms that we have reason to value not only makes our lives richer and more enhanced, but also allows us to be fuller social persons, exercising our volitions and interacting with—and influencing—the world in which we live by which economic and social development should be measured. Thus development should focus on, and be judged by, the expansion of people's "entitlements" and "capabilities" that these entitlements generate. In this context, inadequate income is a strong predisposing conditions for an impoverished life but poverty should be seen as a deprivation of basic capability rather than merely as low income. Because for unemployment situations, the deficiencies in income, (that is, the material debasement for the unemployed persons) which are made up through "transfer economy" by the state, a practice common in Countries of the North, cannot compensate the far-reaching debilitating effects on its manifold effect the unemployment creates on individual freedom, initiative and skills. This usefully allays uneasiness about the relevance of conventional economic goals and especially the importance of the goods *per se* and hence the standard income-based poverty assessment, is not a good measure of entitlements. Sen (1999) defines "entitlements" as "the set of alternative commodity bundles that a person can command in a society using the totality of rights and opportunities that he or she faces". For most people, the crucial determinants of their entitlements depend on their ability to sell their labour and on the price of commodities. Employment opportunity, and the level of unemployment, must therefore be included in any meaningful definition of development. Entitlements also depend on such factors as what individuals can extract from the state in the form of social welfare provision, the spatial distribution of resources and opportunities, and power relations in society. Using the concept of entitlement, Sen (1999) in his analysis of major world famines concluded that several famines were not associated with shortage of food, but rather with a lack of entitlements because the food supply got withdrawn from certain parts of the countries or sections of society and food prices rose, so that to concentrate on food production only neglecting food entitlement for people, can be "deeply counterproductive".

Thus the statement that "Economic Growth is a Means Rather than an End" becomes an important step in the evolution of study of changes, growth and development in the post-World War II period. The "point is not the same as saying that economic growth does not matter. It may matter a great deal, but, the answer cometh slowly since if it does, this is because of some associated benefits that are realized in the process of economic growth" (Sen, 1984).[15]

Since, "money doesn't answerth every thing"[16] so for spatial development (of an area/country), the associated social objectives are three: (1) increase in the availability and distribution of basic life-sustaining goods, (2) enablement of conditions of good health, basic education and encouragement and cultivation of initiatives for both material welfare and self-esteem, and (3) expansion in the range of social and economic choice to the people. Hence though it can scarcely be denied that economic growth (through increase in *GNP* which is admirably precise, encompassing changes in the amount, composition, rate of growth, distribution and composition of resources), but it is one aspect of the process of economic development since it revolves "around the appropriateness of using market prices to the material progress which brings an inequity in its distribution in the society rather than to value social welfare"[17] so it is not the same as "development". Again, since *development being a concept,* it does not always have very clearly defined meanings to the level of measurement of growth as there were social and political aspects that were not incorporated in purely economic definitions of development. The fact that economic growth left much of the population in the dust was recognized as early as in the 1950s by a study of development in the West which revealed that early growth had been accompanied by increased inequality, but as modernization gradually encompassed the entire country, the standard of living of the poor would increase and inequality would diminish. But the "dethronement of growth" for its mono-concentration on GNP as the basis for definitions of development occurred in the 1970s as it became evident that not everyone was sharing in the fruits of economic growth. Studies of income distribution revealed that quite often the poorest 40 per cent or more of the population was being left behind and that economic growth often had little relation to the physical quality of life for most of the people.

Thus development began to be perceived as much more complex, a multidimensional process and one that might not be the same for all countries. A country just beginning development may focus on such fundamentals as nation building, establishing a strong state, and creating an infrastructure of roads and communications. In parts of India, for example, progress may be measured in *personal terms* : like electric lights for the village, a health clinic within walking distance, and low price shops for rice and flour.

For countries that have already realized such basics, development might include the following : (first), economic growth that entails a reduction in inequality and rising standards of living, preferably for all population groups but at least the lowest 40 per cent; (second) sufficiently strong and self-sustaining *domestic economy* with at least some manufacture; (third) a balance between export and domestic production; (fourth) effective citizen participation in government (though not necessarily through western-style democracy); (fifth) education available to all sectors and both sexes; (sixth) governmental that is, political and social stability; and (finally) a respect for human rights. As general as these criteria are, few Countries of the South would qualify as anywhere near developed, though some are seriously moving in the right direction.

Again since social and political aspects cannot be incorporated in purely economic definitions, hence according to Gary Rose (1979),[18] it becomes necessary to be careful about moving from the general level of abstract ideas, so that we can use them as "measures". For example, an increase in gross output leads to an increase in output per head so that the man who doesn't find the provision of bread for his living can get the provision for it but again it may be true that man does not live by bread alone, hence so far as growth itself is concerned, question arises how important is its content as distinct from its magnitude? Thus according to Rose (1979), to investigate a proposition like "development" which implies a *trajectory of improvement*, in both relative and absolute terms, we require to have "empirical indicators" for measuring the people-centered development through a scale for building it up as a process and testing its validity—the entire proposition is often called "operationalizing the concept".

Operationalizing the Concept—Indicator Link in Development

Obviously, "development" is a very complex concept—it can be defined in many different ways like the removal of poverty, which robs people of the freedom to satisfy hunger or not remaining deprived of having sufficient nutrition or to obtain remedies for treatable diseases, or the opportunity to be sheltered or adequately clothed. In other cases, development should be removal of social deprivations like the lack of clear water, or sanitary or educational facilities as well as of denial of political and civil liberties by tyranny of repressive states/authoritarian regime. Thus the real world tries to put efforts various aspects of development into practice, depending on which characteristics of individual people's lives or whole societies one considers as important.

The GNP as an index basically revolves round the *appropriateness of using market prices* (*rather than to value of social welfare*) for which traditional development economies—rightly or wrongly concentrates on it. Again this measure of per capita income is a mean average and it says nothing about the "inequality" in distribution of total income (or, income distribution) between rich and poor of a country. One way to find out this inequality/equality is to compare the "Lorentz curves"[19] for various countries. The simplest and most common method of expressing their inequality is to divide the country's population into "fifths" according to income share of the country's population and working out from the Lorentz curve their respective "Gini Coefficient"[20]—thus some countries within the

Countries of South as Kuwait and United Arab Emirates (with per capita income of more than $ 20,000 per annum), they may have one of the highest GNPs per capita in the world, but a very unequal income distribution. In contrast, the People's Republic of China has a much lower average, but a more equal distribution with a low compared to the former countries so that in interpreting the development of these two countries, China may be ranked higher than Kuwait or United Arab Emirates. This suggests that income distribution of a country may have to do more with its history and such characteristics as land tenure and the depth to which local elites are entrenched than with GNP. Thus for the Southern Countries, the developmental emphasis on growth of GNP cannot be justified following Keynesian hypothesis (as stated earlier) : there is apparently no necessary trade-off between equality and growth so that inequalities in income distribution will diminish after a certain level of income is reached.[21] Thus to the extent that per capita income distribution in a country becomes highly unequal, *increases in measured* GNP *may indicate a worsening situation for the country.* Bangladesh is lowly placed in income categories of the world countries but in terms of income distribution, it fared much better than Brazil, a high-income country in the South by a lower Gini Coefficient integral (World Bank, 1992).[22]

Hence the recognition that economic growth, of any and by itself, does not automatically lead to the wider, normatively defined goals of development, and as we have seen earlier even the improvement in standard of living of the poor in absolute terms (but a decline relative to the wealthy over the time) may cause the income distribution as a "development with negative effect". Thus in the post-war period since the 1950s, redistribution as a means of generating more equality within a country in terms of a new economic order becomes an important step in a "sea-level of changes" in thinking about the process of development (Potter, 1999).[23] And this redistribution need not bring the images of an autocratic Robin Hood government in stealing from the rich and giving it to the poor, because this way of transferred funds does not lead to increased production but would bring deterioration. More realistic strategies may include land reform, progressive (but reasonable) income and wealth taxes, minimum wage laws, low-interest loans to farms and small businesses, and investments in education, health and nutrition. Such reforms can greatly raise the level of productivity of the people at the bottom of the economical pyramid and, by putting more money in the hands of the poor, can increase demand for domestically produced goods.

In economic theorizing on development studies, the main focus is on the economic dimensions only and not on content specific related to the social and political aspects of the countries so that this preoccupation only with national aggregates of GNP and their averages (per capita GNP) leads to the ignoring of spatial, interpersonal and intersectoral distribution problems of vital interests to the countries. Thus human development focussing on people has yet to enter in a meaningful way. *Hence since there is no consensus on what should be understood by* "development" *and how development can best be brought about nor why it becomes difficult for bringing any kind of improvement for the large majority of a country's population,* we review below *a multiplicity of conflicting and competing conceptions so that all these issues can be viewed.* In the process, to discuss the theories and strategies into geographical, historical and broader theoretical contexts to which they have been formulated, we are required to be drawn backwards to the theories of earlier times since the World War II, particularly 1950.

Physical Quality of Life Index (PQLI) : Basic Needs Model as a Measure of Development

All countries ostensibly wish to raise the living standards of their poorest citizens, but this is seldom the primary goal of development. For one thing, the political and economic elites are usually those who control the development process, and they are seldom amenable to the drastic redistribution that such a strategy implies. M. D. Morris (1979)[24] attempts to develop an index of the "physical quality of life index" (PQLI) in terms of measuring a country's provision of basic needs, such as health, nutrition as well as education. Morris says that constructing a human development index is difficult since it involves comparing very different societies. According to him, an useful PQLI measure must not assume only one pattern of development—particularly, that the western way of life is the only way and secondly it must not assume that the values of "development" are the same in all societies. He did not take per capita GNP in measuring development since it is of limited use, because it may mainly reflect the wealth of the upper 40 per cent of the population. He constructed PQLI measure in terms of three indicators as its components, like : (*a*) longevity, as measured by life expectancy at birth, (*b*) infant mortality at age one, and (*c*) adult literacy. His PQLI for a country is the mean of any country's scores on these three simple indicators with equal weights —it is based on a scale from 0 to 100. If life expectancy at age 1 is 38 years, the country would score 0; if 77 years, it would score 100. Infant mortality ranges from 229 per thousand (score of 0) to 7 (score of 100). Raw literacy scores are also included, a top score of 100 would signify a very high level of basic needs provision. In this *physical life score index,* most developed countries score above 95. If this score of above 95 is assumed equal to 100 as average for North, one can express the PQLI index for a country of the South: the larger is this deviation, wider is the gap between South-North, the closer this figure is to 100, the smaller is the gap, and on other hand a figure of 100^{+} indicates the country is having a PQLI index higher than that of North.

Morris (1979) argues that the use of indicators for judging performance of physical quality of life for any country/region should concentrate on outputs (or results) rather than inputs. Input measures, he feels, do not measure success in meeting the desired goals, and may lead an ethnocentric bias to the means employed. The PQLI compares with GNP cross-nationally fairly well since higher is the per capita GNP, higher is PQLI but there are numerous exceptions : Sri Lanka with a per capita GNP of $ 470 per annum was having a PQLI of 91 but Saudi Arabia with a per capita GNP higher than Sri Lanka by 42 times is having a PQLI of only 67 in 1995. Hicks and Streeten (1981)[25] do not assume that Morris' PQLI index gives a correct index of progress on human needs, as opposed to some alternative index having different weights or a different selection of component indices. However, since the data is generally available on these three indicators only, so even though somewhat "open-ended" with respect to its theoretical underpinnings, this approach did correctly highlight the need for making a greater provision for social sectors and their use permits the calculation of the PQLI for a wide range of countries and facilitates the examination of changes in the index over time. A recent study finds that in a scatter plotting of a sample of 80 countries GNP per capita against the PQLI , got an overall non-linear relationship. According to Holloway and Pandit (1992),[26] once GNP per capita reaches about US$ 4,000, increments to basic human welfare tend to be marginal. Thus countries like Cuba, North Korea and China were having a large number of extremely poor people ruled by a small wealthy elite till the 1950s. The strategy these countries required a radical restructuring of their society which they adopted successfully through the Marxist strategy by divesting the elite of the power through violent revolution. A strong central government, deriving the authority and legitimacy through the lower classes, formed around a socialist ideology, controls almost every aspect of the economy, and choose to concentrate immediately on supplying all the basic needs to all sectors of the society in these three countries: that is by aims of providing adequate food (caloric value), sanitary and drinking water, health services, education and full employment. In these countries ideally, medical treatment and education are free, so that even the most remote rural family have access to a clinic than that expensive modern hospitals be built in the cities and a few years of grade school for everyone (100 per cent literacy), the latter is also seen preferable to high levels of illiteracy accompanied by a few college graduates; and food and housing are subsidized to make them accessible to even the poorest. Thus, at the outset, though their per capita GNPs remained even low or moderate relative to other countries among the South, they have attained higher scores 100^+ on the PQLI—(or, "positive deviations"), compared to the PQLI scores of long-affluent nations of USA, Canada, Western Europe and Australia whose levels of human welfare are *higher* on account of their level of economic development.

Thus these three countries, Cuba, North Korea and China have been effective in providing a low but adequate standard of living for the larger part of their peoples, but they have found it difficult to evolve beyond the basic needs approach. In that sense, there may be a sort of built-in-self-destruct mechanism within the socialist economies. If the state is successful in providing basic needs to all or most of its citizens, the second or third generation after the revolution will no longer see this as a priority: having grown up with their basic needs supplied, their dreams will tend toward more consumer goods, expressive freedom, and political participation. China, which by 1994 had attained the fastest growing large economy in the world, has been experimenting with *limited capitalism* and greatly expanded consumerism but seemed unable to provide social and political reforms.

Several other countries that have chosen the Marxist model like Vietnam, Cambodia, Chile, Nicaragua, Ethiopia, Tanzania, South Yemen, Angola, Mozambique have failed on the other hand miserably even at their basic needs goals. Failures of socialism are usually ascribed to structural weakness of the model itself, such as the lack of incentives to produce (because of funnelling capital to basic needs), putting off the industrialization, the emphasis is on redistribution over production, productivity and efficiency, overly rigid centralized planning, and the necessity of repression (coercion) to maintain the system. "However, it would be difficult to find a Marxist government in Countries of the South that was not relentlessly attacked by the US imperialism, either overtly through CIA destabilization or overtly through embargoes and open warfare, so, it is difficult to say how a socialist economy would have fared if it were left to peace and allowed to trade freely in the international marketplace".[27]

Again, the largest negative deviations are recorded by two very different groups of countries: oil-rich Middle Eastern countries (including Iran, Iraq and Saudi Arabia) and many African countries (including Algeria, Angola, Libya, Namibia and Niger).

Similarly motivated in "informational broadening" for making valid statements about comparative deprivations of people is the United Nations Development Programme (UNDP) which from 1991 starts bringing out the *Human Development Report* annually for the world's countries based on two aggregate indices of well-being (or, of the deprivations, reversely) of nations: the *Human Development Index* (HDI) and *Human Poverty Index* (HPI), the latter one is available from 1997. Each of these two indices are depicted in a single measure, but based on the *three* critical dimensions of social attainments as basic means of well-being : (one) *health attainment* as measured by *life expectancy at birth,* that is : *longevity*; (two) *educational attainment,* as measured by a *combination of adult literacy* (two-third weight) as well as the *combined first-, second- and third-level gross enrolment ratio* (one-third weight); and (three) *material standard of living,* as measured by *real* GNP *per capita*, that is, *purchasing-power-parity* (PPP) of countries' currencies (in US$), in terms of its long-term relationship with the exchange rate. The inclusion of income in UNDP's human development index (instead of food) is justified on the ground that it enables the exercise of choice with respect to the other two indicators: educational and health indicators. Education as an indicator is stressed for its instrumental role and for its contribution to equity rather than merely as a means to raise productivity and incomes. Similar considerations in the health sector draw greater public attention to the expenditure on basic health services carried out towards nutrition, water supply, sanitation and housing in poverty-ridden economies of the developing and underdeveloped Countries of the South in a more holistic way than to a narrow focus given by the income-centered view of GNP per head as the only curative care for the income poverty of working population in developed countries of the North.

The exact measurement of these dimensions in terms of the specific indicators of UNDP's human development concept cannot be value-neutral to reflect the specific socio-cultural conditions that prevail in a particular country at a specific period of time. For construction of the index, for each of these three indicators, fixed minimum and maximum values have been established, like:

1. Life expectancy at birth : minimum 25 years and maximum 65 years,
2. Educational attainment measured by a combination of Adult Literacy : 0 per cent and 100 per cent for the underdeveloped countries as well as the Combined 1st, 2nd and 3rd level (*i.e.*, the combined primary, secondary and tertiary school enrolment rates) : 0 per cent and 100 per cent for the Global Countries of the North,
3. And, the Standard of Living as measured by Real GDP per capita (in PPP$) : $ 100 as minimum and $ 40,000 as maximum (in PPP$).

For any component of the HDI with reference to a country, the index for each of the above three indicators is computed according to the following ratio formula :

Index = (Actual x_i / Value – Given Minimum Value of the Indicator) / (Difference in the Given Maximum and Minimum Value).[28]

Focussing on *human development of people* in a country (or, region), HDI provides a measure of capabilities and opportunities of the people, for example, India was having an HDI of 0.448 in 1995 which gave it a rank of 139 among 174 countries of the world whereas for China, the HDI index was 0.669 and its rank among the world's countries was 106; thus China which assumes a basic needs model of development, it scores 20 points higher than India.[29] But this HDI index compiled by UNDP shows that economic well-being of nations do not always correlate well with per capita income—the same growth rate and per capita income of countries can be associated with very different levels of achievement in other spheres such as life expectancy, mortality rates, literacy and education. As the UNDP says in its 1997 Report that "although GNP growth is absolutely necessary to meet all essential human objectives, countries differ in the way that they translate growth into human development". Thus oil-rich countries have much lower HDI rankings compared to their high per capita income while poor countries like Sri Lanka, Cuba and Vietnam have deliberately devoted scarce resources to human development. Thus HDI measure is having an underlying assumption that every country can reach the higher level of performance set in the West or have access to the "purchasing power" of the "upper income" world.

While the HDI measures overall progress in a country in achieving human development, the UNDP's Human Poverty Index, HPI in a composite index, reflects the distribution of progress and measures the backlog of deprivations that still exists in the total population. This is also based on deprivation in same three dimensions of basic development as the HDI, like the per centage of people expected to die before the age of 40, the per centage of adults who are illiterate and deprivation in overall economic provisioning, public and private—reflected by the per centage of population without access to (One) safe water, (Two) health services, and the (Three) per centage of children under the age of 5 years who are underweight through malnourishment.

Introduced in 1997, this HPI index attempts to provide by an aggregate human measure of the prevalence of poverty through different dimensions of human deprivity in a country/in a community. If HPI measures development in terms of progress which is seen as a process of expanding the real freedoms that people can enjoy, poverty can be taken as a deprivation from economic opportunities (to earn an income), social facilities (of health care and education) as well as political and civil rights. And because of the different dimensions of human deprivation, the concept of human poverty is much bigger than it can bring together in this single quantifiable HPI composite index. For example, lack of political freedom, lack of personal security and safety, inability to participate freely in the life of a community and threats to long-term sustainability can hardly be measured and quantified.

The ranking of countries by their HPI[30] also shows some striking contrasts with their ranking by per capita income—income poverty and capability deprivation which together lead poverty to remain "self-perpetuating" in Countries of the South in general is damned annoying but is also a disgrace for Countries of the North : on the basis of an income poverty line is taken as 50 per cent of the median personal disposable income, according to UNDP in 1995, 100 million people in OECD (which total 29) countries were income-poor. This relative deprivation in incomes can yield *absolute* deprivation in terms of capabilities (which lead to perpetuation of poverty), for example, in a generally opulent country income poverty imposes a strain on a relatively poor person in form of social exclusion. The UNDP has calculated that the cost of eradicating poverty across the world is relatively small compared to global income—not more than 0.3 per cent of the world GDP— and that political commitment, not financial resources, is the real obstacle to poverty eradication" (Desai *et al.*, 2002).[31]

Development, then, is clearly a "normative" concept, *i.e.*, it involves values, goals and standards that make it possible to compare a particular situation against a preferred one. It follows that development can properly be evaluated only in the context of human needs and values as perceived by the very societies undergoing change. It also follows that although development implies economic, social, political and cultural transformations, these should be seen not as ends in themselves but as means for enhancing social well-being and the quality of human life. In this book, we shall be concerned with both the means and the ends of many-sided approach to development, addressing our subject matter from a broad perspective so that development is understood in geography not by any single overarching remedy like growth in per capita GNP (which has been held on the professional thinking till in the recent past) at its core but with an integrated and multifaceted approach alongwith the functioning of markets —so that simultaneous transformations are achieved in terms of progress on different social and political institutions, which reinforce each other as net outcomes.

Conceptions of Development

Thus development in simplest sense means "good change" and in everyday parlance it is synonymous with "progress". In these two words, "good change and progress" development is primarily "forward looking", imagining a better world; in societal terms it implies a vision of a desirable society, "well-being for all", a state of being with respect to a set of values which can be measured so that we can talk of "more development" or "less development"—these have been used so commonly, including in such manifestations of the world countries as "developed, less developed (or, developing) and underdeveloped".[32] *Development is a zero-sum game: for some countries to become developed, other countries must become underdeveloped ("dependent")*. Initially developing countries are "undeveloped," but immediately they come into contact with developed (capitalism) they become "underdeveloped". *Development is considered with society and in this context development itself is a social process, a social construct* : "Development is economic growth plus social change".

To this economic assumption that economic growth leads underdevelopment to development through the creation of a dynamic domestic sector capable of generating both self-sustained growth and the transfer of the "decision-making centre," so that there is a sociological interpretation of the transition from traditional to modern societies. Thus to analyze development properly, it has to be considered in its totality: the "historic specificities," both economic and social, underlying the development processes at the national and international levels. Within given structural situations, we must understand the conflict between social movements that are "set in motion" by social classes in developing societies. Our approach must examine not only structural conditions and the ideologies of the social movements, but also their relations and their reciprocal determination. Development always alters the social system of domination as it changes the organization of production and consumption. It cannot be reduced to changes at the institutional level or to the analysis of actors' value orientations. This view of the problem encourages us to analyze the political behaviour of social classes that maintain control at the structural level and those that oppose such control. Moreover, it moves us to consider the value orientations that give the action its framework of reference, not at the individual level, but at the cultural one, as ideologies.

Again it is important that dimensions of this process of development are numerous, particularly for the developing Countries of the South. An array of competing political aims and social values and contrasting policies of social change parade under the single banner of "development". Thus the case is for taking a broad and many-sided approach, a "comprehensive development framework" rejecting a compartmentalized view of development as a single all-purpose remedy, the latter had much hold on professional thinking till in the recent past, but it is replaced by the former which is having much relevance now.

Development strategy in terms of *enhancing human well-being* originally relied on drawing down natural assets out of the land-related environmental resources in the primary sector, but gradually capitalism brings a materialist dialectics in which capital as a process creates a social asset which through its network and trusts on capital brings human-made assets in the secondary sector by production of commodities for human well-being—this form of productive activity through commodities long back replaced the natural assets in their contribution to human well-being, and the present-day focus of development has been on labour and capital resources. Since human beings are both a resource for, and beneficiaries of development, labour and human resources have received a great deal of attention. But in the 1950s most of the Southern Countries' growth strategy continued to focus largely on natural assets. Again since 1970s the cost and time of moving commodities, people and particularly information ratcheted downwards which brought "dematerialization of space" that is, *globalization* which brought a new form of production strategy in terms of "human asset" to increase the human well-being further. Thus while the limits to focussing on physical capital alone is found in low physical capital-labour ratios whose experience of a rise in aggregate GNP growth rate tend to a limit in the Southern Countries because of the diminishing marginal productivity of physical capital, whereas increasing growth in GNP goes on taking place because of the direct channelling of increasing human assets getting coupled with human capital-increasing effects of education (that is, human-made assets). Below we discuss the different senses in which the term "development" is used and how it relates to capitalism.

Development in relation to Capitalism

Development is a concept which goes just beyond simply "good change." *First,* development implies an all-compassing change, or transformation from backward to modern economies, not just an improvement in one aspect, like growth in per capita of GNP income. *Second,* it is not just a question of a one-off process of change to something better, but implies *a process which builds from within on itself, where change once started is continuous one* and where improvements build on previous improvements. *Third,* it is a matter of *qualitative structural changes* occurring at societal level as well as at one individual level. *Finally,* development is not always seen positively—if social change is all encompassing and continuous then the implication is that previous ways of life may be swept away, with the loss of positive such as traditional values and forms of livelihood as well as negative features.

Conceptualization, theorization and practice of development particularly in relation to economic, are innovations of the West, practiced and elaborated over two hundred years in a vast literature from Adam Smith's (1723-1790) analysis of absolute advantage through division of labour (1776)[33] and David Ricardo's (1772-1823) increased specialization through trade[34] to the contemporary post-Keynesian macro-economic theory beyond the welfare-state,[35] all of which appears appropriate for the industrial countries of Global North. The first theoretical framework for understanding the working and ramification of capitalist system was devised by Adam Smith, the Scottish philosopher, historian and father of *classical economics* who started the first revolution in *growth theory.* In his *Wealth of the Nations* he talked about the liberal economic society in which regulation was by competition and the market and not by the state, and in which each man, thrown on his own resources, laboured effectively for the enrichment of the society. Adam Smith's historical perspective regarding the way modern system of production had developed on the strength of private ownership of capital resource as an *institution of private property* was a simplification of reality but it introduced a terminology of "labour productivity as well as its interaction with exchange and trade" as a key to economic growth which are still familiar to us today. The growth of *industrial capitalism* (or, capitalist mode of production) originated in a transition from *feudalism* through *mercantilism* in England and North-Western Europe where accumulated wealth was available to be invested in new ways, and where the nature of social relationship enabled this to be successful. *Capitalism* means the economic and social system is based on *private ownership and the ownership is by individual.* The *archetypal capitalist* owns land, buildings, tools and equipment, hires labour and buys raw materials as his endeavour in order to produce goods and services competitively so that when they are put to the market for pricing, for sale they earn profit. Thus capitalism involves the use of two naturally given resources, *land* (by its endowment factors) and *labour* (by its qualities and potentials), to be used in the most efficient and productive manner possible so that surplus commodity values are produced through the *market* for which labour gets a fair basis in leading his economic life as well as the capitalist who gets the surplus value (that is, the profit). This profit produced by the labour gets reinvested which alongwith mobilizing the (human-made) *capital* (or, money value) lead to *mass production of goods* and *services* that people consume in order to accumulate wealth and to expand the scale of activities. And this went alongside the creation of new forms of livelihood giving a new kind of people with new motivations so that the available accumulated wealth was to be invested in new ways, and the new nature of social relationship enabled this to be successful. Integral to the system is the determination of prices, which, in a free market, according to Adam Smith (1776), are determined by the supply of, and demand for, the factors of production. In order to maximize their profits, capitalist will always seek to minimize their costs of production, including the wages they pay their workforce, so as to outcompete other producers by having lower prices. One way of doing this was through the *division of labour*, dividing manufacturing tasks into simple, repetitive operations which could be performed by unskilled, and therefore by cheap labour.

Another important way in which manufacturers could seek control over their fellows was to be first in the use of new *technology* by following *competitive advantage* of David Ricardo (1772-1823), according to whom, *competition is the instrument of change* making it the "central tradition" in any economy of a nation-state. The significance of the invention and adoption of new ways of doing things, whether through science and technology, or ideas and organization, to sort out such intricate issues as repositing (or, even relocating) the production structures which has often been crucial to firms, industries and regions in *capitalism* for getting ahead of competition according to what each can produce *best* (at least *cost*), ensuring an equitable sharing of gains from trade and investment and apportioning factor of rewards between labour and capital (as two traditional focus of development) in the fairest possible manner. The way these two social Darwinists, Adam Smith and David Ricardo, could break decisively the restraints of feudalist and mercantilist society of Western Europe and able to take them to the era of *industrial capitalism* which raised the economic level of the whole populace, so that the entire era can also be described as the era of the "modern doctrine of development".[36]

In this context, according to Weber (1922, 1967)[37] who in "The Protestant Ethic and the Spirit of Capitalism" (1905), as a economic historian was preoccupied with the nature of capitalism, wrote in a *Capitalistic Order of Society*, there is a pursuit of production for profit (through sale of produce more than the consumption), and forever renewed profit, by means of continuous, rational endeavour to take advantage of its opportunities for profit-making. He noted that capitalism as the economic and social system embodying land, labour and capital requires *self-regulating markets* (markets as capitalist management of economic activity) for those to be bought (and sold) as well as exchanged via *commodization* and *competition,* and not state planning or intervention; he emphasizes capitalism encourages *laissez-faire*, and it works best, when the dominant belief emphasizes planning, efficiency and careful accounting, in particular, when people try to account for possible future effects of present actions. Indeed, an absolutely crucial aspect of capitalism is that it is *intrinsically dynamic*: it tends to built on itself and grow or "develop" from within spontaneously and naturally towards fulfilment of a "potential" through *interaction* and *intervention*, which may entail destruction of the old in order to achieve the new —so that the previous ways of life may be swept away, with the loss of positive as well as negative features. He found the civilizations of the East and West as fundamentally different—in the East he found the quietest and myst-ridden culture[38] in the peasantry alongwith a vested interest of the privileged ones through casteism (as in India) which together became a brake on economic developments in particular. On the other hand, like Marx, among others, Weber noted that one of the most important historical transformations of many European societies was that from their relatively static and contained rural communities they gave way to rapidly changing urban conglomerates over the physical space and the traditional patterns of deference that governed their feudal society, got changed by the rationalization of economic structures: all these made their life more efficient as a whole : thus Weber noted, in the West there had been demystification and the culturally pervasive extension of rationality in all aspects of modern life. Weber characterized rationalization and more particularly "rational capitalism" as the most fateful force in modern life due to which people began thinking of human beings as rational, autonomous individuals. He noted that the capitalism's "pursuit of wealth, stripped of its religious and ethical meaning" produced "specialists without spirit, sensualist without heart". Weber claimed the "rational, sober bourgeois capitalism" that the West has developed in types, forms and directions which had been not only totally dichotomous with the traditional society of the East but this European modernity never existed before elsewhere. Along this line of reasoning, it is suggested that capitalism itself only comes out of certain cultures, and thus it cannot be expected to develop, to be understood, to be theorized about it elsewhere, at least not in a form recognizable to those in the heartland of capitalism in the West.

By the early nineteenth century there was a "structural change" so that industrial capitalism had developed into what could be described as a *market society*—that is, not only the economic aspects but the whole of existence of the society got governed by the principles of "market's for self-regulating system of pricing". Markets have always been a part of all human societies and all societies have been limited by economic factors. But what is characteristic of self-regulating market (which includes shops, stock exchanges) that not only the hidden hand of the market set prices for commodities (as the case may be in large grain markets in North West India) but the factors of production—land, labour, capital alongwith the productive organization itself (as mentioned earlier) are treated as commodities as well to be exchanged each other or for cash, so that capitalism becomes *generalized commodity production.* So long the capitalist growth is an *immanent development,* the human society might be moving *inexorably* to a higher and higher stage of progress, but in this *immanent process* the movement towards commodification of the factors of production may give rise to tremendous negative consequences for "man", "nature" and productive organization. Hence there might be unemployment, poverty and human misery to a group of "dispossessed people" which cause threat of social disorder—a conflict would arise between the *dispossessed people* who will try to correct these social miseries and those who tried to promote capitalism. The world events in the first-half of the twentieth century (like the two world wars, the Depression of 1930s, the growth of Fascism and authoritarian communism) showed that the capitalist market system was indeed "creative self-destruction" making for extreme volatility in local, regional and national economic prospects (so that this year's boom town becomes next year's ghost town), but history shows in fact it has regained and increased its strength through extraordinary technological innovations and become more fully a "world-economic system". For controlling this (that is, the negative propensities of capitalism), there would be requirement of room for deliberate policy and incentive actions of states (through intrinsic and constitutive importance of democracy) and development agencies not only to grow and innovate leading to increased labour productivity, increasing economies of scale as well as improved production processes but also "an increasing propensity to convert the gain of individualist motivation into productive investment through associating *social capital* which may eventually benefit society in general". If capital in general is thought of in terms of accumulated resources which allow for productive activity, then social capital is an aspect of "human capital": unlike capital, it is less embodied in land, factories and buildings but more in human beings invest in themselves in their knowledge and skills through education, learning, skill formation, in-service training on one hand and to their nutrition and health for longer life in their personal context as well to play their organized role in productivity capital of the country. In recent studies of economic growth, there is a much greater emphasis on this intangible resource of "human capital" than used to be the case not long ago.

Naqvi (2002)[39] has mentioned associating human resource capital (or, skilled labour, by recognizing that labour in different economies possess different levels of education and skills) with productivity as *total factor productivity.* Because of diminishing returns to capital, the growth rate of output is only sustained by appropriate doses of technological change. This argument of extending *total factor productivity* is the reinforcing cycle of reinvestment of "social capital" and increasing productivity which get going through organization of large corporations and their networks. Perhaps the most important impact of this type of success is in the countries of East Asia also, where there is a logical sequel to the pursuit of this social capital (or, human resource approach): beginning with Japan in 1850s and extending to Thailand, the priority attached to human resource development in the initial period is being regarded as an important factor for the type of success that these countries have achieved in their economies.

The creation of social opportunities makes a direct contribution to the expansion of human capabilities in terms of health-care, education which again lead to the quality of life and to its flourishing. The rewards of these instrumental achievements (or, the instrumental role of these social capability expansion) bring the social changes in terms of increasing productive abilities of people and thus on economic growth on a widely shared basis which all go beyond the economic change by the production of conventionally defined commodities. It appears that demand for human capital resources increases more and more as higher level of development is reached. Again, in early stages of development, when labour costs are low, the highly labour-intensive nature of mass literacy, basic education and health care—makes human development expenditure comparatively cheap. Hence the income poverty of a country does not severely limits the ability "to invest in themselves", social returns to investment in primary education may be more than twice the 10 per cent norm normally applied to "opportunity cost of capital", higher for general education and for female education, the latter is effective in reducing fertility rates. Thus priority to human resource development only can bring the possibility of "convergence", namely, the possibility of Countries of the South closing the gap in their per capita incomes and this, according to Sen (1999), should be the "first and foremost an *ally* of the poor" for the Countries of the South. According to IMF's "World Economic Outlook" (2000)[40] 24 per cent of Countries of the South in terms of their long-run steady-state growth annually have managed to converge to mainly the USA's per capita income levels. This seems to be strange in view of the enormous differences in levels of income per person around the world: a typical person in USA earns in less than 10 days the annual income of the typical person in Ethiopia, the country with the lowest per capita income. If the per capita, GDP of countries in the initial year is plotted against their intervening growth rate over a period, Baumol (1986)[41] showed that their relationship is a strong negative one, indicating that the countries which were rich initially, they grow slowly, while the countries which were poor relatively grow most rapidly. Robert, J. Barro (1992)[42] who worked on "convergence", that is, closing the gap betwccn rich and poor economies, found that during the period 1960 to 1997, the growth rate of GDP per worker plotted against the deviation of GDP per worker relative to that of USA from its own steady-state values in log scale (following an equation which is not given here), shows countries that are poor relative to their own steady states do tend to grow more rapidly. Thus, in 1960, good examples of these countries were Korea, Japan, Singapore and Hongkong.

Finally, once again moving with the spirit of the times when logical purity demands the reduction of plurality of causes to *one* original with the *benthamite metric of utility*, of utility alone to the exclusion of all else—there has to be an understanding of the *development process* in terms of growth and distribution forming an irreducible set of objectives of development policy and these two can together form a mutually *virtuous circle* if proper development policies are implemented

Notes and References

1. Fei, John, C.H. and Gustav, Ranis, 1964, *Development of the Labour Surplus Economy : Theory and Policy*, Yale University Press, Yale.
2. By giving the value to the *public park* from the environmental point of view which satisfy *aesthetic* rather than immediate material needs to increase the land use value.

3. Abler, Ronald Abler, John S. Adams and Peter Gould, 1971, *Spatial Organization : The Geographer's View of the World*, Prentice-Hall, N.J.
4. Knox, Paul and John Agnew, 1994, *The Geography of the World Economy*, Arnold, London. Search for a materialist framework for analysis and thereby integrating space-time-environment into theories of social process was also the aim of David Harvey (1996) in his book, *Justice, Nature and the Geography*, Blackwell, Oxford.
5. "Development" in terms of escaping from underdevelopment became an *euphesim* in United States' hegemony over the world— the term *development* was first used by Harry, S. Truman in his inaugural address on January 20, 1949 as President of United States : "The old imperialism—exploitation for foreign profit – has no place in our plans. What we envisage is a programme of development based on concepts of democratic fair dealing".
6. The total stock of wealth (or, gross product) of a country is measured by the aggregate value of all goods and services produced by the size of all economic activities) in the country—this is the income equivalent to the total valuation of GDP. Since GDP includes also the net income to the domestic economy from all factors of production: goods and services including those *from abroad* for factor services, like capital investment and labour provided plus the export-import balances, a part of which has to be repatriated back to donor country, the latter is known as *capital depreciation*. Thus GDP minus capital depreciation is the *gross national product*, GNP, which is, the value of economic output resulting from the use of resources—land, labour and capital—owned (or, claimed) by national members (residents) of the country in a year plus what they earn outside the country in terms of remittances from abroad, returns on investments abroad. What they own of "claim" is also their income in the year, thus GNP is a measure of *national income*. GNP is always less than GDP and for countries dependent on foreign capital, their national income, GNP may be less than GDP by even 20 per cent. So for countries dependent on foreign capital (for development) there has to be an outflow of profits by companies and capital invested by foreign countries, otherwise also deficit in export-import leads to outflow.
7. In 1997, external debt of Brazil was 193 billion US$, the highest for any country in the world.
8. Hayami, Yujiro, 1997, *Development Economics : From the Poverty to the Wealth of Nations*, 2nd Edition, Oxford University Press, Oxford.
9. *Cited from page* 80-82, Lewellen, Ted, C., 1995, "*Dependency and Development* : *An Introduction to the Third World* ", Bergin & Garvey, Westport.
10. *Cited* from R. J. Johnston, 1984, "The World is Our Oyster", *Transactions, Institute of British Geographers*, Vol. 9, as mentioned, in Rapley, John, 1996, "*Understanding Development : Theory and Practice in the Third World* ", Lynne Rienner Publisher, Boulder.
11. Like the "overall need for political freedoms in the form of open participation (or, participatory opportunities as well as civil rights) and protective security in the form of social safety nets" for East Asian countries, where the country like South Korea which has been growing at 5 or 10 per cent per year for decades did have 5 to 10 per cent fall in GNP in recent years, the disastrous

effect of which was not shared by the population but heaped on the poorest part of the population. In the context of social and welfare consequences of the capitalist growth of the NICs, political repression is cited as a common feature of NIC governments.

12. Marsh, William, M. and Grossa, John M. Jr., 2002, *Environmental Geography : Science, Land Use and Earth Systems*, Wiley, New York.
13. Sen, Amartya (1999, 2002, 2nd Edition), *Development as Freedom : Human Capability and Deprivation*, Clarendon Press, Oxford.
14. Goulet, D., 1971, *The Cruel Choice : A New Concept on the Theory of Development*, Atheneum, New York.
15. Sen, Amartya, 1984, "Development Which Way Now?" *In,* Sen, Amartya, (1984), *Resources, Values and Development*, Chapter 19, pp. 485-508, Harvard University Press, Cambridge, Ma.
16. *Cited in* Colman, David and Fredrick, Nisson, 1986, 2nd Ed., *Economics of Change in Less Developed Countries,* Philip Allan Publishers Limited, Oxford.
17. *Social Welfare* : understanding of the various diversities of individuals in society, giving value to their inherent strengths for bringing out their resources towards a better quality of life.
18. Rose, Gary, 1979, *Deciphering Sociological Research*, Macmillan.
19. The simplest and most common method of expressing inequality is to divide the country's population into *fifths* (or, quintiles) according to income share and to plot them as a per centage cumulative frequency, which provides Lorentz curve.
20. Greater is the Lorentz curve, greater is the degree of inequality of the distribution: the measured inequality of the curve is the *Gini Coefficient.* Since a larger proportion of population in developing countries tends to be in the younger age groups, not expected to be in the labour force, so inequality comparisons between the Countries of Global North and that of the Global South are often distorted by demographies.
21. Classical political economy of John H. Keynes who in his *General Theory of Employment, Interest and Money* in 1936 argued that fiscal policy in command economy by state can manage well capitalism's *boom-and-bust cycles.*
22. World Bank, 1992, *World Development Report*, Oxford University Press, 1992. *Note* Income distribution statistics for Countries of the South is difficult to have, World Bank list such statistics for only a couple of such countries. And, income comparisons between Countries of the South with those of the North are often distorted by demographies since a much larger population in the Southern Countries are in the younger age group of below 15.
23. Potter, R. B. and Llyod-Evans, S., 1999, *Geographies of Development*, Prentice-Hall, Harlow.
24. Morris, M. D., 1979, *Measuring the Condition of the World's Poor : The Physical Quality of Life Index*, Pergamon Press, New York. The problem in Morris' PQLI is that life expectancy and infant mortality are significantly correlated.
25. Hicks, N. and Streeten, P.,1981, "Indicators of Development : The Search for a Basic Needs Yardstick", *in,* Streeten, P. and Jolly (eds.), *Recent Issues in World Development*, Pergamon.

26. *Cited from* World Bank's "World Development Report, Attacking Poverty", Oxford University Press, New York.
27. *Cited from UNDP's Human Development Report 1998*. UNDP publishes Yearly Report of World's countries on "*Human Development*" since 1991 as the process of enlarging people's choices.
28. For the construction of HDI, we require to fix the minimum and the maximum values are established for all the three indicators as follows:
    - Life expectancy at birth : minimum 25 years, maximum 85 years,
    - Adult literacy (or, Combined gross enrolment ratio) : minimum 0%, maximum 100%,
    - Real GDP per capita (PPP$) : minimum $ 100, maximum $ 40,000.

    The construction of the income index is a little more complex—the world average income of $ 5,990 in 1995 was taken as the threshold value (y^*). When the actual income of a country is within $ 5,990, its PPP$ is : $W(y) = y^*$ for $0 < y < y^*$ with $y^* = \$\ 5{,}990$ (PPP$).
    For y between $ 5,990 and $ 11,980, it is : $W(y) = y^* + 2(y - y^*)^{1/2}$
    For y between $ 11,980 and $ 17,970, it is : $W(y) = y^* + 2(y^{*.1/2}) + 3[(y - 2y^*)]$
    For the maximum income of $ 40,000 which is between $6y^*$ and $7y^*$, the $W(y) = y^* + 2(y^{*.1/2}) + 3(y^{*.1/3}) + 4(y^{*.1/4}) + 5(y^{*.1/5}) + 6(y^{*.1/6}) + 7[(40{,}000 - 6y^*)]^{1/7}$.
    With the above formula, the maximum income of $ 40,000 becomes $ 6,311 (PPP$). Life expectancy at birth : minimum 25 years and maximum 65 years Adult literacy : 0% and 100% as well as the Combined first-, second- and third-level gross enrolment ratio : 0% and 100%, Real GDP per capita (PPP$) : $ 100 and $ 40,000 (PPP$).
29. Following UNDP's HDI index, its construction is illustrated with example of three countries below : Greece, an industrial as well as two developing countries, China and India.

Country	*Life expectancy (in years)*	*Adult literacy rate (in %)*	*Enrolment Ratio*	*Real GDP per capita (PPP$)*
Greece	77.9	96.7	82	11,636
China	69.2	81.5	64	2,935
India	61.6	52.0	55	1,422

Life Expectancy Index : Greece = (77.9 – 25)/(85 – 25) = 52.9/60 = 0.882,
China = (69.2 – 25)/(85 – 25) = 44.2/60 = 0.737,
India = (61.6 – 25)/(85 – 25) = 36.6/60 = 0.610.

Adult Literacy Index : Greece = (96.7 – 0)/(100 – 0) = 96.7/100 = 0.967,
China = (81.5 – 0)/(100 – 0) = 81.5/100 = 0.815,
India = (52 – 0)/(100 – 0) = 52.0/100 = 0.520.

Adjusted Real GDP *per capita* (PPP$) *Index* : Greece – for Greece its per capita income being $ 11,636 (PPP), it is above the threhsold level of $ 5,990 (PPP) being taken as the world per capita average income in 1995, but it is less than twice the threshold. So, for Greece, the adjusted real GDP per capita would be equal to 5,990 + 2(11.636 – 5,990) $^{0.5}$ = 5,990 + 2(11,636 – 5,990) $^{0..5}$ = 5,990 + 150 = $ 6,140.

Now, for calculating the discounted value of the maximum income of $ 40,000, which is between 6y^* and 7y^*, we use the following form of Atkinson's formula, which is : $W(y) = y^* + 2(y^{*.1/2}) + 3(y^{*.1/3}) + 4(y^{*.1/4}) + 5(y^{*.1/5}) + 6(y^{*.1/6}) + 7[(40{,}000 - 6y^*)^{1/7}$ = $ 6,311.

For China and India, since their per capita is less than the threshold of $ 5,990, so they need no adjustment.

Now the Adjusted Real GDP per capita (PPP$) Index for Greece : (6,140 –100)/(6,311 – 100) = 0.972, for China : (2,935 – 100)/(6,311 – 100) = 0.457 and for India : (1,422 – 100) / (6,311 – 100) = 0.213.

Now, since HDI being the simple averages of the above three indices, so we have HDI for Greece = 0.924, for China = 0.669 and for India = 0.448.

30. The human poverty index, HPI is calculated for the developing and underdeveloped countries (HPI-1) as well as for developed countries (HPI-2). The HPI–1 concentrates on three essential dimensions of human life reflected by deprivations in—(*i*) longevity, (*ii*) knowledge, and (*iii*) decent standard of living. Deprivation in longevity is represented by the per centage of people not expected to survive to age 40 years, (p_1); deprivation in knowledge is by per centage of adults who are illiterate (p_2), and the deprivation in decent standard of living in terms of overall economic provisioning is represented by a composite (p_3) of the simple average of per centage of people without access to : (*a*) safe water, p_w; (*b*) health services, p_h and (*c*) per centage of moderately and severely undernourished children below 5 years. p_c. This composite variable p_3 is constructed by taking a simple average of the three variables (that is, equal to $p_w + p_h + p_c$) divided by 3. Now, the HPI-1 for underdeveloped and developing countries is computed as follows :

HPI–1 = $[1/3\,(p_1^{\,3} + p_2^{\,3} + p_3^{\,3}]^{1/3}$.

The human poverty index for developed countries, HPI–2 concentrates on deprivations in four dimensions of life, quite similar to HPI–1 for the underdeveloped and developing countries–like in longevity, knowledge, a decent standard of living plus the additional variable of "social exclusion". The first deprivation relates to survival—the vulnerability to death at a relatively early, that is, the per centage of people not expected to survive to age 60, that is, p_1. The second deprivation relates to knowledge—being deprived of the world of reading and communication that is, those who are functionally illiterate, as defined by OECD countries, p_2. The deprivation in a decent living standard in terms of overall economic provisioning is represented by the per centage of people below the income poverty line set at 50% of the median disposable personal income, p_3. And the fourth deprivation, is non-participation (or exclusion), measured by the rate of long-term (12 months or more) unemployment, p_4 , of the labour force. Now, like HPI–1, the HPI–2 is calculated as :—

HPI–2 = $[1/4\,(p_1^{\,3} + p_2^{\,3} + p_3^{\,3} + p_4^{\,3})]^{1/3}$.

Following UNDP's human poverty index HPI –1 and HPI–2 as in above, its construction is illustrated with example of three countries below : Greece, an industrial as well as two developing countries, China and India.

Country	*People not expected to survive to Age 60 (as % to total pop. p_1)*	*Adults who are functionally illiterate (% of age 16 – 65) (p_2)*	*Population below Income poverty (in % to total) (p_3)*	*Long-term Unemployment Line as % of total labour force (p_4)*
Sweden	8.0	7.5	6.7	1.5
Canada	9.0	16.6	11.7	1.3

	People not expected to survive to Age 40 (as % to total population)	*Adult Literacy Rate (in %) (as % to total population)*	*Population without Access to* Safe Water, p_w	*Population without Access to* Health Services, p_h	*Underweight Children below age of 5, p_c*
China	7.0	18.5	33.0	12.0	8
India	33.0	48.0	19.0	15.0	53

For Sweden and Canada, we have HPI–2 equal to $[1/4(8.0^3 + 7.5^3 + 6.7^3 + 1.5^3)]^{1/3} = 6.8$ and $[1/4(9.0^3 + 16.6^3 + 11.7^3 + 1.3^3]^{1/3} = 12.0$. Similarly, for China and India, we have HPI–1 equal to $[1/3\ \{(7.0^3 + 18.5^3) + \{(33.0 + 12.0 + 8.0)/3\}^3\ ]^{1/3} = 17.1$ and $[1/3\{(33.0^3 + 48.0^3) + \{(19.0 + 15.0 + 53)/3\}^3\ ]^{1/3} = 35.9$.

31. Desai, Vandana and Robert, B. Potter (eds. 2002), "*The Companion to Development Studies*", Arnold, London.
32. In the post-war western development discourse, there is a polemical mode of analysis in expressing state of the world countries in terms of divided between "developed world" on one hand and "underdeveloped and developing world" on the other, which highlight the profound economic, social and political differences. The World Bank is fond of monetary criteria like per capita income, literacy, life expectancy, levels of primary school enrolment or volumes of export trade which categorize the countries of the world according to groups of this bipolar world. But these indicators often provide only a very superficial insight into global patterns of impoverishment and inequality. World Bank and United Nations Development Reports regularly carry bland statistical sections which often seem loosely related to "on the ground" dynamics. World Bank divides countries into: high, middle, and low income. Developing countries are in middle and low income bands. Developing countries fall in tropical and sub-tropical climatic areas. According to World Bank, 2002, the poorest country in the world is Ethiopia and the richest Switzerland with US$ 102 and US$ 43,600 per capita annual income. The World Bank sorted 113 countries of the world into six groups in ascending order as per their per capita annual income.
    High Income Countries: N. America, W. Europe, E. Europe, Antipodes, and Japan.
    Middle Income: S./C. America, S.E Asia, N. Africa, Middle East.
    Low Income: Sub-Saharan Africa, S. Asia, China.
33. Adam Smith, THE FATHER OF MODERN ECONOMICS, is credited with both defining the nature of incipient capitalism and for helping establishing

the policies that would speed up the growth of capitalism. In his book, the *Wealth of Nations : An Inquiry into the Nature and Causes* (1776), he argued that it is the labour of the people which measures the wealth of a nation, not the country's land and others. According to him, labour productivity would increase as specialization increase, leading to mass production which would raise the economic level of the entire populace.

34. Alongwith Adam Smith, David Ricardo, the doyen of political economy in the 18th century, in his "*Law of Comparative Advantage,*" (1817), showed the way to economic prosperity through trade.
35. Post-war Free-market economies enabling individuals to pursue their self-interest to the benefit of society through overriding Keynesian statist command economy.
36. *Cited from* Lewellen, Ted. C. (1994), *Dependency and Development : An Introduction to the Third World,* Bergin & Garvey, Westport.
37. According to Weber, culture, which refers to people's intellectual, spiritual, and moral endowment, produce people to a greater degree than people produce cultures, and that therefore what emerges from or functions in one culture cannot be transposed to another. Along this line of reasoning, Weber suggested that capitalism itself which emerged in Protestant societies of Western Europe could emerge only out of it, and thus it could not be expected to develop elsewhere, "at least not in a form recognizable to those in the heartland of capitalism in the West" : in "*The Protestant Ethic and the Spirit of Capitalism*", Max Weber, 1922, 1967, originally in German, Unwin, London. *However, this hypothesis that some cultures are prone to develop while others are prone to lag behind, is questionable.*
38. Cultures are shared meanings embedded in community, ethnic group or nation through language, religion, custom and tradition.
39. Naqvi, Syed Nawab Haider (2002), *Development Economics—Nature and Significance*, Sage Publications, New Delhi.
40. IMF (2000), *World Economic Outlook*, Oxford University Press, Washington D.C.
41. *Cited from* Jones, Charles, I. (2002), *Introduction to Economic Growth*, Norton & Co., New York.
42. *Ibid.*, 41.

2

A PERSPECTIVE OF CHANGING WORLD

The Global Countries of North and South

The aftermath of the World War II became a *political watershed*[1], the former imperial powers of European countries, England, France, Germany, Spain, Portugal , Belgium, The Netherlands who colonized a majority of territories of the world through their overseas expansion from the late fifteenth century up to 1945, that is, for some part of the past 500 years, were undermined, the colonies became "millstones around necks" of European powers: in India there ushered ideas of the nationalists who had been clamouring for independence since long which crystallized the response against alien rule and, England exhausted by the five years of World War II, relinquished India—the Indian Independence Act got partitioned India into two Dominions: India and Pakistan in 1947. The partition resulted in massacres of religious minorities in both countries; half a million died in riots while quite a number of millions fled their own borders for safety. The liberation of India, the *jewel of the imperial space of British Empire* started decolonization and political independence of the colonial countries in Asia and Africa followed in short order though independence often was granted only after violent nationalist uprisings. Thus within a decade or two after the World War II, over 140 nations in Asia and Africa became independent one after another and took their places on the international scene, so that while in 1939 at the time of the outbreak of World War II, the colonial empires in Africa, Asia and the Pacific Islands were at their peak, but by 1965 they had virtually ceased to exist.

For this rapid decolonization, there is no simple explanation apart from the rise of nationalism—got initiated by Kamal Ataturk in 1920s[2] and led by Indian nationalist forces, this got spreaded among the colonial population depending on the territories. However, the demand for independence came both from the colonies themselves as well as also from external influences which eventually militated against the continuation of the colonial empires. There had been a drive to *economic autarky* led by the nationalist forces from beginning of twentieth century in India. The Russian Revolution of 1918 gave impetus to the notion of mass movements and class struggle—it generated anti-imperialist sentiments which had their impact on West European thinking, encouraging the view that colonial domination was essentially immoral. Finally, after the World War II, as the world hegemonic power, the USA also began to exert pressure on the European powers as it pursued its own expansionist goals in the mid-twentieth century. However, the World War II had the most direct impact upon the colonial relationship and the imperialist powers. During the war France, Belgium and the Netherlands were cut off from their dependencies and it was difficult for them to return to full colonial rule after the war. Other colonies were less affected, and in the long-established Caribbean colonies, for example, the notion of Britain as the "mother country" was greatly encouraged by the sense of allegiance that the war fostered.

Yet the World War II brought fundamental changes to the colonial world and during the 20 years after 1945, substantial decolonization which occurred, took place in two main waves: (first), before the 1950s when independence came in the Middle East and in South Asia, and in the (second) post 1950-phase, the majority were in Africa and the Caribbean. Despite the changes in the wider international climate, in the white settler colonies in Africa and in Asia, the colonial powers were loathed to give up its territories without a fight which led to bloody and contracted struggles in Algeria, Indonesia, Rhodesia, Kenya and Mozambique. France was the lone colonial power that determined to maintain its holdouts : Indo-China in Asia and Algeria in Africa. But the French were forcibly expelled by the communist Vietminh, the country was temporarily divided, with communists to the north and the former French supporters to the south. Algeria also got freedom with relatively little bloodshed. The colonialism came to an end during the 60s when Portugal and Belgium relinquished their control over major parts of the African continent and the curtain had been drawn finally except a few holdouts around the globe on colonial rule when South Africa gave up its hold on Namibia in 1990 and on

July 1, 1997, Hongkong became a Special Administrative Region, SAR of China. The colonies which remained finally were mostly islands, like Gibraltar, Falkland under Britain either too small in the remote oceans with population of 150,000 to 500 to be viable as independent states.

The end of the World War II gave rise to a socialist block, consisting of the Soviet Union, USSR alongwith the industrialized countries of Eastern Europe and soon because of the attempts of the two superpowers, USA and USSR to force all nations into a capitalist or communist block, the term "Second World" appeared since 1949, identifying these countries of Eastern Europe as well as the countries like China, North Korea, Vietnam.

In the post-war western development discourse, there is a polemical mode of analysis in expressing state of the world countries in terms of divided between "developed world" on one hand and "underdeveloped and developing world" on the other, which highlight the profound economic, social and political differences. The World Bank is fond of monetary criteria like per capita income, literacy, life expectancy, levels of primary school enrolment or volumes of export trade which categorize the countries of the world according to groups of this bipolar world. But these indicators often provide only a very superficial insight into global patterns of impoverishment and inequality. World Bank and United Nations Development Reports regularly carry bland statistical sections which often seem loosely related to "on the ground" dynamics. The maps produced from these statistics is that they have allowed some observers to label whole areas as a homogeneous group "Third World" or "Global South" as if the same could be said for all its 176 constituents.[3] The problem with labelling such large number of countries with vast regions of the world as "low income" or "Third World" is that a set of negative assumptions are often made and a "pejorative uniformity" is imposed. Thus evidence of unevenness and injustice in the contemporary world is not always effectively conveyed by these statistical measures which are taken as indices of what constitutes "development". There is always a politics of labelling which is necessary to understand and be critically aware of. All too often the term "developing world" has been defined as a problem.

The meaning of the term "Third World" has been in a continual state of evolution since the phrase was first coined in 1952. On August 14 of that year, the decolonized countries of Asia (except Japan), Africa and Latin and Caribbean America were given a pedigree given in France *Tiers Etat* (or *Monde*) or the "Third World" by a French demographer, Alfred Sauvy in an article in *L'Observateur*—the poor, powerless, marginalized people—in France prior to the 1789 French Revolution. Such a useful concept would not be confined long to the people in the history of one country. This term of the "Third World" was quickly widened to include countries that were decolonized as well as marginalized in the international system, compared to the First World which is a fairly coherent unity consisting of the industrialized countries, including the United States and Canada of North America, the countries of Western Europe, Australia, New Zealand, Israel and Japan. All these are having democracies, per capita income is relatively high, and although the distribution of wealth in terms of per capita income between these countries ranging between more than $ 40,000 to lower than $ 10,000 (for Portugal and Greece),[4] the larger part of their citizenry are middle class. Thus definitions of the "Third World" have been contested, as have the origins of the phrase. Yet the concept of three worlds can hardly be said to convey a precise meaning. The geographical boundaries of the "Third World" vary enormously according to which statistical indicators are used and the idea has different meanings in different cultures. The idea of "three worlds" has provided many observers with a convenient shorthand expression and is undoubtedly linked to the formation of "mental maps" and "imagined geographies" of inequality. Given that definitions of development have often been contested, these "pejorative terms of imagined geographies" often overstate a set of (tragic) stereotyped common features of the so-called "Third World" such as poverty, famine or political instability. In discussing unevenness and inequality therefore a representation of the imagination of the people and places subject to the development process is constructed which is only partial and provisional.

It is extremely difficult to summarize the lives of people who are in some way subject to development and the variety of struggles that they encounter. Thus relations between "worlds" are imagined and simplified by geographical expressions like "North" and "South", "rich" and "poor", "developed" and "underdeveloped". The Brandt Commission,[5] sponsored by the United Nations, depicted a *world divided between the Countries of North and the Countries of South* rather than First, Second and Third Worlds. This cleavage of the economic world into two interdependent but highly unequal camps, cast in terms of "North" and "South" came into common use with the earlier publication of the Brandt Report (1980)[6] "North-South : A Programme for Survival Report of the Independent Commission on International Development Issues". The North includes the industrialized countries of USA and Canada in North America, Europe (West and East), The Commonwealth of Independent States, CIS (former Soviet Union in Asia), Japan, Australia and New Zealand; the South includes China and all of the countries of Latin America, Africa, the Middle East, South Asia and South East Asia. By this definition, the South has over 3.75 billion people—three-quarters of the world's population—living on one-fifth of the world's income. This international cleavage not only reflects differences in prosperity but also different forms of economic organization, different kinds of resource bases, different demographic characteristics, different political systems, and different roles in the system of international specialization and trade. Defining and measuring "economic development" is therefore problematic, as we shall find, there are strong grounds for thinking in terms of underdevelopment, rather than development for countries of the South, since the term "development" as has been mentioned in Chapter 1, implies a trajectory of improvement, in both relative and absolute terms.

This North-South divide of the "Brandt Line" if based on GNP per capita, the most widely used international economic indicator, there is a very rough geographical correspondence in it, at least for Europe, Africa and the Americas. When one reaches Asia and the Pacific, however, the rich-North/poor-South distinction falls apart. Australia (rich) is below the equator and China (poor) extends well north of Japan (very rich). The North/South distinction also suggests a *geographical determinism* that is, *tropics* (having mean annual temperature, more than 25 degree celsius), create poverty while *temperate* and *cold* climates are conducive to development but that view is difficult to defend historically since the great civilizations have thrived in most of the regions now considered the Countries of the South—like in Mesopotamia, Egypt, Peru, Mesoamerica, China and India.

Again since the collapse of the USSR and the pro-Soviet regimes in Central and Eastern Europe in 1989 followed by declaring sovereignty of 27 million non-Russians as "Commonwealth Independent Socialist" (CIS) countries within their local government units in Asiatic part of Russia, the sensation of geopolitical complexity of an ideological war between the "capitalist" West, led by USA and the "communist" East, led by Soviet Union, got lessened. In terms of metaphor, this hostility better known internationally as a Cold War (*i.e.* "poor, frozen and frosty relationships, but short of all-out "hot" war) was over and the term "Second World" (which was geo-politically a coherent group under Soviet regime until about 1989), became flatly misleading and it should not have many ardent defenders. But on other hand the myriad complexity of geopolitics of the world remains reduced to a simple black and white representation, be it numerical or cartographic: a developed First World and a developing or underdeveloped Third World, because both have gained a place in the common vocabulary of nations and the conditions they describe are profoundly entrenched in the political, social and economic structure of the contemporary world. In both the First and Third Worlds, many like Worsley (1984) [7] found the term Third World to be derogatory, a "pastiche of political and economic meanings that never made others may accept it as simply descriptive". Mao Zedong, the President of Republic of China used a variant of the three-worlds classification, seeing the First World as the superpowers—the United States and the Soviet Union, the Second World as the clients of the superpowers, and the Third World as the rest. Brandt Commission (1980) in its report, "North-South: A Programme for Survival", not only called into question the notion of separate worlds but also maintained there should be increased levels of economic activity and interaction would benefit all nations—regardless of their geographical and ideological locations.

Thus what it amounts to is that no term is truly adequate, but all are useful, and, to some extent, unavoidable, depending on what aspect one is emphasizing. Instead of Third World, we are having the terms like the "South", or the "Southern Countries" or the "Global Countries of the South" (or the countries of Global South) and the variations of "developing or underdeveloped" interchangeably. They all describe the less developed countries of Central and Latin America, Africa (except South Africa), Asia (except the present sovereign countries of Asiatic Russia and Japan) and Australasia (except Australia and

New Zealand). The North is generally taken to mean the major "Western" capitalist powers of Europe, USA, Canada, Japan, Australia, New Zealand and probably South Africa. The countries of socialist economies of Eastern Europe are not included in this North-South context. So for the bipolar world, these two terms, the "North" and the "South" will be used henceforth instead of First (or developed) World and the Third (or, developing/underdeveloped) World respectively. We are adopting this terminology of the North and the South hereafter to refer to different parts of the world to avoid the more value-laden distinctions such as "developed" or "undeveloped," "modern" or "traditional," "First," "Second," or "Third World."

As a single indicator, the value-loaded nature of GNP makes it the better measure of productive capacity of the national economy of a country as well as its employment potential. The per capita measure of GNP of a country is often taken as a real division between the "industrialized and developed" *i.e.,* the Countries of the North and "other than developed" *i.e.*, the Countries of the South by the international institutions like the World Bank, International Monetary Fund. In their working out of the "South-North Gap", United Nation's Human Development Report 1998 made a distinction between 226 countries listed for the entire world, in characterizing 50 countries as industrialized which they placed in the North while the remaining 176 countries were identified as belonging to the South.[8] Though the Countries of the South do possess certain common traits characterized by their large-size of population in gross poverty, a largely deficient agricultural sector internally and an unequal dependence on the First World for their trade and prosperity, but they are too vast to encompass a neatly coherent collection of traits. However, the major feature which they all share commonly are the constraints imposed on them in their past and in the contemporary world economy their dependence on the economic conditions and social and political processes dominated by the Countries of the North. Hence the generalizations made about development context of these large number of small and large nations which we like to mention them as Countries of the South, or the Southern Countries, or the South in this book, there can be many exceptions, many disclaims.

The idea of "development thinking in sum total" by terms of *equality* or "catching up" with the wealthy through economic activity, has often been discussed in relation to "developing and under-developed countries" (which we term as Countries of the South, or the Southern Countries, or the Global South, or just "the South"), but the South also came in its present form only after World War II. During the long period of European and American expansionism, there was little real concept of development: Colonies and protectorates were appendages of their masters and as such, they had their particular goals of supplying raw materials and providing market. Development was not one of these goals and there was little concept of development even in the writing of "dialectical materialism"[9] of Frederich Hegel and "labour theory of value" by Karl Marx [10] in the nineteenth century though development has taken numerous forms in numerous contexts. Again though no single theorist was having such an enormous effect on development theory as Hegel and Marx, who inspired generations of dependency theorists, but nowhere Karl Marx suggest that Western capitalism was the cause of underdevelopment. It was Vladimir Lenin who popularized the idea of the expansionist and destructive nature of capitalism, who wrote that as capitalism spreads in search of raw materials and market, it leads to colonialism where capital accumulation is by class domination and subordination by politico-economic means. This stage dominated by *finance capital*, is named by Lenin as "imperialism". This perspective of colonialism and imperialism, articulated by Lenin in 1916, inspired the "dependency theorists" in the 1960s and 1970s. In a famous speech in 1949, President Truman spoke of an "underdeveloped world":

> the "handicap and the threat this underdevelopment poses to the developed world" .

Born of the ashes of colonialism and the fires of anti-communism in the United States, thus "development" occupied centre-stage as an ideal in the global arena after the World War II. Truman goes on to explain the need for "modern, scientific and technical knowledge" and announces the beginning of a "bold new programme" to resolve "underdevelopment" and "poverty" in poor countries.[11] Since then development in terms of escaping from underdevelopment has become "an amoeba-like concept, shapeless but ineradicable which spreads everywhere because it connotes the best of intentions for the best of all".[12]

Empires and Trade : The World on the Eve of European Expansion

Geography, by its deterministic factors of arid and lifeless deserts, rugged mountains, large stretches of jungles, often cause a lack of integration resulting in a spatio-temporal unevenness not only of regions but also of countries. In that way, geography is often cited as a reason for underdevelopment. However, these geographical barriers are no obstacle to the indigenous cultures of a region—to its people whose cultures evolve in symbiotic interaction with deserts, mountains, and jungles. It is only with the *alien* development patterns transported across the sea that geography became an "enemy to be conquered". This conflict between.the *organic* development of indigenous peoples with their traditional societies and the *imposed* development of capitalist economy by the European conquerors is a universal theme in history of the Southern Countries.

By the time European expansion began, the continents which now comprise the Southern Countries had nurtured highly complex civilizations with advanced levels of cultural, technical and social development. Some of these areas were not "undeveloped" or "underdeveloped". Rather they were *pre-developed* in the sense of having a non-European conceptualization of being "developed", with societies which had achieved a high degree of cultural, economic and political development before the arrival of Europeans, and in some cases in advance of progress compared to that in Europe itself.

In the populous regions of Central America, Asia and parts of Africa, great empires rose up and expanded. The continent of Asia boasts two *nuclear areas* —"the regions where archeologists believe civilizations arose spontaneously, without being influenced by previous civilizations". These two were the Indus Valley of India and Western China in fourth and fifth millennia BC. The four other nuclear areas of the world are Iraq, Egypt, Mexico and Peru. Indian civilization would experience an incursion of Aryans from the west and was invaded by the Persians and, later, the Greeks under Alexander, the Great. In the east in China, Chinese civilization, isolated by the Himalayan Mountains, the Gobi Desert, and the vast Manchurian Plain, would escape such influences and would continue, relatively untouched, into the twentieth century. Its vast Ming Dynasty of the fifteenth and sixteenth centuries inherited a long Confucianism tradition in philosophy, art and literature. The Manchus who were at the peak of power, considered themselves superior to the outsiders and wanted little from them. Few technological achievements elsewhere could compare with China's abacus, fine silks, lacquers, porcelains and explosives. China carried on an overland *silk trade route* with Persia (present Iran) and Syria but imported little, thus the exports of China had to be paid not with imports but with cash. Though dynasty followed dynasty, the everyday life of the Chinese remained relatively unchanged over centuries and millennia.

The Middle East, which stretches from Israel to the Indian sub-continent is having a physical environment which is hostile to human habitation except in their narrow strips of fertile and irrigate riverine stretches and so apparently the area would suppose to be little contested. Yet this arid region—has been a "cauldron of conflict for six thousand years". Here the earliest civilization emerged in Mesopotamia (in what is now Iraq), between the fourth and third millennia BC. Theories of "environmental circumspection" suggest that the first city-states (like present day Singapore) developed here because people were confined by desert within narrow riverine valleys and deltas. With nowhere to spread, natural population growth forced greater elaboration of social systems and technology. The creation of complex irrigation systems led to distinct social classes, differentiated between those few who controlled the water supply and those who laboured, those who ruled and those who served. Egypt experienced a similar pattern of early development. Over the centuries, few areas in the world would be the focus of so much warfare and social change, as kingdom followed kingdom and empire built upon empire in this part of the world. In this hostile area the vast Byzantine empire centered in Constantinople, had been conquered by the expanding Ottoman Turks. Though Muslims had to retreat from Spain as well as from Europe by seventh century AD, but they were on the move elsewhere, spreading south across the Sahara desert to connect with Mali and Songhai and expanding north into Turkistan and east into India. Muslim culture, centered on the Middle East, made important contributions to learning in the fields of mathematics and astronomy. In medicine, considerable progress was also made. Knowledge of chemistry and other natural sciences was advanced and their knowledge in cartography influenced medieval European map-making in charting the ocean route.

Throughout much of South-East Asia, Hindu trading empires were contemporaneous with the European Dark Age. At a time when Europe was a collection of small feudal principalities, Indian traders settled along the coast of the Indian Ocean countries, bringing the Hindu and Buddhist religions, particularly on the east.[13] In the early seventh century AD, the powerful Khmer kingdom, with its capital at the huge city of Ankor, was founded in Cambodia and this kingdom would last for five hundred years. Much of South East Asia including the large island complexes of Indonesia, Borneo and the Philippines developed no common culture or language but was integrated into a loosely structured, culturally and commercially sophisticated unity.

Early development in Africa, as in Latin America and Asia, varied greatly. For example, there were marked contrasts between the societies of the Egyptian, Ethiopian and Sudanese empires on the one hand, and the food/hunter-gatherers of the Congo or the Kalahari Desert on the other. Some states had well-established trading and agricultural systems, as in the kingdom of West Africa and Rhodesia (present Zimbabwe) in south.

In the New World, the mighty "Aztecs" who dominated central Mexico and ruled over the subject nations from the Atlantic to the Pacific coasts and as far south as Guatemala, had developed a complex, highly stratified society, with a nobility which held military, secular and priestly authority together. Skilled craftsmen worked on gold and other precious metals. Intensive agriculture supported a dense rural population and, a large urban system, culminating in the capital city Tenochtitlan, built on an island in a lake where Mexico City (the present capital of Mexico), was one of the most beautiful cities in the world: this city with a population of over a quarter million, was larger than any city in Europe at the time which impressed the Spanish conquerors on their arrival in 1519. In Western South America, the "Incas" administered a century-old territory, 3,000 miles in length along the Andes and the western coast from present-day Equador into Chile. The Incas were brilliant administrators and engineers who ruled with forceful beneficence. Their agricultural techniques included irrigation, terracing and use of the foot plough. Contact between the widely scattered rural communities and the imperial centre at the City of Cuzco was maintained by a well-developed system of roads. Impressive monuments and fortresses further demonstrated the cultural level achieved by the Incas.

During the fifteenth century Europe was relatively marginal in the affairs of the wider world. "But that all these would fall, soon or late, under European domination attests to some remarkable changes in the world history and economy from the latter fifteenth century onwards" (Knox and Agnew, 1994).[14]

Europe on the Eve of its Expansion

The year 1492 is remembered for Christopher Columbus "discovery of America" and by the time of Columbus's first voyage, the Portuguese mapped out the entire coast of Africa: they took the full advantage of such invention of the "Sextant", used to fix the ship position at sea. Meanwhile the coastal navigation was being transformed into true *seagoing shipping* so that within five years of discovery of America, in the other parts of the world oceans, Vasco da Gama reached west coast of India by sea in 1497.

The year 1492 is significant also from the point of view that the Muslims who occupied Spain for almost nine centuries, were expelled, and the country got united under King Ferdinand and Queen Isabella. Although this was still incipient, but the emergence of Spain as an unified nation-state would come to have a profound impact in Europe which was till late in the fifteenth century, remained divided into feudal lords or city-states, only nebulously united as countries under weak monarchs. In much of the rest of the world, on the other hand in Middle East in Mesopotamia (present Iraq), Egpyt, India, China, Mexico (the Aztecs) and Peru (the Incas), there were *empire-states* which administered over large territories.

The near millennium-long Medieval Period in Europe, characterized by great religious architecture but relatively little economic or technological progress, slowly gave way to a period of artistic florescence, religious factionalism, scientific enquiry, nation building, and maritime exploration. In the political sphere, a new breed of rulers declare themselves free of papacy and made united larger and larger areas through warfare, intimidation, and political weddings, creating "true states" out of areas previously united by little more than common language. Economically, Europe remained largely agricultural, but during the later Middle Ages, towns became increasingly important, nascent capitalism emerged in these towns, where merchant elite developed around craft guilds and commercial fairs. Fortunes built by burgeoning trade with the New World on other side of the Atlantic, provided the wealth for large banking houses that flourished in the Italian republics and, later in Holland and England. Feudalism—in which serfs, bound to their estates, produced for the benefit of landlords—began to decay. Originally, "rent" was paid by labour but by the fifteenth century, money-rents were already well-established, in some areas, providing taxes for state development and for investment capital in mercantile trade. State bureaucracies emerged where previously there had been only royal courts or personalistic patron-client rule.

Despite these increases in accessibility to previously uncharted areas of the world through increasing number of voyages by the sea-faring nations of Europe for mercantile trade, the first region to be brought into the Western European orbit as a *dependency*—and thus, in retrospect, the *earliest to assume the status of Third World*—was the landlocked countries of Eastern Europe. This trade, dominated by Holland (presently, The Netherlands), established a system in which an underdeveloped *periphery* exported raw materials to a wealthy *core* at low cost and imported finished goods at high cost—Poland and Lithuania provided rye, cattle, furs, timber; Hungary exported cattle and copper. In return, these Eastern European countries received textiles, hardware, and other manufactures. The Americas, both North and the South would be the next to assume this pattern, as Spain among the European nations was the earliest one to plunder its newfound southern continent of gold and silver and then turned indigenous economies to producing agricultural goods for the mother country.

But as the European economic and cultural world was gradually transformed, sea-faring nations of Europe began to voyage beyond the shores of Europe. They travelled overland into Asia, in search of scarce commodities for European consumption. These included spices, sugar, silk, muslins, porcelain and the like. The Portuguese, funded by Italian bankers, were the first to navigate around Africa into the Indian Ocean, with its Arab-dominated sea-based trading networks, while soon after Christopher Columbus (as already stated) sailed westwards across the Atlantic in 1492 to open the Americas to European exploration.

For Spain and Portugal, this is the heroic "Age of Navigators", as the Spanish, in particular, moved rapidly to take control of these new maritime trade routes looking out to the Atlantic while the Portuguese established a far-flung network of fortified port bases in present Mozambique on the Indian Ocean to have trading with India. Equally significant was their use of African slaves to produce sugar in plantation farms in their Brazil territory. The Spanish founded a more militaristic, oppressive and exploitative empire in Central and South America which saw the decimation of local population through warfare and the catastrophic effects of European diseases, especially small-pox and measles. It also brought enormous inflow of gold and silver bullion into Spain and ultimately onwards into European economy which more than any other is said to kickstart the European *capitalist world system*. Fundamental to mercantilist philosophy was the simple principle: Collect wealth, but don't give it out, prosperous yeoman farmers and craft-manufacturers began to accumulate capital and merchants began to invest capital, all founded the new class of capitalists. Significantly, it was the state, not individuals, that got most of the money and that also controlled expenditures, through, for example, passing laws to limit imports and to prohibit the departure of precious metals. In this period of *merchant capitalism*, money had to be spent, of course, but at home, on armies and on ships that would become the potential for gaining wealth abroad. Previously, overseas trade had been in spices and fine silks for the noble classes, now the emphasis was on agricultural commodities and industrial raw materials for manufactures.

The Rise of Capitalism

Along with the emergence of the European nation-state as the primary unit of international economic power, there grew the "doctrine of mercantilism" or *merchant capitalism* (or, the *proto-industrialization*), there has been substantial arguments about the processes of transformation involved. For the nation-states of Europe the overriding political objectives of maintaining law and order at home and national strength abroad involved clear economic objectives. They implied the need to maximize government revenues, to increase national wealth and that through the increasing national share of world trade and monetary resources. The locational advantage of England lying astride the intercontinental sea-routes (of what was then the most dynamic region of the world economy—north-western Europe). Textile manufacture was the most studied production process of this period which got mechanized with water-powered machinery by the early eighteenth century, the industry was characterized by some "family" type of "domestic" production system located in the countryside, merchants were both the providers of capital and principal traders in a regionally-specialized, complex. The textile industry had a relentless tendency to geographical specialization as well as concentration with the result that by 1770s in England, west Yorkshire was beginning to dominate light worsted woolen cloths, Lancashire in cotton, Wales in flannel and east Midlands in hosiery. The complex interaction of resources, labour skills, technological innovation, and capital circulation in particular regional economies, which increased production and lowered costs, meant that new markets had to be developed.

The mechanization of textile industries in England only in the eighteenth century, brought the "emergence of capitalist mode of production in England" (Wolf, 1982).[15] Because of its efficiency and technological superiority, this mode spread rapidly throughout the world, virtually replacing the *kin-ordered mode* by the *tributary mode* and radically transforming the cultures it encompassed. The *kin-ordered mode*, to the extent that it had survived mercantilism, was obliterated. Everywhere capitalism created a system in which labour was bought and sold as a commodity in return by the wage for the goods produced. Rigid class stratification emerged where none had existed before, and in many places, the results were cultural collapse, detribalization resulting in mass immigration in search of jobs in the New Continent.

Market opportunities for cotton textile products were to be found both in the Europe-wide trading as well as among the colonial population overseas. Thus, cotton textiles were first imported to England from India, the techniques of production were learnt, manufacturing commenced, factory production lowered costs, and the factory-made cotton textiles were exported to India, there they undercut and destroyed the Indian artisan cotton and silk textile industries.

The Munducucu, a horticultural tribe in Brazil, was brought into capitalist economy by the development of rubber tapping in the area's highland region. Highland villages disappeared as workers moved to the riverbank groves of rubber trees. The rubber trader replaced the Chief, who became caught up in a web of debt bondage to village stores that they could not escape in their life time.

In the nineteenth century, European agriculture continued to increase production but was unable to keep pace with the demand from the growing urban population. In 1840, Britain repealed Corn Laws, which had protected local grain producers, and opened its markets to colonial and American producers. New methods of extensive grain farming were used in mid-western North America, and in the new British colonies of Australasia. The invention of steam ship, allowing more rapid transport, and of refrigeration and meat canning from the 1870s, led to similarly extensive ranch grazing of sheep and cattle in these countries. Together with the Pampas grasslands of Argentina, all became an important part of the "semi-periphery" of Britain's global economy.

It was accompanied by new waves of migration to the colonies and USA, especially from Ireland and Scotland, where the Great Potato Famine and the Highland Clearances, respectively were other manifestations of capitalist agricultural change in Britain. The United States absorbed enormous numbers of European migrants throughout the nineteenth century; during the period of years between 1881 to 1910, about 25 million migrants from Europe particularly from England, followed by Italy, Germany, Hungary, Russia, Scandinavia, Spain, Ireland and Portugal colonized United States to characterize the latter country as a "moving frontier of settlement" changing the frontier settlers to self-reliant, opportunistic, individualistic and democratic.

The nineteenth century England was thus characterized not only by an enormous variety in scale of production, but also by an equally enormous variety of manufactured products. "The Workshop of the World" was as true of nineteenth century England in general as it was in several of the country's manufacturing cities like London.

The market mechanism *per se* in this period provided a basic arrangement in providing passion to the people for interacting with each other and undertaking mutually advantageous activities. Thus *Capitalism*—the competitive drive for profit by individuals or privately owned companies—grew up alongside and, in many ways, out of mercantilism (or, merchant capitalism). Although such an economy had been developing among town merchants, under mercantilism, the state remained the focus of investment and profit internationally. The Spanish conquerors of Central and Latin Americas functioned under state authority and claimed lands for the crown, a "royal fifth" of all gold and silver went directly into state coffers. However, state bureaucracies had neither the wealth nor the competence for the degree of expansion that was inevitable once the progress of conquest and colonization were set in motion. Although states made every attempt to control overseas expansion and trade, they increasingly did so by granting monopolistic royal charters to *joint stock companies* formed by independent associations of merchants, such as England's East India Company (EIC) for trade in India. These companies were a logical development of the "wedding of mercantilism and international capitalism". In many places, mercantilism blended into a sort of state capitalism. Though there were constant attempts by the various governments to control them, their political and economic strength at home and the distance of their operations from their governments gave them a high degree of independence.

In the overseas dominions, both mercantilism and early capitalism implied *tributary mode of production*; that is, the natives were compelled to produce for the benefit of the Europeans but did not work for wages. The means by which the natives were obliged to produce, varied: in some cases, it was a matter of military force or of building up a dependency on European manufactured goods. The pattern in Latin America introduced by Spain was based on the *encomienda,* a large land grant that included the right to the labour of the natives who lived there. In many areas in India, like in Bengal, the British introduced a "landlord system" based on English law that brought peasant under more complete domination, elsewhere the peasants were taxed a substantial part of their production. The slave trade, which decimated large parts of Africa for more than four centuries, provided mass labour for single-crop production of cotton and corn in USA and sugar in the West Indies. Though "tributary system"[16] took many forms, the result was invariably that production for domestic use was turned to production for export to Europe, and this had a profound effect on every aspect of the cultures that were touched.

A fully capitalist mode of production—based on bu⁻ing and selling of human labour by elites who owned the means of production—that is, dividing the society into two *social classes*, the elite class as the producer and the labour as the production agent, did not emerge until the state was transferred *from a manager and collector of tributary production to a production of support for private enterprise*. From this point of view, capitalism's first real break with mercantilism emerged only in the latter half of the eighteenth century with the development of mechanized industry in Britain. Though wage labour had long existed on a small scale, with the "industrial revolution", it gradually became normative throughout the world. European factories had the capacity to turn out hundreds of times the goods of individual craftsmen. These factories needed to be supplied with raw materials, such as cotton for textiles, and the end products required markets. Factory production gave rise to vicious competition among entrepreneurs, breaking the system of state-sanctioned monopolies. The state remained no longer the focus of economic progress. Its function became the protection and encouragement of private capitalism. Now the state's job was to create the infrastructure that capitalists would use to make and move goods and to provide armies to safeguard markets, displace recalcitrant natives, and coerce labours to produce at subsistence wages. New laws had to be written to enforce labour contracts and protect private property.

A second industrial revolution, starting in the mid-nineteenth century, greatly amplified world trade and investment, which increased over 400 per cent in the century and a half after 1850. The replacement of coal by steel, new sources of power—including electricity—and major advances in medicine and chemicals helped spread industry at a hitherto unknown rate. Among the global effect was the destruction of handicrafts, which were replaced by machine-made goods manufactured in Europe or in local industrial enclaves. Steam transportation and the opening of Suez Canal provided faster and cheaper access to markets. The pace of agricultural transformation also increased, with the introduction of Latin American crops, such as cocoa in Africa and rubber, into Asia and the conversion of large areas of subsistence lands to plantation production for European consumption.

Early European Expansion

But over the long period of European colonialism, the colonial episode played out its global course: the era of West European overseas expansion which lasted from the late fifteenth century to the mid-twentieth century falls into two distinct phases: the (first) phase between 1450 and 1830 and the (second) from the nineteenth century up to 1945. The first phase was characterized by the discovery of both the Americas, their possession and finally Europe's loss of most of the colonial Americas. But the industrial revolution in the meanwhile stimulated a new spate of European expansion, particularly in Africa.

South America : The first phase was largely located in the North and South Americas and is associated with considerable settlement by the European migrants. Within 70 years of Christopher Columbus' arrival in the Caribbean, when discovered by Spain, Latin America was not only fully occupied, but it was also home to two *nuclear areas, the two empires*, Aztec of Mexico and Incas of Peru which were at the apex and could be the rivals of any empire in the world at the time. Surprisingly, perhaps, it was these two societies, which were organized into powerful states, that were easiest to conquer. Tribals in the Amazonian rain forest or in the mountains were either inaccessible in the jungle or sufficiently mobile that they could escape from the Spanish "conquistadors" (conquerors) like Hernan Cortes European armies; farmers, however, could not run away, and the states, with their cities and top-heavy bureaucracies, could be decapacitated. In 1519, Hernan Cortes with 550 men and a few cannon, sailed from Cuba for Mexico, he marched inland and as he went he made allies with enemies of Aztec, and the conquest of Tenochtitlan (the capital of Mexico) was achieved through a bloodiest warfare in which 240,000 people died in 1521.

The Incas were conquerors —in less than one hundred years, the Incas empire had spread from the central Andes to span the 5,000 kilometres between Equador and Chile. This empire growth was forged by a professional army, consolidating its gains with the construction of highways and control points. The Inca economy was also no less sophisticated—a strict dynastic hierarchy organized the extraction of labour and goods from its subjects through a centralized state, military might and the legitimizing force of religion. Great irrigation systems were built to ensure a regular surplus of potatoes and maize on communal plots of land which were traded. In 1532, eleven years after the conquest of Mexico, Francisco Pizarro with 180 men set foot in Peru and ascended into the Andes. Taking the advantage of the civil war between the two sons of the recently deceased emperor, through brutal tactics, Pizarro kidnapped the Incas emperor and strangled the emperor—the manner the emperor met his death continues to shock to this day. The Incas were overthrown, and most of central and western South America were occupied. Thus, in a mere two decades, a region of 9,000 kilometres long—from the Rio Grande to the alluvial deltas along the Pacific Coast in present day Peru—had been effectively conquered by the fierce battle tactics, rapacious will and brutality of men "with nothing to lose and a world to gain". It is significant for this process of European conquest that the lands of the New World, discovered or undiscovered, were divided between Spain and Portugal by the Pope in 1494.

Portuguese occupation of Brazil was less rapid, and because of the Amazonian rain forest which stretches for thousands of miles Atlantic Ocean, initially remained close to the seaboard. After initial small-scale extraction of forest products, a plantation sugar economy was established in the north-east coastlands. The large landholdings were worked initially by the local people, but after they fled or died, slaves imported from Africa provided a major element in the Brazilian economy until after 1850. Portuguese, envy of the mineral wealth of Spanish America, prompted exploration of the interior, and resulted in the discovery of gold in the late seventeenth century. This led to the first significant settlement in Brazil, and also contributed to the westward push of Portuguese territory beyond the 1494 demarcation.

The other European powers came later in the region, securing smaller territories in the Caribbeans and Guianas. Sugar was also a major element in the colonies of the other European powers, in the Caribbean islands and the Guianas, again based on plantations and African slave labour. The wealth generated made these possessions the most valuable parts of the emerging empires of eighteenth century Britain and France. However, sugar monoculture had a "stultifying effect" on other sectors of rural economy, limiting production of food crops and fostering the need to import essential foodstuffs. The plantation system had a pervasive influence on both the economy and society in the Caribbeans. Plantation society was dominated by Europeans, and African slaves forming the lowest echelon of society so that even after emancipation, their dependents found themselves at the base of a highly stratified and status-conscious society.

These invasions were motivated for plundering: the *conquistadores* came to plunder and convert the local people to Catholicism. The Spaniards in particular sought precious metals, exploiting gold and pearls in the Caribbean before 1510, and major sources of silver were discovered in northern Peru. Mining for this silver was the key to the early colonial economy, providing wealth for export to Spain, the impetus to urban growth, and fostering extensive development of ancillary activities to serve the mines—producing food, pack animals, timber, mercury and salt for use in silver processing. This economy was exploitative not only of land but also of labour. The mines required substantial labour forces, the Spanish colonists forced the surviving subjugated Amerindan clans to work in their mines and their farms at *starvation wages*, through a system of drafting labour. Peasant families had to pay taxes to the state and women were forced to spin and weave cloth for export to Europe. Peasant agriculture became commercialized, accentuating division within Amerindian society. By the eighteenth century, some member of pre-conquest elites, like *chiefs* participated freely in the European colonial economy. However, the majority were so poor and exploited that even the Catholic Church urged reform but the colonial conscience rarely stretched that far.

The first two decades of Spanish rule decimated the Inca polity. In less than one hundred years, 75 to 90 per cent of the native American population had died from disease, overwork and the destruction of their way of life in what is referred to history as *great dying*. Later the local population was also induced or forced to work on the substantial landholdings created (called *encomiendas*, the large land grants) for those who married locally and settled. This created, over centuries, not only a mixed race but also a mixed culture: variously termed *mestizo* (or, *ladino*), this blending of Spanish and local population became the racial and cultural norm in Latin America subsequently. The colonial territories were exploited in the interests of metropolitan Spain, which exercised close administrative control. Urban development and trade took place but as everywhere production for domestic use became secondary to supplying Europe, so most major cities were deliberately located at natural coastal ports, and restriction in the development of manufacturing continued. Where people produced for their own use, now they labour to supply the wealthy citizens of countries thousands of miles away; thus where populations were relatively small and widely distributed earlier, now millions crowd the shanty towns of a handful of coastal port cities; where tribals practiced small-scale slash-and-burn horticulture, now ranchers pennanently destroy hundreds of thousands of acres of forest every year.

The effect of the great outflow of Latin American wealth was profoundly paradoxical. Spanish rulers did little to develop their backward agricultural economy. Instead, they used their bullion to build spectacular cathedrals or purchase arms and luxury goods from England. Between 1503 and 1600, Spanish gold and silver flooded out of Spain to which the price of bullion declined and this contributed towards the decline of Spain itself, which fell into debt and stagnation, while other parts of Western Europe were undergoing a transition towards industrial capitalism. Spain remained a case of late industrialization in Europe and its industry was technologically lagging well behind the other European countries even at the end of twentieth century.

During the colonial period, Spanish America was divided into four large viceroyalties: the viceroyalty of New Spain, 1521 (former Aztec empire and present Mexico); of New Grenada, 1715; of Peru, 1543; of Rio de la Plata, 1776; in addition there was Viceroyalty of Brazil, 1760 (of Portugal). Over the centuries, there grew up a conflict between the Spanish-born administrators and the "mestizo" or local-born Europeans; from 1810 to 1825, revolution erupted throughout Central and South America. Gradually US interest in Latin America became manifested when in the Monroe Doctrine of 1823, which announced to Europe that there would be no recolonization of the Americas. The five viceroyalties were broken up into mutually hostile states. The period from the end of Spanish-American revolutionary war to about 1898, when USA assumed a much more active role with three decades of gunboat imperialism, during which US forces invaded and occupied the islands and countries of the Caribbean, Cuba, Nicaragua, and Panama became economic and political satellites of USA. The more overt form of domination ceased with President Franklin Roosevelt's *Good Neighbour Policy*, but in the post-World War II period, the Caracas Declaration of 1954 would reiterate the Monroe Declaration of 1823 warning this time not the Europeans but the Communists to stay out of Latin America. This period was the age of the *caudillo,* the local strong men,[17] whose fragile power was based on personal charisma and they carried on the command from generation to generation to become the legally titled *haciendas* of the post-colonial period.

The 1880s saw a return to export-led economic growth, under an ideology of *laissez-faire* international trade. The plantations owned by the caudillos in the process became to be called as "rationalization" of agriculture in which the peasant-farmer is a wage labourer, with no property and no security guarantee whatsoever—the peasant may lose his small plot of subsistence land, so that he is forced to survive on the scant wages that can be earned during the three or four months of planting and harvest.[18] Efficiency and profit became the sole goals and export production, once more, brought the landed gentry into the national economy and politics, creating a modernizing elite.

The period until the Great Depression of 1930s was one of continued export-import growth that created something of a middle class and saw the development of trade unions but had little effect on the masses. Relative to the other continents of the South, Asia and Africa, Latin America is well off: its southern cone countries of Uruguay, Argentina, and Chile with their large European populations, are modern industrial nations with substantial middle classes; Brazil and Mexico are major industrial powers, though they are also the most debt-ridden countries of the South, each owning more than a hundred billion dollars. Although democracy became the norm, few countries are more than a decade or two away from repressive dictatorship, and in some countries, democracy is a little more than a facade on military or oligarchic rule. Thus despite a great deal of progress in the twentieth century, the conditions that can bring about revolution or dictatorship are still very much present.

A Continent of Contrasts : Throughout Latin America three centuries of European rule brought profound change. Vast areas of land had been brought into cultivation, major mineral resources were exploited, cities and ports had been built, and trade albeit closely controlled and narrowly based, had been developed. Concentrations of land, power and wealth that were established over five centuries ago continue today—out of 20 countries with only one or two exceptions, wealth is extremely maldistributed, and the process of modernization seem to be amplifying the problem rather than closing the gap between the rich and the poor. Because Latin America escaped the social and economic disruptions in both the World War I and II, no other region on earth has such a long continuity of elite rule. In those few countries such as Peru, that have enacted land redistribution programmes, the old money maintained its power by shifting from agriculture to industry, mining, or fishing. Neither the emergence of new elites nor the rise of populism has much affected the extreme concentration of wealth. Because of the Depression of 1930s when countries in Latin America went by the Import-Substituted-Industrialization (ISI) policy (the policy of emphasizing export and minimizing import), the local elites would seldom have much problem in accessing foreign luxury goods. In El Salvador—a country with a population of over 5 million—one can speak of the "fourteen families" (now actually about a hundred families closely interrelated by kinship and marriage, holding the wealth of the country). Thus most of the benefits accrued to the North, or to the settlers of European origin in the region; consequences for the local population were limited and often adverse: European lifestyles, language and religion were, with varying degrees of success, impose on the local populations. In many areas the demographic consequences of conquest were considerable. As a result of warfare, enslavement and disease the local populations of Central America, the Andes and Brazil were substantially reduced or, in the case of Caribbeans eradicated. Substitution of the indigenous labour force by black slaves also profoundly modified the demographic pattern. Slavery was most important in Brazil, the West Indies, and the Guianas, which received 7-8 million slaves during the period of three centuries from 1550 to 1850. In consequence, in much of the Caribbean and in the sugarcane plantations, goldfields and early coffee plantations of Brazil, blacks became numerically dominant in the population. In addition the *miscegenation* between whites and blacks, and with the surviving local population elsewhere, give rise to substantial proportion of people of mixed race which exists in much of the continent today.

Shortly afterwards political independence came to the mainland Latin American countries including Mexico from Spain and Brazil from Portuguese ruler much earlier by the first half of the nineteenth century.

European Expansion in the Nineteenth Century

Untill the mid-1750s, the industry in countries of Western Europe was little more than a by-product of overseas trade rather they were motivated by political rivalry, militarism, and a desire for markets and for sources of cheap raw materials, seeking control of territories in "tropics" of Asia and Africa. But the previous era of mercantile trade laid the foundation for later expansion and, by the end of eighteenth century, the relative power in military, scientific and financial powers were such that the establishment of a "core-periphery pattern at a global scale" could begin. In these new colonies Europeans were important in various administrative and commercial capacities but, with some exceptions such as Algeria, Angola, Mozambique, Rhodesia, and Russian Central Asia and Siberia, these were not colonies of settlement. The countries which had never been under formal European government were Iran, Afghanistan, parts of Arabia, Liberia, China and Thailand. Nevertheless European expansion affected them all, so that even when European empires came to an end, the Countries of South still bore the legacies of the preceding five centuries. It inherited political, economic and social structures that, even after the dissolution of formal empires, were still influenced by the economic forces of modern imperialism.

European colonization in the Americas reached its peak by the late eighteenth century. Shortly afterwards as the European empires in the Americas began to crumble by the Revolutionary Wars of the 1770s, marking the foundation of an Independent United States, provide a convenient termination for the first phase of the colonial expansion of the core European states. In this first phase of colonial adventure, the core European states had utilized their ill-gotten gains to develop a series of specialized industrial regions supplying their colonial markets as well as the New Continent. Food production, industrial manufacture, service provision, transportation were all radically transformed by the capitalist enterprise. Cultures, societies, governments were necessarily impelled to change, too. States became more centralized and powerful. The differences between rich and poor at the level of both individuals and between countries became more marked. For some individuals and groups of countries, capitalism brought prosperity, improved living conditions, but for the people in the colonies, it brought destruction of their individual cultures, impoverishment, degradation and slavery and all went to scar them more deeply in the twentieth century.

The second phase of colonization started, through the nineteenth century which was more rapid than the first. It was geographically distinct in that it was focussed upon Asia, and especially after 1880 on Africa and the Pacific, and within 120 years Western Europe had claimed most of the rest of the world. The period from 1800 to 1914 was the age of West European imperialism when the expansion of European empires was faster than at any previous time. In the early 1700s, the possessions of Europe included only 10 per cent of the land surface of the world and 2 per cent of the world's people. In 1800, Europe and its possessions (including ex-colonies) covered by land about 55 per cent, in 1878, 67 per cent, in 1914, 84 per cent and yet more by the outbreak of the World War II in 1939. The area under possession of the European powers and the total population in their colonies in 1939 were: under England 4.25 million sq. miles and 470 million people, under France 4.3 million sq.miles and 26 million, under Germany 900,000 sq. miles and 13 million people; continent-wise 56.5 per cent of Asia, 91.4 per cent of Africa, 27.2 per cent of Americas and 98.9 per cent of Polynesian islands (Hobson, 1948).[19]

By the end of nineteenth century the "scramble for Africa" had resulted in virtually all of the continent, except Liberia and Ethiopia being divided between England and France who as the dominant powers took the major share, and rather smaller sections went to Germany, Portugal, Belgium, Spain and Italy. Territories in the Pacific had been claimed by England, France and Germany, the USA and Spain. Russia occupied Central Asia as far as the Afghanistan, Tibet border and the British increased their control over the Indian sub-continent as far as the north-west frontier. The latter two powers agreed to preserve the neutrality of Afghanistan, Tibet and Persia (now, Iran) as non-colonized buffer states. Similarly, Siam (now Thailand) lost its empire and was subjected to unequal economic treaties, but remained independent as a buffer between British and France possessions in Indo-China. With this one exception, South East Asia was completely divided between the Netherlands, Britain, France and the USA by 1914. The Chinese empire survived expansion of the European empires as well as the expansionist designs of Russia, Japan and USA. While diplomatic deadlock preserved China's political independence, economic imperialism prevailed through informal control over the government. The colonial powers also secured extra-territorial rights in the Treaty Ports like in Hongkong.

Africa: In Africa, the precise type and impact of the colonial expansions varied; by 1875, less than one-tenth of Africa turned into European colonies, but by 1895 only one-tenth remained unappropriated so that by 1898, Africa remained a "patchwork of quilt" . After 1869, the opening of Suez Canal, European countries sought to expand their economic and political influence. In cases such as Kenya and Algeria, settler colonies were established. In Ivory Coast and Cameroons, plantation colonies controlled by metropolis export-import companies were created. Elsewhere, mineral resources provided the basis for exploitation in Rhodesia and the Congo. In some cases, essentially negative influences prevailed—in Uganda and Chad political control was taken to prevent the expansion of other political powers.

However, common to all was the establishment of a *dual economy*, with a modern, European, export-oriented sector alongside (and often undermining) the traditional self-sufficient economy. Internal conditions and local peoples were no longer the determinants of events. The African territories became part of an international economic and political system controlled from Europe.

In the Islamic states of north-west Africa the French, and later the Italians and Spanish, acquired land primarily by conquest. The French settlers engaged in the production of three Mediterranean crops for export: olive, wheat and vine. These occupied the best land and used machinery, though the pattern of exploitation varied. In the earliest colony, Algeria, the French encouraged large-scale immigration of small landowners, in Tunisia and Morocco the creation of very large commercial enterprises was more common. Throughout these territories though, modern, export-oriented agriculture stood in contrast to the traditional cultivation and extensive pastoralism of the native peoples.

In contrast, European activities in Egypt were primarily commercial. Financial indebtedness brought Anglo-French control over governmental and financial affairs, and then British occupation in 1882. Exports of cotton drew Egypt into the world economy, with Europe providing the industrial and financial skills and Egypt the raw materials. The British encouraged the development of irrigation and extension of the cultivated area. Cotton output increased substantially, but it continued to be an export crop, rather than the basis for a local textile industry, despite the fact that Egypt's cotton textiles constituted one-third of imports.

In West Africa impact was markedly different. It occurred initially through missionaries and explorers whose activities aroused considerable public interest in their home countries. Until the 1860s European activity was associated with suppressing the slave trade, and there was little expansion inland. However, a flourishing commerce begun to develop in vegetable oils and groundnuts. These products were grown and collected by Africans, with little intervention by Europeans, but the trade required extensive commercial contacts and led to rivalries between British, French and German trading companies, and prompted increased European control over the region. Gradually treaties made with local rulers became more binding until they led to the virtual loss of indigenous control. In both British and French West Africa economic policy subordinated African interests to those of the metropolis. Though there was less alienation of land than in other parts of Africa, and peasant farmers produced most of the groundnuts and palm oil, the colonial system forced them to concentrate on export crops, as well as on producing subsistence foods.

East Africa was drawn inevitably into the network of world trade after the opening of Suez Canal in 1869. Rivalry between the European powers ensued, with Britain eventually emerging as the dominant influence. Early interest was in the suppression of slave trade, but by the early twentieth century efforts were made to make the colonies profitable additions to the empire, mainly by agricultural development. New systems of land use were introduced in Kenya. Large-scale immigrants were encouraged and substantial areas of land were alienated. The pattern was not uniform, however, for in Uganda export cotton was primarily a peasant crop. In those areas where land was sold, leased or given to Europeans by the authorities, traditional African economic activities were disrupted. The pastoral and nomadic tribes were severely affected and some put up considerable resistance, which the British quelled, frequently by force. Labour shortages were overcome by a system of taxation which forced the African to work for wages, while at the same time preventing the accumulation of capital within the African reserves.

In southern Africa, the local economy and social systems were irreversibly altered with the penetration of Dutch settlers from the seventeenth century onwards. Traditional seasonal grazing lands were appropriated, and disease, particularly smallpox epidemics, decimated large numbers of population. The Africans had no option but to work for the Europeans or to retreat into less hospitable desert areas, for their cattle were seized or killed and a system of taxation forcibly imposed. The Rural Reserve Policy, generally applied in the second half of the nineteenth century (except in Rhodesia, now Zimbabwe) where it came much later, introduced a new element into the territorial relations between Europeans who settled and the local Africans. The best land went to the Europeans, with the native reserves generally on the more marginal areas, and suffering from overcrowding.

Discoveries of gold and diamonds after 1860 brought even more fundamental changes in the political economy of southern Africa. A system of migratory labour evolved and men left the reserves for varying periods of time to work in the mines. This gave the mining magnates full control over the workers. It also greatly influenced the character of urban areas which developed and the social and economic relations between Europeans and Africans, and between the European-dominated commercial farming areas, styled the "native reserves" in British colonies.

There were major contrasts between colonial regimes that developed in Africa and in the Indian sub-continent, and as well as within the latter. These differences reflected the varying conditions within which the colonial systems operated, as well as the diverse purposes the colonies were intended to serve.

The Indian Subcontinent : In India, by 1450, Islam extended through the Muslim Turks under Babur (who claimed a direct descent from the Mongol Chenghis Khan), invaded the Indian plains through Afghanistan using the new technology of musket and artillery. The Mughal empire would ultimately cover most of central India from the Arabian Sea to the Bay of Bengal. Under the Mughals, the imposition of Muslim rule on a predominantly Hindu country (less than a fifth of the people would be converted to Muslim) meant that majority of subjects felt little common cause with the state or allegiance to it. Again the caste system in Hinduism—of four main castes and hundreds of subcastes—assured that interests and loyalties of the subjects remained focussed on local groups rather than on nation.

In 1497, Vasco da Gama's discovery of a sea route to India opened the way for extensive trade. For a while, Portugal which was more interested in commerce than empire, monopolized trade with India. But after they were decisively defeated at sea by British, the latter consolidated their earlier "trading posts" through the East Indian Company (EIC): this Company, formed by the independent associations of merchants and chartered as monopolies (for trading in India) by the mother country, was a logical development of the wedding of "mercantilism and international capitalism". The colonization of India which was started by Britain by establishing their trading posts well entrenched in Bombay, Calcutta and Madras where they established permanent warehouses and port settlements known as "factories". The Company officials developed strong alliances with indigenous commercial elites, who formed a new and powerful capitalist class dependent on EIC trade. Together with the native elites, the British were able to play off their "Hindu card" against Muslims and Moghul against local "subedars". As the Moghul Empire in India collapsed finally in the 1700s and led to an anarchy, the "factories" became increasingly fortified and defended by native troops, called "sepoys". It was from these fortified settlements and with the use of the sepoys that the EIC gradually extended its rule over the heavily populated regions north along the Ganges from Calcutta and in southern India. Thus British Rule in India was accomplished piecemeal over long period and this got extended until, by 1849, the British controlled the whole country. The British ruled much of the rest of the country indirectly through treaties made with "princely states". The Indian subcontinent in the British Empire became an important source of agricultural exports, particularly of cotton, jute, hides, oilseed and tea. Development of irrigation works began, for example, in Punjab. In developing an export economy in India's large size of some 4 million sq. kilometres, a good transport system was essential and this prompted the construction of a major railway network. The British set out to develop a substantial railway network: the first line was opened in 1853, and by 1870 there were 6,400 km of railroad and up to 1914 construction averaged some 1,200 km a year, the nodes of the rail networking system became Bombay, Calcutta, Madras and Lahore and it evolved to integrate the principal ports and cities, and the most productive areas, particularly the Ganges and Indus plains. The Indian railways remain a significant component in the contemporary transport network. Conversely, easier access to the Indian market for manufactured cotton goods destroyed the existing artisan cotton textile industry.

English became the medium of official communication and of education, and India eventually adopted British political institutions. Acceptance of alien rule was fostered principally by making very few changes in the basic structure of Indian life. Moreover, members of the established social groups, the aristocracy, the merchants, bankers and educated professionals, were absorbed into the expanding colonial civil service, which gave them the illusion of power : In 1900 the British, with less than one per cent of the civil service, ruled over half a million Indians. The advantages of British rule probably lay most in its imposition of law and order over a vast, hitherto divided sub-continent, the provision of centralized government; and the construction of a comprehensive system of roads, railways and canals which could facilitate subsequent large-scale industrial and agricultural growth. India—"the jewel in the crown"—became the base for colonization of Burma (now Myanmar) and many of the South East countries.

Sri Lanka's colonial experience (as Ceylon) was dominated by the development of a plantation economy. While rubber plantations were able to attract peasant labour as the work was intermittent, local people refused to work on the coffee and tea estates. Tea production was largely carried out by Tamil immigrants imported from southern India. The system of compulsory labour for maintenance of existing irrigation system was effectively destroyed by the British administration, so that self-sufficiency in food production was lost, and rice, the staple food, had to be imported, especially in the dry zone.

The expatriate plantation sector grew rapidly and created great wealth but little of this wealth ever reached the Ceylonese, though there were some exceptions. An upper class of indigenous landowners had existed long before the arrival of the British, making their profits by renting out rather than cultivating land. The rise of the coffee industry gave them an impetus for engaging in production, and later many began to produce tea, coconuts and rubber. Because the owners of such estates were local, the profits remained in Ceylon.

East Asia: After 1858, Japan which was formerly an isolated country, set out to copy Western technology. They proved their success, beyond anyone's wildest dreams—with their decisive defeat of the Russians in the sea battle of Tsushima in 1905, Japan would take advantage of the European powers' preoccupation with fighting each other in World War I to usurp the trade with China and India. In its attempts to build a Great East Asian Co-Prosperity Sphere, Japan overran Manchuria, invaded China, and occupied much of the South East Asia. These would not be displaced until the defeat of Japan and subsequently the end of World War II.

China being geographically isolated from the rest of Asia, escaped occupations by any European power (like Japan). Unlike India, the Chinese monarch resisted further dealings with Europeans. Thus for the European countries eager to trade with China for its tea, silk, porcelain and cotton goods, the Europeans were allowed to trade only at Canton.

Since the Chinese traded only in exports (and not in import), so exports from China had to be paid with precious gold and silver. Forced in this circumstance, The British solved the problem by growing opium in Bengal and importing the same to China—this trade in opium systematically built up a huge trade based on increased addiction to this powerful drug. When the Manchu emperor made an attempt to whip out this odious trade by destroying 20,000 chests of British opium, the British responded by seizing Chinese ports. The two-year opium war ended in 1842 with the Treaty of Nanking, which gave Britain the Hong Kong port (taken under a lease of 99 years) and that also opened five Chinese ports to foreign trade. Thus though there was the difficulty of any attempt to colonize or directly rule China, but as a result of the two-year war, an "open door policy" was enacted throwing China's borders wide to the full impact of Western capitalism. The Chinese local crafts like those in case of India got destroyed by cheap manufactured imports, and standards of living dropped. By the turn of the twentieth century, Western influence and internal weakness had brought down in 1911 the 260-year old Manchu Dynasty which was at its peak once, and the country, though formally became a republic but in reality, collapsed into virtual monarchy, with local warlords pillaging the countryside. The Koumintang (or, National) Party was formed to unite the country and they had to turn against their rivals, the communists for power. The communists led by Mao Zedong organized the peasants and took over mainland China as Peoples' Republic of China in 1949. The Koumintangs escaped to the island of Formosa where they established a one party regime to form Taiwan.

Although the "dominion theory" of communist expansion proved false, but the end of the Cold War found three of the few remaining communist governments situated in Asia: they are China, Vietnam and North Korea. Both China and Vietnam, two of the poorest countries of the world as measured by per capita income, began making overtures for increased economic relations with the West and showed tendencies to introduce *limited capitalism.*[20] North Korea remained as isolated and hostile holdout of Stalinism.

South East Asia: The countries of South East Asia were once basically self-sufficient, but were changed into suppliers of primary commodities for Western markets by the impact of colonialism. They were thus brought into almost total dependence on the fluctuations of a world market over which they had no control. Where a labour supply was not forthcoming, immigrants were brought in or came spontaneously from Indian provinces to occupy positions as middlemen in colonial commercial activities. Meanwhile, the majority of the region's indigenous inhabitants became increasingly marginalized within their own society, and dependent upon the new monetary economy into which they were forced.

In Indonesia Dutch control under the "culture system" established a planned economy on a vast scale and, especially in Java, which resulted in a large-scale production of coffee, sugar, tea and other foodstuffs for the Dutch market. This system led to the diversification of crops, the introduction of scientific methods of production, including modern irrigation works and a great improvement and extension of the communication networks. The culture system involved gross exploitation of the Javanese cultivators, and forced labour when voluntary workers could not be recruited. Less intensive colonial contact in the outer islands of the archipelago meant that people were not so involved in the production of export crops and remained relatively unaffected by the rising demands and expectations of a monetary economy.

The experience of British colonies in the South East Asia differed only in detail from that of Dutch Indonesia. The wealth of Burma (now Myanmar), which had developed through the production of rice, tea, lead, rubber, copper and other export goods, was largely in alien hands. *Laissez-faire* colonial policies permitted free trade, British control of resources, and the immigration of Indians who acted as commercial middlemen and money-lenders. However, for the peasantry whose self-contained subsistence economy had focussed upon the social life of the village, circumstances deteriorated and by 1940 at least 50 per cent of them became landless agricultural tenants and labourers. British rule had liberated the Burmese from some of the restrictions of traditional society, and there had been some improvements in health conditions. Balanced against such benefits was the fact that the Burmese had a disproportionately small share in the prosperity of their country and, for the majority, life had become marginalized as they strove to survive within the framework, though on the fringe, of the new international monetary economy.

Malaya's transformation into a leading producer of tin and rubber was associated with large-scale Chinese and Indian immigration, so that in its peninsular state a multi-racial society emerged where the immigrants outnumbered the indigenes. The port of Singapore became the centre of British financial and commercial interests in the region, and virtually a city of Chinese traders. While development on the peninsula was rapid in the regions of Chinese-dominated tin mining, the rest of the country was little changed and the social imbalance between the immigrants and indigenous Malaya became very clear. The development of the west coast states was further increased by the British construction of roads and railways linking the mining areas, plantations and ports.

In contrast to the situation in most of South East Asia, in Malaya there existed the preconditions for a balanced agrarian policy. There was an abundance of land relative to population size and a highly prosperous tin industry that financed public works and other developments without resort to a rigid taxation system. Furthermore, the high level of Chinese and Indian immigration obviated the need for coercing Malays into working on the plantations. Smallholders were not seen in competition with the estates, so that in contrast to other colonial situations the British authorities in Malaya also encouraged the small-scale farming.

In Sarawak, Brunei and Borneo, European enterprise was minimal. The countries were poor in resources and, in the case of Sarawak, the paternalistic administration deliberately discouraged modern commercial exploitation. The result was a far slower rate of economic growth or development of an infrastructure than was the case elsewhere in the region. In 1940 Sarawak was little changed from what it had been a century before, with much of the land still covered in virgin forest and the greater part of the population engaged in subsistence farming. Of its known resources only oil was of any real significance for the outside world. The advantage was that large number of people had not become marginal within their own society and economy.

Contrasts in colonial activity also existed among the French South East Asian territories of Vietnam, Cambodia and Laos. However, the differences were more of degree than of kind. The pre-existing differences in the development of the northern and southern parts of Vietnam were further exaggerated by the colonial regime. The densely populated north was largely left to its traditional self-sufficient subsistence economy based on rice production. The south bore all the imprints of classic colonial situation: large-scale commercial enterprises and the creation of a distinct dual economy. Development, which was concentrated in the more sparsely populated southern region, was associated with the appalling poverty of the peasants, their continual indebtedness to money-lenders and other middlemen and an ever growing population. Immigrants to Vietnam were mainly Chinese, who had been going there long before the beginning of colonial commercial activities, but who went in increased numbers during the colonial period. They played their typical role of middlemen operating between the Europeans and the Vietnamese peasants.

Cambodia and Laos, like the neglected eastern territories of Malaya, were perceived to be countries of low economic opportunity and were, therefore, omitted from the mainstream of colonial activities. There remained areas where a simple subsistence pattern of life prevailed. Cambodia contributed small amounts of rice, rubber, maize and pepper to international market, and Laos produced some teak, tin and coffee. As elsewhere in the colonial world, even the small amount of commercial activity that developed was monopolized by outsiders —in this case, French, Vietnamese and Chinese. The simple, though highly refined, cultural life of the Khmer people of Cambodia and likewise the traditional world of the Laotians were, as a consequence, only minimally affected by the direct influence of colonialism. However, they were not to escape the many indirect implications of being poor, unsophisticated societies in an increasingly prosperous and technological world.

The Philippines shared many of the features of colonialism in the region. Typically, there was a dual economy: foreign-owned plantations produced a few primary goods for the world market, while the small-scale subsistence sector was characterized by poverty and indebtedness to a group of middlemen comprised mainly of Chinese immigrants. On the other hand, Philippine society differed from any other in South East Asia in that at the end of the nineteenth century its history of colonization by Europeans—the Spanish—had already lasted for three centuries. By the time the island societies came under US colonial rule in 1898 they were in many respects westernized in terms of cultures and attitudes. This factor was of major importance to the pattern of US penetration in the twentieth century. The traditional Asian self-contained subsistence economy had already been influenced by one oriented towards the world markets. A basic infrastructure existed and the range of commercial agencies and credit facilities characteristic of a monetary economy were an accepted part of Philippine life. But the major economic change under US rule was the spectacular rise of the sugar industry which turned the economy into a virtual monoculture.

The Polynesian Islands : Traders, whalers and missionaries accounted for most of the Europeans in the Pacific in the nineteenth century. They were mainly British and some French nationals. By the end of century, the US also had significant commercial interests in the region. The implications of colonization for the Pacific islands were broadly similar to those of Asian countries affected by the spread of plantation economy. The earlier small-scale trade in sandalwood and seaslugs for the Chinese market was replaced by a much larger export trade in coconut oil and guano. Both of these products were controlled by small settlements of Europeans in the islands. Demand for plantation labour led to many social and political disturbances, and, as a consequence, by the 1860s the political systems in Fiji, Samoa and other island groups were beginning to disintegrate.

European countries were not the first to create empires, nor were they the only imperial powers between the sixteenth and the twentieth centuries. Hindus of India colonized South East Asia, the Muslim influence based in Arabia, extended from southern Spain and North Africa to India by the thirteenth century. The Turkish (Ottoman) empire, stretching from the Mediterranean to the Indian Ocean, survived until the World War I.

Not until the nineteenth century did European powers come to affect the entire globe but unlike their predecessors, European expansion created a "world economic order" from which virtually no country could exclude itself. A new global system evolved with Europe emerged as the centre of the military power, wealth and technological invention, and, with few exceptions, the rest of the world constituting a relatively weaker and less advanced periphery.

The exceptions included the former European colonies in North America and Australia, where large-scale European settlement produced a different form of colonization. They became self-governing European societies overseas and were effectively extensions of Europe. Subsequently they evolved with their own identities, though remaining still part of the global centre.

A further group of colonized territories that do not form part of the South are the Asian territories of Tsarist Russia, which were incorporated into the USSR.

Japan, which was also never colonized, became an expansionist power in the late nineteenth century, and it has achieved a technological pre-eminence which currently rival the industrialized countries in the West.

The historical processes associated with the changes that led to the emergence of the gap between the Countries of South and of the

North are the subject of diverse interpretation. The question that many have raised is: Did European expansion and influence caused the economies and societies of the Southern Countries to retain traditional and outdated methods of production and social institutions? If so, then it would be pertinent to ask whether that was due to European exploitation of overseas territories, or whether traditional methods of Southern Countries' production were so deeply entrenched that they were unable to respond promptly and fully to the modernizing influence of Western development. An alternative view questions whether expansion of the world capitalist economy drew the countries which lay outside the central core of global developments into an international system of capitalism, within which they were unable to cope.

The impact of European expansion was not uniform throughout the world. The pattern of expansion varied, so did the motivation for it. And the countries affected were themselves varied in character, historical background and exploitable resources. However, one thing is clear: the gap in wealth and technology which currently exists between Europe and European settler colonies on the one hand, and the Countries of South on the other, is relatively new.

The Impact of Colonization

Colonial status was thus primarily a political phenomenon but the economic aspects of European dominance in the world, which colonialism heightened, did not necessarily depend on formal relationships. In this context, distinctions must be made between three overlapping concepts: *colonization*, *colonialism*, and *imperialism*. "Colonization" is the creation of permanent communities in foreign trade, for example, "frontier economies of white-settled colonies of North America, Australia and South Africa".

"Colonialism" is the establishment of full state sovereignty over another country or region by a foreign country. Apart from the search for raw materials and markets, the colonialism got stimulated by multiple motives, like for religious and cultural ideologies of bringing Christianity or civilization to backward peoples, the need for strategic control of trade routes or critical waterways, and the desire to gain national prestige. Often the process of expansion began with trade and ended with full colonialism. On the other hand, much colonialism in Central Africa and parts of Asia were administered through native officials under European control. Economically, therefore, all the world outside of the settler colonies of North America and Australasia came under the dominion of Europe and, later under the influence of these former white-settler colonies as well.

England led the way in colonialism; it was said that "the sun never sets on the British Empire" that came to include multiple African colonies, India, Hong Kong and Australia. By the late 1800s, the French empire encompassed Indo-China, Algeria, Tunisia, and much of the Sub-Saharan Africa. Other countries in the competition for empire were the Portuguese, Dutch, Spanish, Belgians, Germans, Russians, Japanese and Americans.

Imperialism is the act of extending a state's political domination over another territory, either directly or indirectly, in order to establish military bases, to protect trade, or for other interests. Imperialism, without colonization or colonialism, became the policy of the United States in the early twentieth century, for example, in Central America and the Islands of Caribbean where US troops invaded and occupied at will, without establishing permanent settlements or creating colonial governments.

Imperialism is not a stage, not even the highest stage, of capitalism: from the beginning, it is inherent in capitalism's expansion. The imperialist conquest of the planet by the Europeans and their descendant colonists in North America was carried out in two phases and is perhaps entering a third. The first phase of this devastating enterprise was during the expansion of mercantile capital, organized around the conquest of the Americas, in the framework of the mercantilist system of Atlantic Europe at the time. The net result was simply the total genocide on which the United States was built. The fundamental racism of the Anglo-Saxon colonists explains why this model was reproduced elsewhere, in Australasia by their Biblical right to wipe out the "infidels", this was quickly followed by the infamous slavery of the Blacks, so that the colonists could briskly take over to ensure that the useful parts of the continent were "turned to account."

The second phase of imperialist devastation was based on the industrial revolution and manifested itself in the colonial subjugation of Asia and Africa. "To open the markets"—like the market for opium forced on the Chinese by the Puritans of England—and to seize the natural resources of the globe were the real motives in this phase which established the new legitimizing discourse of capital. This second phase of imperialism is at the origin of the greatest problem with which mankind has ever been confronted: the overwhelming polarization that has increased the inequality between peoples from a maximum ratio of two to one around 1800, to sixty to one today in per capita income with only 20 per cent of the earth's population being included in the centres that benefit from the system. At the same time, these prodigious achievements of capitalist civilization gave rise to the most violent confrontations between the imperialist powers that the world has ever seen. Imperialist aggression again produced the forces that resisted its project: the socialist revolutions that took place in Russia and China (not accidentally all occurred within the peripheries that were victims of the polarizing expansion of really existing capitalism) and the revolutions of national liberation. Their victory brought about a half-century of respite, the period after the Second World War, which nourished the illusion that capitalism, compelled to adjust to the new situation, had at last managed to become civilized.

The question of imperialism (and behind it the question of its opposite—liberation and development) has continued to weigh on the history of capitalism up to the present. Thus the victory of the liberation movements that just after the Second World War won the political independence of the Asian and African nations not only put an end to the system of colonialism but also, in a way, brought to a close the era of European expansion that had opened in 1492. For four and a half centuries, from 1500 to 1950, that expansion had been the form taken by the development of historical capitalism, to the point where these two aspects of the same reality had become inseparable. To be sure, the "world system of 1492" had already been breached at the end of the eighteenth century and the beginning of the nineteenth century by the independence of the Americas. But the breach was only apparent, because the independence in question had been won not by the indigenous peoples and the slaves imported by the colonists (except in Haiti) but by the colonists themselves, who thereby transformed America into a second Europe. The independence reconquered by the peoples of Asia and Africa took on a different meaning.

The ruling classes of the colonialist countries of Europe did not fail to understand that a new page of history had been turned. They realized that they had to give up the traditional view that the growth of their domestic capitalist economy was tied to the success of their imperial expansion. For that view was held not only by the old colonial powers—primarily England, France, and Holland—but also by the new capitalist centres formed in the nineteenth century—Germany, the United States and Japan. Accordingly, the intra-European and international conflicts were primarily struggles over the colonies in the imperialist system of 1492. It being understood that the United States reserved to itself exclusive rights to the whole new continent.

The construction of a great European space—developed, rich, having a first-class technological and scientific potential, and strong military traditions—seemed to constitute a solid alternative on which to found a new resurgence of capitalist accumulation, without "colonies"—that is, on the basis of a new type of globalization, different from that of the system of 1492. The question remained how this new world system could differ from the old, if it would still be as polarizing as the old one, even if on a new basis, or if it would cease to be so.

No doubt this construction, which is not only far from finished but is going through a crisis that could call into question its long-term significance, will remain a difficult task. No formulas have yet been found that would make it possible to reconcile the historical realities of each nation, which weigh so heavily, with the formation of a politically united Europe. In addition, the vision of how this European economic and political space would fit into the new global system, which is also not yet constructed, has so far remained ambiguous, not to say foggy. Is this economic space to be the rival of the other great space, the one created in the second Europe by the United States? If so, how will this rivalry affect the relations of Europe and the United States with the rest of the world? Will the rivals confront each other like the imperialist powers of the earlier period? Or will they act in concert? In that case, will the Europeans choose to participate by proxy in this new version of the imperialist system of 1492, keeping their political choices in conformity with those of Washington? On what conditions could the construction of Europe become part of a globalization that would put a definitive end to the system of 1492?

Today we see the beginnings of a *third wave of devastation* of the world by imperialist expansion, encouraged by the collapse of the Soviet system and of the regimes of populist nationalism in Countries of the South. The objectives of dominant capital are still the same—the control of the expansion of markets, the looting of the earth's natural resources, the super exploitation of the labour reserves in the periphery—although they are being pursued in conditions that are new and in some respects very different from those that characterized the preceding phase of imperialism. The ideological discourse designed to secure the assent of the peoples of the central Triad, the OECD countries (the United States, Western Europe, and Japan) has been refurbished and is now founded on a "duty to intervene" that is supposedly justified by the defence of "democracy," the "rights of peoples," and "humanitarianism." The examples of the double standard are so flagrant that it seems obvious to the Asians and Africans how cynically this language is used. Western opinion, however, has responded to it with as much enthusiasm as it did to the justifications of earlier phases of imperialism.

Furthermore, to this end the United States is carrying out a systematic strategy designed to ensure its absolute hegemony by a show of military might that will consolidate behind it all the other partners in the Triad. From this point of view, the war in Kosovo fulfilled a crucial function, witness the total capitulation of the European states, which supported the American position on the "new strategic concept" adopted by North Atlantic Treaty Organisation, (NATO) immediately after the "victory" in Yugoslavia on April 23-25, 1999. In this "new concept" (referred to more bluntly on the other side of the Atlantic as the "Clinton Doctrine"), NATO's mission is, for practical purposes, extended to all of Asia and Africa (the United States, ever since the Monroe Doctrine, reserving the sole right to intervene in the Americas), an admission that NATO is not a defensive alliance but an offensive weapon of the United States. At the same time, this mission is redefined in terms as vague as one could wish that include new "threats" (international crime, "terrorism," the "dangerous" arming of countries outside NATO, etc.), which plainly makes it possible to justify almost any aggression useful to the United States. In addition, NATO is freed from the obligation of acting only on mandate from the United Nations (UN), which is treated with a contempt equal to that which the fascist powers showed for the League of Nations (there is a striking similarity in the terms used).

American ideology is careful to package its merchandise, the imperialist project, in the ineffable language of the "historic mission of the United States." A tradition handed down from the beginning by the "founding fathers," sure of their divine inspiration. American liberals—in the political sense of the term, who consider themselves as the "left" in their society—share this ideology. Accordingly, they present American hegemony as necessarily "benign," the source of progress in moral scruples and in democratic practice, which will necessarily be to the advantage of those who, in their eyes, are not victims of this project but beneficiaries. American hegemony, universal peace, democracy, and material progress are joined together as inseparable terms. Reality, of course, is located elsewhere.

People in the West are persuaded that because the United States and the countries of the European Union are "democratic," their governments are incapable of "ill will," which is reserved for the bloody "dictators" of the East. They are so blinded by this conviction that they forget the decisive influence of the interests of dominant capital. Thus once again people in the imperialist countries give themselves a clear conscience (Samir, 2001).[21]

The process of colonization within any country or group of countries was conditioned by the pre-existing historical circumstances and the attractiveness of its resources to the colonizers. In turn, these affected the degree of exploitation and the nature of the developments that took place and consequently the level of marginalization that occurred within the economy and society. The communities least attractive to European settlement, such as Laos, Borneo, Brunei and Sarawak in South East Asia, remoter islands of the Pacific and the more isolated areas of Central Africa, were among the least exploited. Yet, their greater retention of social traditions and economic practices did not enhance their possibilities for future development. On the contrary, they have remained among the poorest of the Countries of South. In so far as they gained from their isolation, it has been in that Western material expectations have been minimally aroused. On the other hand, Malaysia, well-endowed with resources for which there is international demand, or the Philippines with its cultural sophistication, are the examples of countries which though exploited by the colonial system, have nevertheless, emerged with the facility to compete favourably by comparison with other Southern Countries.

The net balance in the advantages *versus* the disadvantages of colonial penetration was not, in the last analysis, a measure of the total influcncc that European expansion had upon the countries on the periphery. It was being on the periphery of global events and, concomitantly, the internal marginalization of large sectors of the population with which this was associated, that made the vital difference to future developments.

It is not surprising that the preservation of political sovereignty in China, Afghanistan, Thailand and Iran was insufficient to isolate these countries from the effects of an evolving global power balance, which they were in no position to counter. They were all the victims of unequal trade treaties with European countries, and undoubtedly part of the colonial sphere of influence in the Middle and Far East. The *Open Door Policy* imposed upon China by Britain, France, the USA and Japan virtually made China an economic satellite of the imperial powers and the Communist revolution of 1949.

Other countries which maintained political autonomy also bore significant hallmarks of colonialism. Thailand's economy, for example, depended on the export of four primary products—rice, teak, rubber and tin—of which only rice was controlled by Thais. In return, manufactured goods were all imported. A typical colonial economy developed, with a foreign-owned export sector and a traditional agricultural sector, and even rice was transformed from a subsistence crop to a commercial one. As in the former colonies, peasants were forced into a marginal way of life in relation to the country's mainstream of production. While Thailand gained none of the apparent advantage of a former colony, such as European medical and educational services or a communication system, its slower piece of change might be viewed as a net advantage. Thailand escaped many of the socially disruptive effects of colonialism that were experienced by neighbouring Vietnam and Myanmar.

One of the crucial results of colonialism was the creation of new countries. Few of the colonies at the advent of colonial expansion were nation-states. Europeans needed clearly defined borders, even if those borders made little cultural sense. One of unanticipated effects was to create the basis for nationalism that would ultimately become a primary factor in the demise of the colonial system after World War

II. Throughout the colonial period, ethnic uprisings and wars of liberation erupted throughout Africa and Asia.

The invasive process of European capitalist system left certain legacies on patterns of economic, social, political and cultural life which continue to shape the thought and action of the present—the newly independent countries which include all countries of Africa and Asia except Japan—acted as "an adjunct to the imperial countries", supplying them with raw materials, food crops for their industrial labour and a surplus for investment in the "frontier economies of white-settled colonies of north America, Australia and South Africa. Immediately after their independence when the bonds of "Global Colonialism" was over, these ex-colonial countries turned their attention from problems of political emancipation to domestic problems of national unification, but class- and religion-based social struggles were seldom easily resolved. Moreover, after political independence, they did not become rich and industrialized to achieve the elusive goals of economic independence and economic development, so to the questions of economic regeneration they remain rather "in the periphery of the system produced by the expansion of capitalism". Inheriting neglected administrations (from the colonial days), orientated mainly to the extraction of "staple" crops for export and exportable natural resources to feed the consumer culture of Europe and North America, and faced with the smug intransigence of the ex-imperial governments, the colonies of the yesteryears could not enter the "exclusive club of industrial nations" but became known as the Underdeveloped and the Developing countries of the South. The economic, political, and cultural domination of these new sovereign states of the South by developed countries of the North continued as *neo-colonialism.* This new system would be maintained *less* by the deliberate efforts of conquering governments (through military power would, as always, remain a crucial factor) than by the inherent logic of a world marketplace in which the rich could control the terms of trade for goods and services.

However, in the decades since independence, many countries of the South have found their own voice. The first leaders of the new states, who emerged from elites created by colonizers, have died off or been pushed out of power. Several formerly impoverished countries have established themselves as strong participants, rather than passive pawns, in international affairs. Far from being totally subservient to the dynamics of international trade, most countries have developed unique internal structure that must be taken into account. Thus, it is questionable whether neo-colonialism or better the American imperialism—in its simplistic form as "an extension of colonialism" —is as useful an explanation of the Southern Countries' poverty as it once was. Nevertheless, the system in which one quarter of the world's population utilizes over 80 per cent of the world's resources—a system with its origin in the fifteenth century—has not only persisted but also intensified: in terms of relative poverty, it is estimated that in 1750 there was virtually no economic discrepancy between Europe and the rest of the world, however, by 1977 the Countries of North were, on the average, between eight and fourteen times richer than the Countries of the South.

Whatever the blessings or curses of European expansion, it cannot be doubted that indigenous systems were hugely and permanently changed. Prior to European expansion, virtually all regions were "developing" in the sense that they were moving in the direction of more complex and articulated political and economic systems, larger towns or cities, and greater social differentiation. In the Americas, in Africa, in the Middle East, and in Asia, we see the rise, maturity, climax and fall of civilization after civilization, each new one building on the edifice of those gone before. Development, in this sense, was never something that belonged to the West.

Such development was brought to the abrupt halt by European imperialism. Natural processes of economic and political evolution were abruptly shifted in entire new and unforeseen directions, which were usually out of the sync with the cultural traditions of people. Development and progress would no longer be defined by the values of tribal and peasant peoples or by the native states but would be determined by the local overlords of far-away powers. The countries of the South can no more repudiate their colonial past than they can abjure their role in the present world system. What they can do, and many are trying to do, is to find a place in that system that fits their own cultures and aspirations.

Changes and Continuities as a State of Being : The Global Countries of the North and the South in the 1950s

Since the mid-fifties of the twentieth century the affluent countries of the capitalist core, which were already having higher levels of income and prosperity for their people, were having a passionate preoccupation with growth: the post-war industrial boom had been almost without precedent, the US economy performed exceptionally well from the late 1940s right through to the early 1970s. Real disposable income per capita in USA rose from just over US$ 2,200 (in constant 1972 terms) in 1947 to US$ 3,800 in 1972 (OECD, 1983).[22] The 1960s were particularly prosperous, with economic growth averaging over 4 per cent per year, thus expanding the GNP by 50 per cent over the decade. Meanwhile, the average family obtained a real increase of 30 per cent in its disposable income and this has led to the propagation of a new gospel of growthmanship which Galbraith marked as the years of "Affluent Society" (Galbraith, 1977).[23] The concept of more growth, higher incomes and all through higher production has evolved a fair coherent entity in the core countries of the capitalist world, including United States, Canada, the countries of Western Europe, Japan, Australia and New Zealand. These countries, besides capitalism, all are democracies, practicing welfare nation-states; they have already built up what is known as *infrastructure* in form of an adequate system of transport, communication and power. Furthermore, they are all having a well-knit and fairly well-integrated society—if the national income goes up, its benefits are transmitted to all the sections of society and to all parts of the country in a manner and in a measure which is accepted as equitable. And so they can judge their performance from year to year by increasing the earning power of their nation, that is, by the *actual rate of growth achieved in their economy and attaining development.* And hence to the extent that the mercantalists developed a distinctive common doctrine and it was the doctrine that the prime movers in national economic development were the industrial, commercial and the maritime branches of economy whose advantage were laid in their ability to increase the national share of international trade, by following Ricardo's "Theory of Comparative Advantage" among nations.

In contrast, the process of development which emerged in underdeveloped economies of ex-colonial countries, got characterized by a "dichotomous" or "dualistic" nature where the capitalist sector exists with the traditional sector: in the latter, surplus labour existed through a number of sources in these countries, including farmers, petty traders and retailers so that the labour is perfectly elastic in the subsistence sector. Thus labour in this dualistic economy was employed at a swage and this surplus at subsistence wage which became available to the capitalist sector, remains unlimited. The income distribution became skewed and most of the gains in economy were going to the traditional landlord class that did little saving but this capitalist class grew at the expense of the large subsistence sector. This concept of dualism also became apparent in some spatial contexts where certain regions moved ahead of others by virtue of some initial advantages.

While this spatial manifestation of unbalanced growth is argued inimical to the growth of the economy as a whole, Gunnar Myrdal and Albert Hirschman advocated strategies of "unbalanced or inequality growth" essential for the economy to lift itself to higher income levels so that the less dynamic areas may benefit from "centrifugal spread and backwash effects". This "Regional Income Inequality Model" of Myrdal (1957) [24] is based on the contention that in a market economy changes in the location of economic activities produce cumulative advantages for one region rather than a straightforward equalization of growth across all regions. In the short-term, therefore, growth is concentrated in one region rather than spread across all regions. In the long term "spread effects" can counter the "backwash effects" of regional concentration as demand in backward regions is met locally rather than supplied from the advantaged region. The main point of this model, however, is that a process of *cumulative causation*—the build-up of advantages due to external economies producing further advantage—lead in the short-term to geographical concentration of growth and higher regional income.

Myrdal's model was followed by others based on the same logic. Hirschman's (1958)[25] model assumes "polarization" and the "trickling-down effects". But this alternative strategy of experiment with "growth poles" manifestly failed: the uneven development is claimed in part to be a product of orthodox capitalist development process which unproportionately privileges the urban areas and industrial sector and leave behind in the process a host of areas, sectors and social groups of the periphery whose resourcefulness became "sequestered" by the core.

At the very start of this work, we noted that the idea of development in the sense of *good changes* and continuities lead to different meanings of development as a "state-of-being" and as a "historical process". If these two aspects are taken together, the changes and their continuities leading to development, do not draw towards closer conformity with one another. Indeed, many argue that the very historical processes which led to a state of development of capitalist economy in the Countries of Global North, were also the cause of underdevelopment in the Countries of Global South: the former constantly trying to achieve the commoditization of land, labour and economy not in terms of *immanent capitalism* but through *imperial capitalism* and the latter through attempting to "protect" these elements. The European expansionism of the last 500 years has been a complex process, with different phases, varying characteristics, and consequences that have been diverse and complicated in the Southern Countries affected by the commercial, financial, industrial, and political forces spurred outward from Europe throughout to what become a GLOBAL ECONOMY. Thus when for analytical purposes we identify certain broad patterns and effects, we inevitably oversimplify matters. Still, it is nevertheless true that trade pressures have been a constant of the European expansionism—there have been specific targets of such trade, and how these targets were reached would shape the social and political institutions of the Southern Countries.

In contrast to the capitalist economies practiced by the modern industrial countries of the North, the Countries of Global South are having a lower size of their economy, reflected in GNP per capita. In sharp difference to the productive capacities of the industrial societies using their labour (available to their society) in a more productive and efficient ways to produce a greater quantity and a more diverse range of goods and services, the Southern Countries which were characterized by their *immanent* (*i.e.*, own internal) systems of economy with the production in fixed supply and in a limited range of secondary and tertiary sectors so that people were largely engaged in agriculture, exports were mainly from the primary sector and their per capita *income poverty* translated into shorter life expectations, higher rates of infant mortality and lower levels of educational attainment. And presently with no exogenously given technological innovation, many of the Southern Countries face problems of income distribution and nutrition, barriers to international trade as well as the surrounding international environment which put structural barriers for application. And they fairly lack the qualities of industrial technology, investment capital, high levels of education, and the entrepreneurial ethic which had evolved in richer nations of the North. This makes the early *modernization analysts* tended to view the underdevelopment of the Southern Countries as a primary condition, an initial stage through which all countries must pass on their journeys towards modernization.

A second look at the Southern Countries, however, reveals that the characteristics that define the Countries of the South as underdeveloped, are in no way *primal* or natural. Poverty certainly has a long history, but the destitution of urban shanty town dwellers and of peasants pushed off their lands by the expansion of *export agriculture* is very different from the feudal poverty of the past. "Neo-colonial economic dependency", extremely rapid population growth, bloated militaries supplied by a vast international arms market of the North (the USA in particular), and governments striving to provide attractive investment climate to huge Multi-national Enterprises (MNEs) —none of these suggests some innate condition that will soon be outgrown. In many ways, Brandt's (1980) term of "North-South" which designated a relationship, specifically, a relationship between the countries of the North with those of the South did not progress far. While in 1750 there was virtually no discrepancy between Europe and the rest of the world; but by 1977, however, in terms of *relative poverty, the Countries of the South had deteriorated steadily* when the developed Countries of the North were, on the average between 8 to 14 times richer per capita than the Countries of the South.

Notes and References

1. It has been the opinion of many that "the major consequences of World War II will be recognized not in terms of alignment of political power .... but in terms of the birth of a large number of political entities in the so-called less developed world" (Fet and Ranis, 1964, *Development of the Labour Surplus Economy*, London).
2. The only country which was freed before the World War II in Asia and Africa was the present day Turkey which came out of the crumbles of the break down of Ottoman Empire in Middle East during the World War I by the bold venture of Mustafa Kamal, who took on the name of *Ataturk* (Father

of Turkey). The creation of the Republic of Turkey in 1921 under the leadership of Kamal Ataturk inspired the nationalist thinkers in the colonies of South Asia and Africa.

3. *See the Classification of Countries in the major world aggregates*, in, Appendix A, UNDP, 1998, *Human Development Report, 1998,* Oxford University Press, New York.
4. Greece and Portugal are usually considered in the First World by dint of their geographic coverage under western Europe, but having their incomes per capita are the lowest in the countries of Global North. Israel and South Africa are two other borderline cases, where the occupant population are in minority; they seem belonging to the North industrially but doubtful from (human) development point of view, if the total habitat is considered.
5. Brandt Commission, 1983, *Our Common Crisis* : *North-South Cooperation for World Recovery,* Pion, London. This Brandt Commission sponsored by United Nations, *depicted a world divided between Countries of North and the Countries of South rather than First, Second and Third Worlds.*
6. Brandt, Willy (1980), *North-South : A Programme for Survival,* Cambridge, Mass. MIT Press. Though dated, this report of an Independent Commission of a group of international statesmen and leaders on International Development Issues, distinguished between the "rich", the "developed" "North" and the "poor", "developing" "South". This report not only provides a good overview of the problems of the South but also offers long-term recommendations which include a new approach to international finance.
7. Worsley, Peter, 1984, *The Three Worlds* : *Culture and World Development* University of Chicago Press, Chicago.
8. *Op. cit* 3. *Note* the list of 50 countries in the North is inclusive of 25 countries in Eastern Europe and Commonwealth of Independent States in Central Asia.
9. Friedrich Hegel talked about the theory of evolution of social system: man in mixing his labour with nature for material production creates society in an on-going dialectic (*i.e.*, related, being two sides of the same coin) process, so that production factors with the technology as its production function relate the people with each other in a society. A particular mode of production over time, determines the very nature and structure of a society, in which capitalism as an evolutionary process of social system, replaces feudalism is a stage for developing highest form of economy.
10. The value of a commodity is determined by the subsistence wage of the labour necessary to produce it plus the profit of the owner of production. If a labour is driven by low wages to work for 10 hours but the production requires 6 hours, then in Marx's *Labour Theory of Value,* the extra 4 hours' work is the owner's profit. If this exploitation of labour is *protected by state and legitimized in the society*, ultimately over time this exploitation of the labour would be recognized and the proletariat (wage labourers) would rise up against the capitalist in violent revolution and establish "socialism".
11. *Op. cit* 5 in Chapter One.
12. *Cited from "Poverty and Development into the 21st Century*", by Allen, Tim and Thomas, Alan, eds. Oxford University Press. *Also op. cit.* 5.
13. *Cited from "Dependency and Development: An Introduction to the Third World"*, by Lewellen, Ted, C. 1995, Bergin & Garvey, Westport.

14. *Ibid,* 4 in Chapter One.
15. Wolf, Eric, R., 1982, *Europe and the People without History*, University of California Press, 1982.
16. *Tributary System* : Extracting "tribute" (both prestige and basic goods) from dependencies under overarching political economy based primarily through the use of political-military force.
17. Private professional standing armies and police forces were established for the protection of these ruling families (the "caudillos") against the disgruntled peasantry and labourers, a function that security forces maintain in present times. Fidel Castro of Cuba carry the *caudillo* mantle in the twentieth century.
18. The feudal agriculture, no matter how exploitative it might be, implies bonds of mutual dependency between master and peasant, and the peasant has his own meagre lands for subsistence crops.
19. Hobson, John, A., 1948, "*Imperialism*", Allen and Unwin, London.
20. *Limited Capitalism* : refers to closed economy of People's Republic of China in which the entrepreneurs being allowed to own small businesses and to hire a small number of workers. Private companies now account for a large proportion of China's industrial output.
21. Amin Samir, 2001, *Imperialism and Globalization*, Monthly Review Press, New York.
22. OECD, 1983, *Industry in Transition*, OECD, Paris.
23. Galbraith, J. K., 1977, *The Affluent Society*, Houghton Mifflin, Boston.
24. Myrdal, G., 1957, *Economic Theory and Underdeveloped Regions*, Duckworth, London.
25. Hirschman, Albert, O., 1958, *The Strategy of Economic Development*, Yale University Press, New Haven.

3

THEORIES OF UNDERDEVELOPMENT

Modernization and Development Paradigms

Development covers both "theory and practice", that is both ideas about how development should or might occur, and real world efforts to put various aspects of development into practice. Thus following the general definition of a *theory* as a set of ostensibly logical propositions, which aim to explain by virtue of exclusion or inclusion of different kinds of information and data regarding how development has occurred in the past, and/or how it should occur in the future. Hence there is the need for a brief survey of development theories to realize how theories as *normative models* (that is, they can generalize about what should happen or be the case in an ideal world) have commanded support and influenced development policy at different times and in different countries. Therefore, in studying the way in which development policy has evolved in different countries and at the international level, it is necessary to have an understanding of the influence of theoretical views about development.

Again different development agendas reflect different goals and objectives reflecting social, economical, political, cultural ideologies which suffuse development theory. For example, both in theory and practice, early perspectives on development were almost exclusively concerned with promoting economic growth but as the economic growth widen income disparities, subsequently the predominant ideology became to emphasize these wider policy imperatives which seek primarily to reduce inequalities. Thus in this overarching concept of development thinking, various theories have been produced which have not commanded attention in a strictly sequential-temporal manner that is, in order to accelerate (or, even to initiate) these theoretical views, they address themselves to certain crucial variables and hence in the process if they have been found wanting because of their simplifying assumptions or the domain in which they sought to relate are strong, then these theories have not been totally replaced. In the process, development theories have tended to stack up, one upon another, co-existing sometimes in what can be described as "partial theories" requiring changes in response to experience and deeper understanding of development processes. By examining both the content and some of the alleged shortcomings of these theories we can gain some idea of what domain development theory should ideally cover and it is instructive to gain some appreciation of the progress which development theory has made (Colman and Nixson,1986).[1]

Thus development theories and their theoretical innovations can be organized in three different ways: (first) chronologically, that is, theories are discussed in terms of when and why they appeared and how they are refined/extended; (second), they can be organized according to the applied context to which they are being directed, and (third), provide an approach to organize development theory in terms of the field into which the theory fits. This third approach leads to organize development theories in relation to economic theories and this has been followed in this chapter.

Developmentalism and Modernization (or, Diffusion) Theory of Development

The political independence of former colonies after the World War II, their push for economic growth by focussing their energies either on transformative or ameliorative development agendas, and the Western societies' economic "evangelization" and neo-colonialism brought a number of writings, all normatively concerned with the specific problem of national development in Countries of Global South which in terms of *modernization paradigm, provide a general framework of understanding that includes certain widely shared assumptions about how best to organize society on "a value system" and how social change proceeds.* This *behavioural framework,* reaching back to Emile Durkheim in 1893/1894[2] is having a deep root within classical sociology: according to Durkheim *developmentalism* consists of two primary assumptions: (One) *Development from within* : this view says that any object, a plant, an animal and a society has within it the tendency to change its form. Thus each nation-state constitutes an autonomous society in the sense that social change within these societies can be adequately understood as processes operating within each society. Secondly (Two) *Development as interaction* : this view says that changes in an element, an animal or society takes place by a combination of the qualities and potentials within the element and the opportunities and resources available in the environment. Thus development in terms of social change operates as a series of parallel paths for all societies. Hence each society or nation-state can be viewed as occupying a position along this common path and this implies three research tasks : (*a*) a theoretical project to define the path, (*b*) an empirical project to allocate countries to the path, and (*c*) a methodology to devise how to speed up their journey along the path to development. And regarding the path "to development", it was conceived that at one end of the path there was a *traditional society* which was irrational and non-progressive, whereas at the other end of the path there was a modern society that was rational and progressive. The general process of reaching this modern idea was, not surprisingly, often termed "modernization" which was seen as a progression from simpler to more complex and interdependent forms of association. Change comes about as a society shakes loose from value system of traditional society with its archaic social and economic arrangements. *Tradition* and *modernity* are thus seen as "polar opposites" as one emerges, the other subsides. In this view, change takes place at different rates in different societies and places through division of labour (in societies), but eventually all societies will undergo such a transformation, converging towards a similar end point. Thus Durkheim's concept of modernization is *evolutionism* plus *functionalism* plus *positivism.* Hence in contrast to traditional society, which embodies the opposite conditions, modernity involves "total" transformation of a traditional (or, pre-modern) society through *evolution* toward an industrial market economy that is characterized by *functions* of continuous economic growth, the proliferation of state bureaucracy, high rates of literacy and formal education, giving a *pŏsitivism* of a reduction in inequality, lowered birth and death rates, a decline of religious influence, meritocracy, democratic political tendency and urbanization.

In its application to the Countries of Global South, this modernization framework sought to identify the industrialization associated with progress, increased productivity, economic independence and quality of living conditions that had developed the Countries of the North and specify where and why these were lacking in those of the South. Modernization theorists, depending on their focus, reached varying conclusions to Rostow's summing up the modernization approach when he wrote about "five stages of economic growth" (1960)[3]: in the Rostovian schema, (derived essentially from the Marshall Aid Plan for European post-war reconstruction), the problem of the South was a mere shortage of capital and technology and development required a rise in the savings rate; the West could help speed up the process of their development by sharing its capital and know-how to bring these countries into the modern age of capitalism and liberal democracy. To others like Apter (1965)[4] it was a question of "value systems" : the people in the South lacked the cultural values like the modernity, the entrepreneurial skills etc. Yet whether from a sociological, political or economic standpoint, modernization theorists generally concurred on one important point that "underdevelopment was an *initial state*". The theorists of this framework, found while the North had progressed beyond it, but the other countries lagged behind.

Presented by the American social scientists as a common doctrine of change particularly during the heydays of Western economy in 1950s and 1960s, this modernization theory thus gives the portrait of the idealism and optimism prevailing during that time of the early 1950s. Counterposed to the early material of growth theory in "traditional economy" as backwardness or primitiveness, including kinship, religion and ruralism which according to *"modernity"* has to be superseded by the more robust formulations of this modernization approach. The expected changes as development, is conceptualized as gradual, qualitative passage from small-scale to more differentiated large-scale social, from simple to complex technology, and from personal to impersonal or anonymous relationships. Thus encapsulated to the idea of development as "following the footsteps of the West" which in effect, was to say to represent the developed world, this paradigm of economic growth meant for the Countries of the South, was "suffused with the intent desire of USA to fill the post-colonial "vacuum", to order the post-war world and to maintain the interests of capitalist business by propagating the idea of state-led developments in restructuring their economy through the Foreign Aid Agencies, which favoured large-scale projects of industrial and agricultural development, since the latter required the guarantee of government involvement. In the 1950s and 1960s, it was fashionable to denounce growth without development through modernization, and call for a complete metamorphosis of the economic, social, political and cultural infrastructure as essential for sustained growth. Some might see development as a set of end states, and modernization as the means of reaching them. However defined, development and modernization are *processes*: economic (industrialization, urbanization and technological transformation of agriculture), social (weakening of ascriptive ties, and advancement by achievement), cultural (the disenchantment of the world, and growth of science, literacy and secularization) and political (rationalization of authority and growth of democracy). These facts of modernity are not only complementary but linked together.

Theorizing on economic dimensions of development (national wealth and poverty) of a country over time can be focussed on two aspects: (one) is the internal causes of the country for which it is assumed that countries are having no international trade sector and (two) their economy suggests a "closed system" with a clear relationship between domestic savings and economic growth. The modernization theory in this case suggests an "open system" for which no such simple relationships exist between national economic aggregates as in the earlier. In this modernization approach, drawing on the theories of eminent sociologists such as Max Weber (who was a near contemporary of Emile Durkheim), when development is referred to societies, it focuses on people who are more prone to be judged by their individual abilities rather than family or ethnic ties and believe in the possibility and desirability of change—it sees economic growth and higher consumption not as ends in themselves but as means to achieve human development. Hence development implies an entwining of numerous and variant threads of separate structures, entities, groups, units and agencies and societal development are viewed upon as "growing systemness". This wholesale reflexivity in modernity in society is possible when a number of disembedding mechanisms create a normativity out of itself when it refuses to accept tradition unreflectively.

Thus this *modernization theory of development*, having its origin in sociology overlaps with "growth theory in economics" and it can be considered as a societal development: this theory is about a development from within and it involves a linear transition from the local ties and parochial perspectives of a rudimentary, traditional (or pre-modern) feudal hierarchies of society of a village and the small-town community (to Weber's *Gemeinschaft* [5] in which community is taken as group of people with overlapping geographic *area* living together in an interdependent fashion based on "natural will" where part fulfils for itself all needs, like a society based on organic solidarity where everyone is either all warrior, or all hunter, or all tool-maker). This idea of people's beliefs and value system (their culture) in which a society works, gives way to a different type of modern relations: "a modern, differentiated and bureaucratically rational, economically prosperous society like a large-scale organization (such as the city or state, *Gesellschaft* [6] of Weber with a rigorous mutual dependence of one with the other, that is a society based on rational solidarity) through spatial diffusion of growth-generating innovations in technology and radical thought. Such a transition is necessary because "natural will-based *Gemeinschaft* type of relationship of traditional or non-rational (pre-modern regulated) societies" which include family, friendship, neighbourhood and community relationships because they are inherently fulfilling, all sharing the acceptance of tradition, of the givenness of some actual or symbolic past event, order, or figure as the major focus of their collective identity, as the delineation of the scope and nature of their social and cultural order and as the ultimate legitimator of change and the limits of innovation, are limited in their imaginativeness and openness to change so that in *Gemeinschaft* "where relationship which emerge naturally, they become ends in themselves", whereas on other hand in *Gesellschaft,* in which relationship emerge rationally to become the "means to particular ends". Thus *Gesellschaft* relationships, are primarily entered into as rational means of acquiring a need unmet in *Gemeinschaft* relationships, examples of the latter include citizen-state, employee-employer, and consumer-producer relationships which can be thought of as exchange relationships in a market-based context that lack the emotional attachment of a *Gemeinschaft* relationship. For example, a grocery owner who keeps all the information about the business in his or her head, and has no clear idea of how much profit is being made, but just keeps the household and business monies mixed up together. In another context, another grocery owner keeps a microcomputer and knows immediately the stock he is having, profit and loss accounts. This latter person carries on his business more *rationalized* than the former. Its procedures are designed in order to achieve the owner's goal—the *maximum profit with the greatest efficiency.* Thus Weber's developmental thinking is having two parts: (one) is the idea of people's beliefs and their cultures in *Gemeinschaft* which was having a value system in their development and the (other) is the idea of "capitalism" as involving a spread of what Weber called "rational behaviour" so that it becomes the norm of everyday life.

Weber pointed out that these structural changes in types, forms, and directions from natural-will based *Gemeinschaft* type of community to a "rational-will based society" which Western societies developed, paralleled the growth of capitalism and urbanization to provide the climate for non-agricultural activities, most notably the manufacturing and the international trade which have never existed before. Employment in jobs in urban settings connects *Gemeinschaft* with *Gesellschaft* through migration to economic expansion and modernization, linking modernity to *geographic mobility.* Development, occupational change and migration appear strongly interrelated, particularly when employment is expanding in modern sectors within large cities, drawing rural labour surplus : cities are the points from which productive innovation emanates and hence cities are seen to be the "engines of modernization". Thus modern society in context of Western world was essentially a product of juxtaposition of three events : the *Renaissance,* the *Reformation* and the discovery of the *New World, so that it could create its "normativity out of itself".*

This above Weberian explanation of development arising from the combination of evolutionism, functionalism and positivism as early as in 1922 can be traced as a primary literature, to be generally referred to as *Modernization Theory of Development*—this Weberian theory is about role of entrepreneurship and inventiveness in social and historical change and it is having an usual conclusion that development will occur only when there is an influx of modern society accompanied with modernization in industry, in education, management etc. and the end result of this development is economic prosperity and relatively politically stable nations of the Western World. The pattern variables of a modern society is thus different—the evaluation of actions in a modern society is based on the principle of universal commitments and cosmopolitan attitudes rather than by qualities ascribed to him at birth, that mastery rather than fatalism orient their attitude toward the material and human environment, that identity be chosen and achieved, not ascribed and affirmed and overall the goals of collective society takes precedence over personal gratification (Lloyd and Rudolph, 1967).[7]

Social Differentiation and Social Mobilization

Most proponents of the modernization theory assumed that all societies are along a continuum so that as a country developed, its traditional institutions would fade away and people would come to identify with the nation-state rather than with their tribal or ethnic groups. According to them, chief obstacles to development which separates a "kin-ordered mode of traditional society" (with tribal and peasant cultures) from a modern one are internal, "rooted in some historical inability in the indigenous population to adapt to the modern world", like their low degree of specialization and a high level of self-sufficiency, emphasis on the importance of family and kinship within a lineage or a clan (which determine the division of labour) all resulting relatively little flow of goods and services from rural to urban areas and lack of development of market and money circulation.

The spread of European mercantilism introduced the "tributary mode of production" (which according to Marx is the combination of "feudal" and "Asiatic" mode of productions) throughout the world. In contrast to the kin-ordered mode of traditional economy, tributary relations are based on "domination and subordination" by which society get organized into two classes, a *ruling class elite of surplus takers* and an *underclass of surplus producers*. This mode encompasses a wide range of variations, from the tight monarchial control to the loose domination of local strongmen. In this tributary mode, the kin-ordered mode could subsist, in modified form within it; a new layer of power would be added, and much formerly subsistence production would be turned to export, but the original "bands" or, societies of hunter/food gatherers, changing to "tribes" practicing agriculture, and living in fixed settlements, give rise to egalitarian society, the latter are larger than in "chiefdoms" which exists in societies *ranked by status and power* as well as by *strata* (or *material wealth*) ruled by the chief on kinship basis, would persist. In the transformation of the kin-mode to tributary mode, many aspects of the society do also get supplanted.

In the "tributary mode", the appearance of agriculture sparks off competition for access to the best land, the rules are worked out whereby such lands are allocated and inherited. In this mode, a social hierarchy in the form of a state emerges which makes it spatially larger and administratively more complex than *chiefdoms*. This mode better known as *regulated (pre-modern) state* implies greater functional specialization and greater probability of the rise of urban forms but the *market is not central to functioning of this tributary mode* so that economic relationship are generally subordinated to political, social, religious or cultural considerations, for instance, land holdings are often tended to be controlled (or, regulated) rather than being determined by the market. In this tributary mode (or, regulated, pre-modern state) most people lived in agricultural societies, their lives were bound to their villages and surrounding regions. Their work consisted essentially in winning the means of subsistence for themselves and their families from their environment, as well as paying the taxes and meeting the other demands which were made upon them. From the time of the appearance of states down until the twentieth century the lives of most people were lived in the context of such peasant communities. Such peasant communities are still characteristic of parts of the world today, though increasingly eroded by commercialism and growing contacts with modernity.

Thus between 1500 and 1900 there was a fundamental change in the way an increasingly large part of the world was organized. Beginning in particular parts of England in eighteenth century and spreading to other parts of Europe, there emerged the "market-based states" whose development was closely linked with the *capitalism* in which the market played a key importance for its functioning. This "capitalist mode" emerged first in the mechanization of the textile industries, it is the elite that owns the means of production. In the regulated, pre-modern states (or, in tributary class), the takers demand a surplus, whereas under capitalism, labours must sell their labour for wages; otherwise lacking access to the means of production, they cannot survive. In addition, capitalist production requires that in order to stay in business, the owner of production must constantly increase output and cut costs. This puts the labour in the precarious position of being threatened with losing his job to new technology (and having no fallback for survival) or having his pay cut. Unlike the kin-ordered mode, both the tributary and capitalist modes of production require an apparatus of coercion: this is, *state*, which provides the law, the courts, the police, and the armies to maintain the system.[8]

Because of its efficiency and technological superiority, the capitalist mode spread rapidly throughout the world, virtually replacing the tributary mode and radically transforming not only the economies, but also the political, social and cultures that the latter encompassed. The kin-ordered mode, to the extent that it had survived mercantilism, was obliterated. Everywhere capitalism created a system in which labour is bought and sold as a commodity. Rigid class stratification emerged where none had existed before, and in many places, the results were cultural collapse, de-tribalization and mass migration in search of jobs. Underdevelopment results from encroachment of capitalism on other modes of production as well as on capitalism's destruction of these other modes. By 1900, capitalism and the *market-based states* become the dominant socio-economic system over a large part of the world thanks to the colonial empires of nation-states of Europe. A number of theorists like sociologist Bernstein (1971)[9] made the theoretical language of Max Weber regarding explanation of the origins of capitalism further elegant and less antiquated.

In one sense, the modernization theory tried to impose a unified capitalist system on the world. Yet, each individual culture is having a different starting point, and each gets transformed in its own way and at its own rate of social progress. According to Bernstein, in absence of any social differentiation and social mobilization being created, disruption of traditional institutions by modernization did not fade away the former (as believed by modernization theorists, mentioned above), rather it is more likely to lead to turmoil and discord than to modernization. Tribes, linkages, ethnic groups, and societies may change goals and functions but often actually increase their cultural cohesiveness.

According to Eisenstadt (1973),[10] the core of modernization is "social differentiation and mobilization". If a society is to be sufficiently adaptive to embrace new technology and new structural forms, there should be a "social differentiation" so that political sphere must get sufficiently differentiated from the religious sphere. In Islamic countries, such as Saudi Arabia, modernization has been retarded by the close association of political power with religion despite the availability of high levels of finance capital.

"Social mobilization" refers to the process by which traditional social and psychological loyalties are broken down and reoriented so that new social and economic arrangements become possible. Traditional structures do not cease to exist, nor is cultural collapse a necessary part of the process. Rather, traditional groups restructure themselves and reorient themselves toward new goals consistent with modernization.

The fact that modernization theory was having its origin in paradigmatic north western Europe and the United States because of the total transformation of traditional societies made possible by their decisive characteristics of advanced technology, investment capital, entrepreneurial skill and high levels of education to characterize these "advanced", economically prosperous and relatively politically stable nations. In that way, modernization theory went by the idea : "there is only one path to development, and that is the EURO-AMERICAN WAY". The basic assumption of this modernization theory in terms of spatial divisions of labour, social structures and the geography of production is relevant to capitalist-dominated economic structure but neither relevant to a non-capitalist modes of production and social formation nor does it have any meaning in relation to feudal mode of economy and society (Massey, 1995).[11]

To this theory which overlaps with economic theory of growth, there is another contribution by Max Weber (1922, 1967). He was much more directly concerned with the question of development, for his central problem was the explanation of the origins of capitalism. He defined *capitalism* as the economic and social system based on the private individual endeavour of those with capital to invest, who produce by employing others, sell the products on a market, and reinvest the profits from their sales in expanding their wealth and the scale of their activities. The important social inventions characteristic of this type of society are the *commodity*—the idea that there have to be things, *land* and *labour* which can be bought (and sold); and there have to be a *market*—a place, shops, the factory, employment offices, the stock exchange—in which commodities can be exchanged for each other or for cash between the sellers and the buyers. There has to be a division of labour in terms of skillness scale so that the exchange can take place through the market. Essential to a market form of distribution are the ideas of price, competition and profit. A market requires some idea of getting the "best" price (through a bidding procedure) for any particular item or "commodity" that is valued both by the sellers and the buyers.

The *capitalism* (or, capitalist mode of production) involves the use of two naturally given resources, land and labour, alongwith mobilizing a (human-made) capital, to produce goods and services that people consume in order to accumulate wealth and enjoy a desired quality of life. Capitalism originated in a transition from *feudalism* in England and Western Europe where accumulated wealth was available to be invested in new ways, and where the nature of social relationship enabled this to be successful. In this context, Weber (1922) endeavoured a *capitalistic order of society*, in which there is a pursuit of production for profit (through sale of produce more than the consumption), and forever renewed profit, by means of continuous, rational endeavour to take advantage of its opportunities for profit-making. He noted that capitalism requires and encourages, and works best, when the dominant belief emphasizes planning, efficiency and careful accounting, in particular, when people try to account for possible future effects of present actions.

Weber (1922) in his study of *capitalism* looked at a set of societies of India, China and Western Europe to find out why capitalist mode of economy became a society-wide system in Europe and not in other places. In these countries, Weber found the social and economic conditions were very similar to those in Western Europe but the "Asiatic mode of production" of these societies were not "integrated" with such firm beliefs (equivalent to those in Western societies) emphasizing individual responsibility, hard work, plain living, saving and planning for the future to give them a strong push forward. He talked about that in capitalist society in which a high proportion of the consumption of services is provided in the form of traded labour and consumer products (for example, food is largely purchased rather than home produced) and most exchanges take place in the market to be recorded by the normal national accounting procedures, but in contrast (to capitalist societies) in traditional societies (in the past and the present) only a small proportion of such services are purchased so that market is not central to the functioning of the economy, "in which many segments of the societies are having ties with people which are non-contractual in appearance and substance, and cannot be reduced to those simply between unattached buyers and sellers" (Bagchi, 1982).[12] Weber identified the former as "rational", market type relations and he contrasted the latter in other than in Western Europe as the "non-rational", non-market, relations. Karl Marx who lived in the nineteenth century (1818-1883) and also studied India and China, believed implicitly capitalism would incorporate all these Asiatic regions and "the dynamism and capacity for expansion of the youthful capitalism would be reproduced in any society ..... and the process would ..... create havoc and hardship within those backward countries it would touch" (Forbes, 1984).[13]

Later, the theorists of modernization theory of development picked up the ideas of Weber (1922)—they equated capitalism with industrialization as well as with the operation of markets for goods and services and they argued that all states follow the same trajectory of development through rapid economic growth and this *market economy* path which is followed in "Core of the World Economy" can be repeated in "periphery" of the world economy. The argument put forward in this context is that theories and models devised from first world locations are deemed to be universal in coverage, hence it follows that "research on the Third World could be highly simplified." According to them, the problems of underdevelopment, poverty and malnutrition are the result of traditional, non-rational thought. All that was necessary was to pluck out a tried and tested First World model and impose it on the periphery with the aim at "increasing the need for achievement of the people of the underdeveloped regions". ....Hence by studying social change in a core country, stages of development can be identified that peripheral country will eventually pass through (Johnston, 1993).[14]

In social science, the idea of developmentalism has been by describing the "demographic transition model" alongwith the ladder-like "STAGES OF GROWTH MODEL" (1960)[15] of the famous

American Economic Historian Walter Rostow (the latter has been represented as the "apogee" of modernization theory). The demographic transition model has its origin in the work of the American demographer, Warren Thompson who in a paper in 1929 identified the initial stage as "the high fluctuating stage" where both birth rates and death rates are high, a sort of natural demographic situation. Rostow in his paper put forward in the 1960s begins with a traditional society and its pre-Newtonian science and hence rudimentary technology. Although seeming to describe quite distinct processes, first stages of these two models actually have much in common: they invoke characteristics of societies which were once universal, but which are now only to be found in poorer countries. This is the crucial step in equating current societies in Countries of the South with the situation which were obtained in the North in the earlier days. Once this leap of imagination is achieved the history of the North can be used as a guide for the Southern Countries navigating the same path. Hence the subsequent stages of both demographic transition model and Rostow's model can be developed as highly abstracted histories of experience of the Countries of North. Universalizing by the sweep of modern economic history, Rostow emphasized that market economy, as an institution and a symbol of freedom, efficiency and prosperity, has been enjoying its heyday universally and hence in its incarnation of global capitalism it forms an overwhelmingly dominant practice and ideology so that to be precise in these cases, particularly stylized histories of social change in core country like England, stages of development can be identified in all societies, in their economic dimensions and they have to pass through the same stages as did the First World countries. This means that England and other societies of Western Europe who have reached the final stage of the "age of mass consumption" and "low fluctuating birth and death rates" are defined as the policy goals for the less fortunate. Hence Rostow wrote, "all global cultures are condemned to modernity" as they integrate not only at a structural level of the nation-state system and the global capitalist market, but also by the human process of endogenous development. According to him, a peripheral country will eventually pass through by bringing cultural and institutional change under state auspices to take place within its society. This reduces to saying, "If only you were like us, your problems would be solved". Clearly the optimism of social knowledge of the early 1960s were also tinged with an arrogance which was only just beginning to be fully appreciated (Taylor, 1989).[16]

Thus the key element in Rostow's thinking was the process of capital formation for which he identified all societies in their economic dimensions on the same historical track of development stages when he wrote about five stages of economic growth leading to development by an "unilinear evolution": a movement by societies through the (1) initial stages of traditional society, is that based on pre-Newtonian science and technology and on pre-Newtonian views of the world, practicising a primitive agricultural economy with limitation on productivity and diversity of product (as prevailed in pre-seventeenth century Britain) due to lack in modern technology and a hierarchical social structure of family and clan connections in which value system is characterized by a *long-run fatalism* that is, people view the possibilities open to their children and grandchildren as no different from those available now and in the past. Therefore, there is little or no conception of progress so that this initial situation has to undergo for a change, and in the process of transition as it begins to utilize the insights of science and technology through borrowing (or innovation) they develop the tendency to fend off the diminishing returns and the second stage starts, that is (2) a *transitional stage*: a "precondition for take-off", that is, a sustained growth towards development stage, this is the stage in which there may be social and economic dislocations because of unemployment and swelling up of urban areas but these are simply inevitable, temporary by-products of the progress they symptomize, since the idea of economic progress, both as a possibility and as something to be desired, takes root in this second stage. Institutions for mobilizing capital—such as banks appear and investment increases, especially in infrastructures like transportation, communications and the extraction of raw materials for export. This transitional stage started in Western Europe in the late seventeenth and early eighteenth centuries as the insights of modern science began to be translated into new production functions in both agriculture and industry when alongwith the dynamism they enjoyed the lateral expansion of mercantile trade with their colonies in the world markets. Political power which was decentralized in the traditional society, took shape further as various regions develop as the true nation-state and localities are united under a single government.

Among the Western European states, England, favoured by geography, increased trading possibilities through world exploration, social and political structures was the first industrial nation in the world to develop fully the pre-conditions for *take-off.* Whilst the pre-conditions for take-off was actually *endogenous in England*, elsewhere they were probably the result of "external intrusion by more advanced societies" (Rostow, 1960). This take-off is the interval, covering a few decades, when the old traditional blocks and resistances to steady growth are finally overcome and growth which becomes the normal condition of economic progress takes root, this is the third stage which is a "watershed stage" when the true nation-state develops (which got initiated to other European states) with the aspirations of making economic progress as a serious goal, the rate of savings and investment rises from 5 to 10 per cent or more of the national income, GNP and this investment spurs the rapid expansion of new industries that yield "limited bursts" and "enclaves of modern activity" which expand and comes to dominate the society. This watershed stage as a "take-off" , occurred in Britain in the early nineteenth century. A surge of technological improvement gets combined in this third stage when the emergence of an economic and political elite that makes economic progress a serious goal and has the power to enforce its aspirations as national policy—thus profits gets plowed back into further development, agriculture becomes increasingly commercialized to provide additional food for the increased population as well as for the increasing number of people who earn their livings in manufacturing and the rising income leads to increases in the demand for commodities produced by the industrial sector. Thus the agriculture becomes increasingly commercialized and the foundations for economic transformations are laid so that the society now moves into a sustained stage for steady state of growth which Rostow terms as the next stage: the fourth *stage of drive-to-maturity*, when the modern sector expands to the whole economy overwhelming the technologically deficient traditional sector. This is a period of "self-sustaining growth" when between 10 to 20 per cent of national income is reinvested, so new industries now forge ahead on the scene, the goods that were formerly imported are now produced at home so that imports fall off, production accompanies with sufficient demand and the later are internally met so that about 60 years after take-off, the country reaches *maturity,* that is, it attains the technological and entrepreneurial skills to produce whatever it chooses.

Finally the last stage, the fifth stage known as the "age of high mass consumerism" arrives, when it attains the technological and entrepreneurial skills to produce whatever it chooses and the leading sectors shift away from heavy industries towards provision of high consumer durables and services in the consumer market place: goods once thought luxuries—automobiles, televisions, suburban homes—become normative throughout the entire society. At this point the social welfare provisions are made and the society accomplishes the complete shift from traditional to modern society. It also exploits the comparative advantages in the international trade—this *end result is prosperity* so that large number of persons gain a command over consumption which transcends basic food, shelter and clothing and "relative" political stability bringing the emergence of the welfare nation-state which chooses to allocate increased resources increasingly getting locked in through (*i*) increased employment, (*ii*) the diffusion of services of social welfare, (*iii*) security for its people, and (*iv*) a mass basic production of consumer durables: this final stage Rostow labels the "high mass-consumption society". According to Rostow, this final stage was not reached in Britain and Western Europe until after the World War II, but in the USA mass production and consumption of consumer goods came earlier during the 1920s and 1930s. Thus rational economic competition may appear to treat the loser roughly in the short-term, but in the end all profit from the economic growth and the overall expansion of economic opportunities. This is a familiar invocation.

Thus, according to this pre-eminent theory of modernization in the early 1960s by Rostow (1960), societies in almost all countries travel along a path over a brief period of 20 to 30 years that sets in place a self-perpetuating acceleration of economic growth through a *ladder like these five stages of growth* and *growth that fundamentally transforms the society and the countries achieved "take-off"*—it is now the turn of the underdeveloped countries if only they could be freed of such constraints to advancement like the lack of capital via aid programmes, financial institutions, trade and multinational corporations. Hence, *the answer seemed to be uncomplicated*, since according to Rostow (1960), increasing interaction and integration between peripheral and core countries will at some stage lead to the onset of development in the former areas.

Rostow's 5-stages model of British economic history were influential in the US government because of the policy implication in the international context: decolonisation by the 1950s brought the Countries of the Global South for whom development became a significant issue which was not just an academic concern nor a matter of altruism. The United States and the industrial democracies of Western Europe needed to formulate policies towards these new countries, and they needed theories on which to rely. The Cold War lent a special urgency to the enterprise. The "losses" of China to the communists in 1949 and Cuba a decade later were seen as the "wake-up calls of a world-wide threat to capitalist economy". Policy makers in the United States became convinced that the Soviet Union, unable to attack the West directly, was trying to take over the world through the newly independent countries. Communism, with its promise of "redistribution of wealth" made its appeal to the underclasses. Development in the Southern Countries was perceived as a matter of national security—if the U.S. allies in the South could be brought to the *take-off stage*, then their own development would become sustaining. But this gradual, non-revolutionary model of development of the modernization theory based on a "belief in progress, rationality and mechanization" through technology and capital transferring to the South for achieving "Take-off" and Rostow's goal of discovering universal stages of development seemed to be an extremely seductive way of seeing the world, overambitious at best.

This completely misses out the overall context in which development occurs. For example, (one) when England was at the bottom rung of Rostow's ladder there was no "high mass consumption going on at the top". Again (secondly), countries that are on the bottom rung today, however, find themselves in a polarized world because the top rungs are occupied, since if "economic maturity" is defined as mass consumption, such maturity would be beyond the reach of many countries and given the ecological implications—in terms of resource depletion and pollution— USA-style of consumption might not even be desirable. This is where the assumption of autonomous societies is crucial to developmentalism. (Third), if social change is indeed pre-eminently a matter of internal dynamics of each society (or, country) then the location of the other countries on the ladder does not matter : each country is an autonomous object of change moving along the same trajectory but starting at different dates and moving at different speeds. (Fourth), this model suggests that all countries progress in linear fashion through these stages towards development in the same sequence as happened in Europe and North America, but the sequence of events has not been straightforward, for example, in the contemporary context rapid changes in the agricultural, industrial and service sectors all happening at the same time, rather than sequentially. And neither the modernization theorists could not easily accept that the Southern Countries differ from the first. Finally, Rostow's model is often wrongly seen as development model, whereas it is actually an "economic growth" model: Rostow was concerned more with economic progress leading to capital accumulation, rather than human welfare and non-economic indicators of development. Some countries have experienced periods of rapid economic growth, yet much of their population has felt little benefit from this—what is called "growth without development". Thus modernization paradigm of the 1960s is not an evolutionary theory, it is an "action programme believed in the all-round adaptability towards progress and it seemed to offer every country an equal chance to develop."

Theories of modernization tended to be pragmatic, that is, their aim was to provide governments with the tools to speed development. One did not have to seek for the causes of underdevelopment: countries started out underdeveloped—that was their initial condition. The crucial questions were: How do the countries become developed? And what are the major barriers to development? The way to answer the first question was to examine the progress of the Northern Countries. To answer the second question, it was necessary to research the internal conditions of the poorer Southern Countries. Rostow saw the problems of underdevelopment an intrinsic to the countries themselves, but the solutions would come within the international marketplace. It became evident fairly early, however, something was wrong. Early theorists tended to equate growth with development, but by the early 1970s, it was obvious that these are two different processes and that they did not necessarily coincide. Even a substantial increase in GNP, growth did not always, or even usually, mean that there would be a rise in the standard of living of the masses, an increase in agricultural productivity, the spread of modern technology throughout the society, or the solidifying of a strong nation-state—all things related to development. One problem with the early models was that they were additive, in the sense that they postulated that by merely adding technology and capital, the country would progress. It was now clear that problems were also, and perhaps were fundamentally, "structural", that is, they had to do with the ways that societies and economies were organized in the Southern Countries.

The Vicious Circle of Poverty

All those critiques were voiced but generally these were not heard in the post-1945 era of social *euphoria,* exhibiting a sublime confidence when modern society came to be celebrated as the ideal social system with traditional society as the problem. All this changed with the end of the post-war economic boom and the onset of a world depression by the 1970s. Modernization theorists had heralded a new dawn for the poor countries and they had a hard time accounting for the growing inequalities within the Southern Countries and for increasing international inequality. Theoretically, infusions of capital and technology (which USA was having in abundance) and quite capable of transferring to the South should have closing the gap between the two, but in reality, the gap was growing at a frightening rate. Few countries were achieving anything resembling Rostow's "take-off" and even for those few, growth was often unbalanced and erratic. The theorists began to look beyond the countries themselves to the structure of the international system. Whereas the cause of underdevelopment has not been a problem for modernization theorists —all countries started underdeveloped—it became a central problem for the new generation of scholars. Within some 20 years, developmentalist models became a sick joke and they were challenged by "dependency, underdevelopment and world-system theories" which totally refute this developmental model of social change—the fact that some countries are rich and others are poor is not merely a matter of timing along some "universal pathway to affluence"; on the contrary, rich and poor are part of one system and that they are experiencing different processes within that system: it is *development and development of the underdevelopment*. The most important fact concerning those countries at the bottom of Rostow's ladder today is that there are countries enjoying the advantage of being above them at the top of the ladder. Perhaps more than anything else world-system analysis is a challenge to modernization: the simplistic world of an "international ladder", *in terms of modernity remains as an ambiguous and chronologically elastic term superseded by the sophisticated concept of the capitalist world economy.* The lack of distinction and explanation drawn by Rostow and others between the processes of "growth" and "development" led to a series of sea-level of changes in thinking about the process of development itself. Academics were less enthusiastic.

Rostow's theory conveys a broad policy message and that thrust is two fold: (first), the implication is that development is a universal process, through which each society moves in its own time, in response mainly to internal factors. But the development of the advanced industrial economies from which Rostow derived his base should instead be recognized as a highly special case. (Secondly), his theory places heavy stress on industrialization but he didn't see the agricultural sector crucial to the take-off stage. He also had a serious problem with identifying a "take-off point", because of its broad, descriptive character as a concept. His stages thus become categories for classification purposes, rather than analytical guides to policy.

Failure of the Modernization Paradigm: Critiques and Crises

Though the Southern Countries has been evolving since the fifteenth century, it did really come into existence in its present form until after World War II, since the 1950s. But over the long period of European colonialism, the colonial episode played out its global course and there was little real concept of underdevelopment: the invasive process of European capitalist system converted the colonies and the protectorates as appendages of the mother countries which left certain legacies on patterns of economic, social, political and cultural life which continue to shape the thought and action of the present—the newly independent countries which include all countries of Africa and Asia except Japan —acted as "an adjunct to the imperial countries", supplying them with raw materials, food crops for their industrial labour, providing themselves as the market for their cheap factory goods and a surplus for investment in the "frontier economies of white-settled colonies of North America, Australia and South Africa. Certainly development was not one of these goals and hence immediately after their independence when the bonds of "Global Colonialism" was over, these ex-colonial countries turned their attention from problems of political emancipation to domestic problems of national unification, but class- and religion-based social struggles were seldom easily resolved. Moreover, after political independence, they did not become rich and industrialized to achieve the elusive goals of economic independence and economic development, so to the questions of economic regeneration they remain rather "in the periphery of the system produced by the expansion of capitalism". The Southern Countries could gain currency to purchase the infrastructure and technology needed for development and the rich countries would help this process through transfers to the South. Logically, this should work since one of the cliches of capitalism is that "it takes money to make money". But when applied to Countries of the South, this is the principle of *vicious circle of poverty*: countries remain poor because they start poor. If the people are too impoverished to save or to pay much in taxes, there will be no money for companies to borrow for investment or for the government to create an infrastructure to promote development. Also, if incomes are low or populations are small, there will be only slight demand for goods, so even if entrepreneurs exist, there is little stimulus to produce for the domestic market. A poor country is faced with pulling itself up by its own bootstraps.

Thus, the process of development which emerged in underdeveloped economies of ex-colonial countries, got characterized by a "dichotomous" or "dualistic" nature where the capitalist sector exists with the traditional sector: in the latter, surplus labour exists through a number of sources in a less developed country, including farmers, petty traders and retailers so that the labour is perfectly elastic in the subsistence sector. Thus labour in this dualistic economy is employed at a swage and this surplus at subsistence wage is available to the capitalist sector remains unlimited. The income distribution remains unequal and most of the gains in economy will go to the traditional landlord class that does little saving but this capitalist class grew at the expense of the large subsistence sector. This concept of dualism mentioned earlier also became apparent in some spatial contexts where certain regions moved ahead of others by virtue of some initial advantages. While this spatial manifestation of unbalanced growth is argued inimical to the growth of the economy as a whole, Gunnar Myrdal and Albert Hirschman advocated strategies of "unbalanced growth" essential for the economy to lift itself to higher income levels so that the less dynamic areas may benefit from "centrifugal spread and backwash effects" of Myrdal (1957) and the "trickle down and polarization effects" of Hirschman (1958). But this alternative strategy of experiment with "growth poles" manifestly failed" : the uneven development is claimed in part to be a product of orthodox capitalist development process which unproportionately privileges the urban areas and industrial sector and leave behind in the process a host of areas, sectors and social groups of the periphery whose resourcefulness became "sequestered" by the core.

However, the economic growth, which had been the main preoccupation of Myrdal, Hirschman and Rostow did not necessarily eliminate poverty and that the so-called "trickle-down" effects of growth failed to benefit the poor in both spatial and social terms and the earlier dualistic, unilinear and essentially Euro-American-centric approaches of modernization (or, diffusion) theory failed in its objectives of bringing a "diffusion of development" in Countries of the South. It became evident by early 1970s, that something was wrong —the modernization theory got collided with the realities of labour-surplus character of Countries of the South. United Nations' "Development Decade" of the 1960s turned out to be a failure[17]—early theorists tended to equate growth with development, but it became obvious by this time that growth and development were two different processes and they did not necessarily coincide—even a substantial increase in GNP growth did not always, or even usually, mean that there would be a rise in the standard of living, spread of technology or solidifying of a strong nation-state—all things related to development.

The modernization paradigm now seems naive and one of its problem was it is "additive" in the sense that it was postulated by merely adding technology and capital, the country would progress. Again this theory was criticized for Euro-American-centrism and for its simplistic initial assumptions based on ethnocentric equation: "modernization = westernization means progress", which dubbed all non-modern societies as "traditional" and this was challenged as being a conceit (Forbes, 1984). But most importantly, its dualistic presentation of the development process in terms of characterization of the traditional-modernity dichotomy of society, culture and personality in terms of swapping a negative set of "pattern variables" for a positive set and diffusion of Western political and cultural values, capital and technology to Countries of the South for a "ladder like 5-stages of Growth Theory" of Rostow failed in practice. To continue to refer to non-Western of Third World as simply "traditional" is argued as misleading: in the sense that on the one hand if the word "tradition" connotes the act of handing down from one generation to another it should "means both process and product" —on the other hand, what is modern for one generation will perforce become part of tradition for the next and hence the product that is passed on (or bequeathed) cannot possibly exclude the modern. Most societies today possess the means for the local production of modernity and hence this paradigmatic modernity of western world can be no more pristine (Appadurai, 1997).[18]

The modernization theory does not consider it relevant to look at the detailed history of the Third World Societies since modernization theory requires resources, capital, infrastructure, education for the influx of the modern society with its achievement, motivation, entrepreneurial initiative or human capital. Initially, it was thought that in the early stages of development, economic growth would be restricted by a "two-way gap" of investment (that is, through internal savings as well as import-purchase capacity by earning foreign exchange from exports). Since this problem would typically exist, so it was thought in the early 1950s that the large-scale aid to the newly independent states in Asia and Africa would help them in correcting these imbalances within the world economy so that they can proceed in an orderly way to economic growth. As a result development was seen as the "diffusion of economic growth to the poor countries, along with the diffusion of *modern values,* such as increased newspaper readership, and various other attributes of western living. Yet it soon became apparent that this was not to be. *This produced something of a crisis in the field of development studies.* Growth and development increasingly came to be separated, the former being used to refer to an economic process, the latter to a series of normative changes associated with the improved well-being of the population as a whole.

As a result in the face of this failure, theoretical reactions grew, and counter-positions gathered force and there emerged a new orthodoxy in development studies.

This modernization approach totally ignored the causes of the complex patterns of world's uneven regional economic development and completely misunderstood the dimensions of effective intervention in problems of underdevelopment. Thus according to Taylor (1989), the concept of Rostow that "all states follow the same trajectory of development" is the "error of developmentalism" to which orthodox social science is prone since this is geographically "state-eccentric" because economic conditions vary immensely throughout the world in levels of income, degree of modernization, use of technology, quality and density of infrastructure, capacity for savings and investment, acquiring skills, property rights etc., so that for example, an integrated transport network if it is developed in a Third World country, it may not be in the pattern of an autonomous network pattern in a core country. Thus in contrast to unilineal theory of modernization, different countries require to develop in different ways—even within a single country, not all elements modernize at the same rate, industry may leave agriculture behind, or *vice versa*.

The Dependency Theory/Paradigm

Decolonization highlighted problems that had been hidden behind colonial patronage in the newly independent countries. Development becomes a new phenomenon over large parts of this world—the ex-colonial countries now regarded development planning as their early priority after political independence. In the process they find their problem is much vaster—declining mortality rates in Third World countries, combined with the persistence of high fairly high birth rates, led to high overall population growth. The ensuing "population explosion" exacerbated the plight of the poor, opened up new problems of unemployment, and intensified pressure on scarce resources. Unlike developed countries, developing countries have to concentrate not only on growth rate but also on building up the potential (in terms of *economic and social infrastructures*) for growth. No doubt the growth rate is important—terribly important, even to ensure that everyone has enough to eat. But they have to fight illiteracy, have to eradicate epidemics, have to built up the structural framework of the economy, of changing attitudes of mind and outlook born of tradition and, above all the question of how to escape from the endemic poverty through development.

But what is that we really mean by *development* ? There is probably less agreement today than ever before on the definition of this term, which is commonly used in a normative sense—"as a condition toward which all people unquestionably aspire".

On the need to shift these countries from a low productivity agricultural base to a high productivity industrial base, the *development economics* that emerged in the 1950s in these newly-independent countries basically experienced the same sequence of events as those experienced by the West a century earlier that is, development is required to be achieved through urbanization, commercialization, industrialization, modernization, and per capita production and consumption.[19] In this largely "indeterminate field of analysis involving economics, sociology and politics", various theories seeking to bring about this shift were propounded. Generally, they were "single-barrier" theories which attributed the failure to achieve economic growth to a single factor, such as *low labour productivity* or *capital shortages.* The *corollary* was that development programmes should promote *external capital inputs* or *encourage the redistribution of labour surpluses from the rural areas to the towns and cities.* Most proponents of the modernization theory assumed that the chief obstacles to development in the Countries of Global South were intrinsic to the countries themselves *i.e.*, in not having the resources, capital, infrastructure, achievement motivation, entrepreneurial initiative, or human capital all being rooted in some historical inability in the indigenous population to adapt to the modern world, and according to them, the solutions would come within the international marketplace. In accordance with Ricardo's "principle of comparative advantage", by selling their raw materials, poor countries could gain currency to purchase the infrastructure and technology needed for development. The Countries of Global North would help the process through loans, foreign aid and multinational investment to the Third World, but these all *exacerbates the gap* both within the former countries as well as in the international arena between the "haves" and the "have-nots". The Countries of Global South were drawn ever deeper and more rapidly into international exchanges, by mid-1980s, more than one quarter of their gross product is being put to the test of the world market ...... their foreign debt reached an astronomical figure of US$ 2.07 trillion by 1995, for a number of countries their debt problems become acute, having debt amounting to 200-250 per cent of their exports and their debt service payment costs them 20 to 40 per cent of their export revenues —the problem leads to depreciation of their exchange rates, declining a current account/GDP ratio, and internal imbalances leading to an increase in their budget deficit/GDP ratio, a sharp decline in government revenue and/or an increase in government expenditure, and a rapid rise in inflation. The necessary corrective measures involve imposition of import cuts on capital goods and raw materials, constricting economic growth the effect of which will not be short-term unless they are followed rapidly by policies that achieve growth of exports or import substitution (Kappagoda, 1997).[20]

And a new feature that has emerged since 1985 has been a sharp increase in foreign direct investment, (FDI) (particularly in *liquid flow* of equities and bonds) which increased from $11.0 billion in 1990 and $ 90.3 billion in 1995. Portfolio flows also become significant during the 1990s, reaching $ 22.0 billion in 1995.

Thus the sustainability of a country's borrowing level depends on the adequacy of the policy environment for achieving growth, while maintaining internal and external balance and using borrowed resources efficiently—which in turn depends largely on economic management in the country. Action by debtor countries, particularly those with more open economies, is often circumscribed by the global economic development and the policy share of Countries of Global North on financial and trade issues.

The formal long-term approach to foreign borrowing implicitly assumes that demand management policies ensure that expenditure is maintained at levels enabling external liabilities to be met. In the short run, an economy often faces various exogenous shocks, such as crop failure, transportation breakdown, recession in the main trading partners and/or adverse movements in the terms of trade, which require appropriate policy responses.

Structuralism and Dependency as A Formulation: ECLA Analysis

Structuralist Tenet : In reaction to what became seen as the failure of modernization theory, a "structuralist" tenet was first advanced by a group of Latin American economists: Working on a theory of development in the Economic Commission for Latin America (ECLA) (an United Nations agency established in 1947 in Santiago, in Chile) for construing the international economic order as polarized between an industrial centre and an agrarian periphery, the Latin Americans predicated on the idea that Latin American countries though became independent in the nineteenth century but the British followed by the American merchant houses took the place of those of the former Spanish and Portuguese rulers which put obstacles : the so-called Latin American bourgeoisie, because of its historic origins and because of its economic connections and even kinship-links with land owners, formed a part of the oligarchies which rule the Latin American countries through the foreign alliances so that the Latin Americans had come to see the structure of their continent's economies remained largely backward but at the same time they found that they are trapped objectively not only with international relationships of dependence by (American) imperialism (as their first contradiction to the process of industrialization in their countries, in common with other ex-colonial countries) but also by the local elite class as an internal structure conditioned by imperialism in form of another (second) contradiction, developed an economy of *dualist* nature, and it became an inescapable situation[21] that is, they had to go on exporting their primary products in return for finished products which deepened their continuing underdevelopment for much of the century. These were worked out during the 1950s in the ECLA by Raul Prebisch: The central theme there, at that time, was "structuralism" that is, the difference of structure of production in each pole, a forerunner of "dependency" theory—on this analysis, the major enemy was identified as *American imperialism*. The basic tenet of the structuralists was that the underdevelopment of Latin American countries were neither inflation nor their foreign exchange problem but their reliance on exports of primary products, which were subject to terms of trade that both fluctuated in the short-term and deteriorated in the long-term. This economic pessimism that Southern Countries required to export more by their primary products to maintain their levels of capital goods import and in the process the incipient national bourgeoisies did gobble up a growing share of countries' resources running their national economy to standstill. For breaking up this syndrome, what was believed that the structural obstacles on the path to development was the free market and the need was to impede the free market. Hence instead of leaving the things to the free market what was needed a "big push" in infrastructure investment as well as planning for stimulating industrialization. But since the capital resources needed for this lay beyond the reach of the private sector, so proposal was for a balanced pattern of public investment in several different industries in a way to kickstart the economy through "concerted state action". This is the "statist approach" that is the states had to push industrialization by creating large statist firms and encouraging private sector to produce goods previously imported so that dependence of their trade with the Northern Countries got decreased so as to shelter themselves from the vicissitudes of global economy. This strategy came to be known subsequently as policy of "import substitution industrialization" (ISI). Nevertheless, it is also in the very same structuralist nature that the limitations of ECLA thought lie; at this level of analysis no consideration is given to the social relations of production which are at the base of the process of ISI, and of the transformation in other structures of society that this brings in its wake. This ECLA theory was totally heretical: it chose to speak of the principal obstacle was located overseas but its conviction was that without a strenuous effort to remove the internal obstacles to development the process of ISI would be greatly impeded.

Because the above approach fundamentally spoke of the "free market" as the structural obstacle blocking the Southern Countries' path to development so that they get structurally disadvantaged. Hence the ISI approach with the aim to produce finished goods, is also known as "structuralism". Drawing on evidence of the historic socio-political institutions and built-in economic constraints shaped partly by the past interaction of Countries of Global North (particularly by USA) with respect to Chile and Argentina during the 1920s and 1930s, the ECLA School advocated this policy of ISI to enable peripheral countries to switch from "outward-looking development" to an inward-directed one. In this context, the state seemed to be a natural instrument for social transformation and social development. The role of the state was crucial in this, because it was also required to raise tariff barriers against imported manufactured goods. From this structuralist base, a "dependency school of theory" got articulated. Latin American economy point the way forward as early as in 1950s. Later on the support for structuralist theory did come from W.A. Lewis (1954) [22] who in a seminal paper on labour and development argued that since traditional family farming by minimum levels of existence keep the wage rate at a constantly subsistence level and ensures a virtually unlimited supply of cheap labour which the state can avail of to harness the built-up of a nation's industries.

Dependence and the Dependency Theory Approach: "Dependence" in any given society of a country is a *continuing situation* in which the economies of one group of countries are conditioned by the development and expansion of others. A relationship of interdependence between one or more economies or between such economies and the world trading system becomes a dependent relationship when some countries can expand through self-impulsion while others, being in a dependent position, can only expand as a reflection of the expansion of the dominant countries, which may have positive or negative effects on their immediate development. Dependence is thus an equal complex set of associations in varying degrees of the internal ones and indeed, internal variables may very well reinforce the pattern of external linkages.

From a *dependency theory* perspective, underdevelopment is not the result of some *lack of their initiatives and vitality* within the underdeveloped countries, but it is caused by continued *exploitation* of the dependent (or, underdeveloped) country by more powerful, wealthy ones as the *undesirable consequence of colonialism.* In the past, the political structure of domination involved the formal apparatus of colonialism, that is, the exploitation of the dependent (or, underdeveloped) country by more powerful, wealthy ones as the *undesirable consequences of colonialism.* Powerful countries have since learned that their economic interests could be served with less trouble and obligation simply by dictating terms of trade to former colonies that have remained economically and technologically dependent. The condition is referred to as *neo-colonialism* or *dependency.*

It is understood that this external structure of underdevelopment is articulated within poor countries by patterns of urbanization that still reflect the political and spatial arrangements of the colonial past. Urban centres are the collection and transhipment points for raw materials destined for the world market. In the hinterland, the labour that produces materials for export receives a small fraction of the market value of its product. Within the domestic economy, the greatest share of the market value produced by rural economic activity is realized as profit in the largest urban centres. Thus cities as centres of wealth and the seat of government, stand in an exploitative relationship to the remainder of the nation, so that dependency involves an association by striking alliances between the national bourgeoisie and international capitalism and leads to increased impoverishment and marginalization. This feature of underdevelopment is aptly termed *internal colonialism*: a process nesting within the notions of core and periphery, that produces certain international forms of patterned socio-economic inequality directly traceable to the extraction of surplus value by exploitative practices through which national and international institutions are linked in the interest of surplus extraction and capital accumulation.

Thus questions were asked about the history of imperialism; *growth and modernization theories* were criticized for such things as their excessive concentration on physical capital, their equating of development and westernization, and the overall simplistic nature of their explanation. The modernization theory, as a body of thought was "distilled from the American and Western experience of economic growth" and it failed in its objectives of a "gradual, non-revolutionary model of development based on a belief in progress, rationality and mechanization or industrialization (Rimmer and Forbes, 1982).[23]

Marxism and its Strand in Development Theory

Faced with these difficulties in understanding both developed and underdeveloped societies, another source of social thought which has been critical of just the conception of capitalist modernity and which seek to moderate or reform the injustices in capitalism was the Marxism, which derives from tradition based on the idea of Karl Marx (1818-1883) and Friedrich Engels as well as Marxist thinker like

Vladimir Lenin. Marx and Engels lived in the nineteenth century and confronted the same problems which the rightists Durkheim and Weber had tried to tackle. In contrast to these earlier thinkers, however, they argued that the processes of social change and development brought out by capitalism were in their nature not gradual and evolutionary. Rather capitalism according to Marx, divides society into two classes: (one) those who own the means of production (that is, the main sources of wealth), the capitalists (or, the bourgeoisie); and (two) those supplying the necessary labour power (or, the proletariat), that is, those who must live by selling their labour to the capitalists. Thus idealizing the society in these two class separations, Marx established the social and institutional (political) framework which enabled the capitalist society to generate unprecedentedly high levels of production and productivity. At the same time, however, by alienating the underprivileged majority from its role in the productive system it had sown the discontent of antagonistic. In his *Das Kapital,*[24] Marx argued that in the capitalist system, profits arise out of the way in which the capitalists (or, the bourgeoisie) dominate labour (the workers) in an unequal class-based relationship. Capitalists, as a class according to him, can only reproduce themselves by accumulating profit, and can continue to expand the base of profit, at the expense of the labour or by exporting exploitation to their colonies. This is because capitalists are in mutual competition, if they do not exploit periphery, they perish, hence modern industry (in the core) not only creates the world market, but that creates the need for a constantly expanding market, Marx and Engels emphasized, "capitalism chases the bourgeoisie over the whole surface of the globe" so that it "must nestle everywhere, settle everywhere, establish connections everywhere" ... "so that wherever it goes all established national industries are dislodged by new industries, whose introduction becomes a life and death question for all civilized nations" (Harvey, 1996).[25]

The core of these classes called *contradictions*, was the lack of fit between what a society could produce with its human and technological potential, and the social relations of production which prevented that potential from being realized. The "double contradiction" in the process of industrialization in Latin America (mentioned earlier) can be seen as a disjunction between the productive potential of a society and the distribution of goods and services among its members which gave rise to *dependency theory*. One way in which this theory is important is in its view that social change arises out of political struggles, radical and sudden breaks in continuity, rather than from gradual evolution. Unequal class-based relationship between and or the workers was perceived by Marx to lead inevitably to class conflict. Tracing this conflict in class relationships in time, Marx showed how groups of people with common interests derived from their position in production (e.g. as wage workers) have to organize to bring about the ends that they want. This conflict is called *class struggle*, and for Marxists it is seen as the motor of social change and development.

From this basic approach many things followed—Marx did not study class relationships in space, so Marx could not conceptualize uneven spatial development due to working of capitalism. In Marxist theories of imperialism, there is no theory of underdevelopment and it is having at its centre a notion of exploitation rather than one of dependency. Lenin developed the idea that *imperialism* was an exploitative system of economic, social and political relations which, while changing the colonized societies economically, socially and culturally, changed them in order that they could provide cheap inputs to production in capitalist societies as well as markets for their products. This arrangement always worked to the advantage of the imperial power. Such an approach to the question of development differs in many ways from those who have looked at it before. The father of this approach is Paul Baran who in his book *The Political Economy of Growth* in 1957 continued the central line of Marxist thought regarding the contradictory character of the needs of *imperialism* and the process of industrialization and the general economic development of the ex-colonial countries. Baran affirmed that since economic development in these countries is inimical to the dominant interests in the advanced capitalist countries so to inhibit in the dependent countries such developments, the advanced nations would form alliances with pro-capitalistic domestic allies for expropriating the surpluses in dependent countries which led to economic stagnation.

However, this movement in terms of popular explanation of the causes of underdevelopment came under attack from a coalition of "dependency" theorists, of whom the most prominent was Andre

Gunder Frank, a German born American economist living in Latin America who was influenced by tradition of Marxist thought. In 1967 A. G. Frank published a *cataclysmic critique of social scientific notions of modernization in the new decolonized countries* and attacked the whole of the then dominant modernization theory. Development is a zero-sum game, according to Frank: for some countries to become developed, other countries must become underdeveloped ("dependent"). Initially developing countries are "undeveloped," but immediately they come into contact with developed (capitalism) they become "underdeveloped" by mechanisms like transfer of value through trade, investment, and profits. Developing countries will remain underdeveloped as long as capitalism exists, and the only way out would be political solution and that through a revolution of an immediately socialist character, for within the context of the capitalist system, there could be no alternative to underdevelopment.[26]

Drawing on the Marxian tradition and on the theories which Latin American economist Ralph had developed earlier in 1950 to explain the problems of their societies, he argued that for ideological reasons Western theorists had incorrectly defined the problem as one of "development". To pose the problem in this way assumed that development could occur in any society if it adopted right economic policies alongwith a work ethic and parliamentary democracy. Frank said this was not possible. Instead of development being possible, what was actually in train was a process of underdevelopment. And by this what he meant that relations between rich and poor nations being beneficial to the latter, they should be seen as positively destructive of them, hindering and distorting their development. Underdevelopment is not a stage which precedes development, rather it now has to be seen as the end result of colonialism and imperialism. Another term, "underdevelopment", was introduced to describe societies which had simple technology but were unaffected by the developed societies. This view of development and underdevelopment sees both these states as the result of interaction between societies.

But from the point of view of dependency theory of Andre Gunder Frank, 1967 *underdevelopment* is not the result of some *lack* within the underdeveloped region, it *is not a condition,* it is caused by the continued *exploitation* of the dependent area by more powerful and wealthy ones so that it is *but a process —the process whereby a country (or, a region)* is *remained* locked into "classical dependence" and their backwardness is not the result of precapitalist structure, but rather a product of capitalism. To those familiar with Marxian analysis it is obvious that the dependency theory has its root in Marxist view that *imperialism* is an exploitative system of economic, social and political relations which, while changing the colonized societies economically, socially and culturally, changed them in order that they could made cheap inputs to production in the capitalist societies as well markets for their products. This arrangement contends that capitalism, by nature is exploitative and must always *subjugate* other areas in order for capitalism to survive and work to the advantage of the imperial power. This Marxists argue, is made necessary primarily because of the accumulation of capital into the hands of the few which threatens the viability of the system unless new sources of capital are constantly found. Such an approach to the question of development differs in many ways from those that we have looked at before.

It can be argued for the purpose of analysis of underdevelopment, we may distinguish four major stages to Frank's model of "dependency theory perspective of development process" (Streeten, 1980).[27] The *first stage*[28] described in the two paragraphs above, involves the obvious analogy that the relationship of exploitation between nations whereby the dependent countries have been historically drained of their resources and wealth by the developed ones, that is, the countries in the periphery have been incorporated into the world-system of economy since the early stage of their colonial periods. For example, in pre-colonial India, where shoots of industrialization began to sprout, the British imperialists rolled it back by swamping the colonial markets with the cheap factory-made manufactures. Karl Marx who inspired generations of dependency theorists, himself vehemently condoned the brutality perpetrated by British colonial rule in India.

The *second stage* of the dependency perspective also builds directly on Marxist analysis. In this context its focus is on the supposed necessity of capitalism to require a large "reserve army" of labour which it can exploit virtually at will. The underdeveloped regions provide not only raw materials, wealth and a market for more developed regions, but also function as a reserve labour supply. Labourers can be drawn into when industry is expanding, and simply laid off when economic conditions get bad. This regionalization of reserve labour force is seen as particularly advantageous to the developed countries (regions), as many unemployed workers return to their old homes. Thus the developed countries rarely have to pay the full economic or social costs of maintaining their reserve labour supply.

However, to this second stage, dependency theory does not differ substantially from traditional Marxist analysis. It is the *third stage* of dependency theory of Frank (1967) which marks the formation of a distinctive body of theory and empirical research. At this level, attention shifts from a focus on wealth and population movement to a direct consideration of the social structures which develop both to facilitate the process of regional dependency and as a consequence of it, the dependency theory took an axiomatic view so that the focus here is on the nature of *social class*, and on the way in which "bridgeheads" are formed between the dominant class in any developed capitalist society (which was the "bourgeoisie" or capitalist class)[29] and the dominant class, the "local elites" of the dependent area. In this way the elite (the so-called *co-opted compradors* of economic interests based in the dominant country) of the dependent region becomes "compromised" into supporting its own domination. At the same time there is a conflict of interest between the elite groups in both developed and dependent areas and even a conflict of interest between the two working class groups, and thus that the foreign policies of Countries of Global North would be concerned primarily with the promotion and protection of capitalist interests so that the articulation between the capitalist and non-capitalist modes of production produces an *economic polarization* but the attention shifts from a focus on wealth and population movement into a direct consideration of the consequences of dependency for the social and economic structures of the dependent country itself. In particular, the focus here is on the nature of social class, and on the way in which *bridgeheads* are formed between the international capitalist [that is, the multinational enterprises (MNEs)] class of Countries of Global North, particularly, USA, and their allies from the ruling elites of the dependent countries.[30] In this way the local oligarchies of the dependent country, becomes compromized into supporting its own domination to facilitate the process of regional dependency in their country and at the same time as "buying off" its own working class with a mixture of material rewards and racist ideology.

It is this *fourth stage* of the dependency theory analysis that attention shifts from a consideration of the links between developed and dependent countries into a direct consideration of the consequences of dependency for the social and economic structure of the dependent country itself. The articulation or integration between the capitalist and non-capitalist modes of production is necessarily achieved through an *economic polarization* which in turn creates two largely independent social structures of interminable metropolis-satellite chain, in which the surplus generated at each stage is successively drawn off towards the core country. The "dependent development" which occurs under conditions of dependency has characteristics which make it distinctive from that found in more developed regions.

It is tempting to dwell more extensively on the details of dependency theory by these *four stages.* In this context, it is relevant that persistence into the post-colonial period of great disparities of wealth and power and the obstacles facing Third World Countries[31] attempting to alter their relationship with the industrialized world has led to a profound action against the dependency of their countries. Frank's (1969) main point was that economic processes operated in different ways in different parts of the world —in his prognosis, the *development of the underdevelopment* is a concept which best characterized the capitalist dynamics by which the Countries of Global North may have experienced development and at the same time caused greater levels of underdevelopment and dependency within the remainder, *i.e.*, the Countries of Global South.

In no uncertain terms, Frank (1969)[32] dismissed the Weberian modernization theory as useless from a policy point of view because they failed to define correctly the kinds of social and economic processes at work in the underdeveloped countries. With detailed historical case studies of Chile and Brazil and drawing on the Mexican tradition as well as on the theories which Latin American economists and sociologists had developed to explain how dependency involved an association between the national bourgeoisie and international capitalism and led increased impoverishment and marginalization of their societies. In the Brazilian context he presented a clearest case of national and regional underdevelopment, the way how the expansion of capitalism, beginning in the sixteenth century, sequentially incorporated vast hinterlands and their cores into the global economy by transforming them into "export nodes". The cities like Rio de Janeiro, Sao Paulo which figured most prominently in this capitalist expansion: they are national metropolises in their own right with their own chain of satellites for their countries but at the same time this process of *development of the underdevelopment* significantly impacted on their country's economy, polity and society in the longer term so that in the world scale they themselves are international satellites for the world metropolis, the USA. That is, taking a photograph of a slice of the world, Frank writes "we get a whole chain of metropolises and satellites, which runs from world metropolis down to the *hacienda* or rural merchant who are satellites of the local commercial metropolitan centre but who in their turn have peasants as their satellites." The problems of their societies, Frank argued that, for ideological reasons Weber's emphasis that "Capitalism is not about greed for gain but a rational enterprise seeking opportunes for profit-making and the enterprise which doesn't seek it would be doomed to extinction" is wrong since Weber assumed that development could occur in any society if it adopted the right economic policies along with a work ethic and parliamentary democracy which was not possible in the Third World countries. Instead of development being possible, what was actually in train was a process of underdevelopment. Instead of all countries following the same path to development, Frank (1969) identified *two complementary but opposite paths.* What he meant that underdevelopment is not an original *condition,* equivalent to traditionalism or backwardness but to the contrary, it is a *process*—the process whereby a condition is created so that the country (or a region) is integrated into the world-wide system of exchange (that is, the world capitalist system) which originated in the sixteenth century for to be *made* dependent. Whereas the rich countries had indeed shown impressive economic growth, this development was countered by an equally impressive process of underdevelopment in poorer countries. They were experiencing the "development of under-development". Peru, for instance, had been a victim of this for several centuries, it was not catching up and it was being continually under-developed. The bottom rung of the ladder was so because of this process of keeping the countries underdeveloped. And this was not a result of the nature of these particular societies: they were kept on the bottom by those at the top. Development and underdevelopment were *twin processes,* both necessary for each other's operation in the modern world. Those at the top of the ladder maintained their position through exploiting those at the bottom. The concentration of poverty is the lack of development, and is not, therefore, a consequence of geographical isolation or the failure of Western technology, capital and values to spread. The benign path to development was in reality an international dogfight which reflected a history of winners and losers : Britain was a winner, Peru was a loser. Thus the easy phase of tariff-protected structuralist model through *import-substituted-industrialization* (ISI), so beloved by ECLA had passed and it has been turned inside out.

The Thesis of Capitalist Underdevelopment: Voluntary Colonialism in Chile

Dependency is a two-way street : The First World is dependent from a position of power on Countries of the South for their primary products and the Southern Countries are forced to depend on the former for carrying on their international trade.

The root of the dependency theory lie in the Marxist analysis of imperialism. This position contends that capitalism, by nature, is exploitative and must constantly subjugate other areas in order for it to survive. This argument is made necessary primarily because of the accumulation of capital into the hands of the few which threatens the viability of the system unless new sources of capital are constantly found. Hence rather than relations between rich and poor nations being beneficial to the latter, the rich nations should be seen as positively destructive of them, hindering and distorting their development. Frank cited underdevelopment in Chile is the necessary product of four centuries of capitalist development and of the internal contradictions of capitalism itself, bringing a highly unequal social structure between the Chilean upper class dependent on foreign capital and the precapitalist relations surviving in agriculture. These contradictions are the "expropriation of economic surplus from the many and its appropriation by the few, the polarisation of the capitalist system into metropolitan centre and peripheral satellites, and the continuity of the fundamental structure of the capitalist system throughout the history of its expansion and transformation, due to the persistence or recreation of these contradictions everywhere and at all times." Frank (1971)[33] asserted that these capitalist contradictions and the historical development of the capitalist system have generated underdevelopment in the peripheral satellites whose economic surplus was expropriated, while generating economic development in the metropolitan centres which appropriate that surplus, accumulate and grow. Consumption patterns, for example, being stemmed from high technology, production of luxury goods, and from the creation of a class in developing nations that identifies the foreign interests than with its own society's needs—and, further as this process continues, internal inequities in the countries develop which Frank viewed "largely a result of the impact of external factors" (Frank, 1971).

The Chilean case illustrates how difficult it is for a small retarded capitalist country to win national control of her own resources in this era of domination by a few advanced capitalist countries, and how the battle for national autonomy, if carried far enough, becomes merged in the battle for social revolution. Two American corporations, Anaconda and Kennecott, acquired ownership rights over the huge low-grade ore deposits in the Norte Chico in the early twentieth century, established the Gran Mineria and came to control the most valuable resource of Chile, viz. *copper, from the mining to the marketing stage. Between 1913 (the year it started) and 1958 when the Anaconda-owned Chiquicamata mine, for example, a total of 6 million tons of copper (over 300 million tons of oxide ore had been removed) and by the 1960s, this mine had a production capacity of 100,000 tons of copper ore per day. By 1970, Anaconda and Kennecott together had taken out of Chile $ 4 billion, whereas their ploughing back of this extraordinary profit was a nominal investment in Chile, at most of $ 800 million, and the real investment was a lot less. During periods of slump in the copper market, USA, one of the biggest buyer of Chilean copper, imposed tariffs on it. The Chileans tried to market their own copper in the 1950s but the experiment was abandoned as soon as it ran into trouble, which was partly deliberately engineered by the monopolistic interests in copper. The Government of Chile tried to bribe the copper companies to invest more in the industry, but the latter simply reaped the higher profits without carrying out investment on the promised scale because such investment did not fit in with their world-wide plans. The regime of President Frei gave* *another bribe in the form of inflated prices for "Chileanization" of part of the copper mines, but this had no better effect on company performance than earlier measures. The nationalization of copper mines by Salvador Allende in 1971 at a price which was fixed after allowing for various kinds of illegal appropriation by the companies was an act of national self-defence, transcending class interests. The American companies fought the Chilean governments in various world courts and in the world copper market. The authoritarian traditional oligarchy of Chile teamed up with American interests and ultimately succeeded in overthrowing the popularly elected Allende regime between 1970 and 1973.* The *coup-di-etat* 1973 destabilized Allende, the new regime *"compen-sated" the copper companies on a generous scale, thus paying back much of its debts to its US patrons.* Thus Chile remained subordinated for ever by a programme of monetarist shock and thereby to be forced into following the ingredients of the neo-classical recipe to deal with its problems.

Modernization and Dependency Theory Compared

Thus just as modernization theory behaving like a liberal model, had been overtly optimistic about the prospects for simultaneous economic and political development, so that there can be a logical possibility of "spatial equilibrium" in the long-term, similarly dependency theory proved to be excessively pessimistic about the prospects for economic growth : the "real" world of development diverges and by 1960s it became clear emphatically that anticipated "spread" or diffusion effects of economic growth did not take place in Countries of the South where spatial inequalities at a range of scales—international, regional and urban—had become more rather than less pronounced.

In this context, a virtual avalanche of critical writings appeared claiming that development and prosperity of the Countries of the North in the world economy were based on development of underdeveloped and squalor in the periphery. The continuing failure of Countries of the North to achieve their own steady economic growth and to cater for the needs of all their citizens led to "rediscovery of poverty" in USA and Europe and an increased concern for welfarism. The concept of *relative deprivation* was used to show how despite apparent affluence, many people remained poor so that while poverty was damned annoying in Countries of the South it became a disgrace elsewhere. The concern got turned from trying to find ways forward to modernization paradigm but to explain the persistence of poverty in a world some people were very rich indeed and others barely survived. The dependency approach offered useful corrections to modernization theory. Most important it highlighted an important influence on societies of Global South that modernization theory had largely neglected: international trade, finance and investment patterns. Dependencistas, or the dependency theorists like Andre Gunder Frank has prognosis that the Southern nations, consigned to producing non-industrial goods and ruled by unrepresentative elites, were doomed to be continued backwardness. The dependencistas' work raised the question of development in a new way. According to this "dependency theory", the problems of the Southern nations could now be explained as the outcome of exploitation by the developed world. To this extent the dependency theory encouraged both a degree of *utopianism*—everything would be all right if dependency was done away with—and also a degree of *pessimism*—if dependency cannot be done away with, then there is nothing that policy makers or planners could do. This theory shifted the focus of research from exclusively all internal factors such as class exploitation, population growth and natural resource shortages, since according to the dependencistas, all these internal factors could be too easily dismissed and they focus to international economic and cultural relations, creating new fields of research including *international political economy*—thus the important point of dependency theory is its "international nature". Very importantly, it made study of the history and the role of the past in understanding the present as well as to distinguish sufficiently between the histories and geographical reality of circumstances of different countries.

Dependency theorists also helped redefine the concept of economic development—whereas early studies of the Global South focussed heavily on economic growth, dependencistas emphasized the importance of economic distribution. When rapid economic growth is accompanied by an increased concentration of wealth and income for a minority, as has frequently occurred, it offers little benefit to the impoverished majority, which may be left worst off. Influenced by the dependency theory, even such pillars of establishment as the World

Bank have oriented their focus toward "growth and redistribution" since 1974. In that way the explanation of dependency theory for the systematic underdevelopment of the non-industrialized or semi-industrialized nations thereby has filled up a lacuna in Marxist theories of imperialism, which have tended to focus more on the contradictory nature of capitalist production and its property relations and less on the politically instituted relations of exchange between nations. Dependency theory has done more than this, however, because it has as one of its corollaries the proposition that as long as actually existing capitalism prevails, the prognosis of Andre Gunder Frank that low-income countries as a group are in principle not going to be the beneficiaries of any structural changes in the international system. But if it is the case that the less-wealthy countries are not going to benefit from *any* structural disposition of the international system as long as actually existing capitalism continues to exist, then it behoves the proponent of dependency theory to provide an account, however stylized, of the current disposition of this system.

Thus though dependency theory appears to provide a simple and powerful model of the origins and nature of underdevelopment, according to the critiques it also exaggerate the role of external influences and consequently downplays the internal obstacles to the development of more equitable domestic, economic, social, and political systems. A lessening of dependence on the industrialized world requires not only the achievement of better international bargains but also improved management of local resources and significant social and economic changes *within* the countries of Global South. Thus dependency theory apart from its contribution to the growing sense of self-reliance prompting Southern nations to seek more influence over their political and economic destinies, but it failed to confront some very important questions which had been central to modernization theory. At a more purely theoretical level, critiques have also been made of the various 'essentialisms' and false 'universalisms' that are said to be devil the development or dependency paradigms. Examples of these include the presumption that industrialization and the possession of industrial capital are the crucial requisites of economic progress (so that an economy is deemed to be developed, developing, or undeveloped depending on whether or not it has traversed an appropriate path to industrialization, and thus to have amassed a commensurate kind and scale of industrial capital); the inability to think beyond the state as the primary and essential vehicle of economic development; the (often unacknowledged) importation of problematic assumptions regarding the role of foreign investment and foreign trade in the so-called less developed economies; a eurocentric bias; and an underestimation of the implications of widespread and haphazard industrial development for the environment. In the face of these and other challenges to their paradigm, the proponents of theories of uneven development or dependency have, according to their critics, done little more than reiterate (*i*) their conviction that some form of socialism can still function as a potential countervailing force to capitalist depredation, and (*ii*) their belief in the efficacy of some kind of strategic decoupling of the less-developed economies from the capitalist world-system, again with the hope that this can function as a protection against capitalist encroachment. But, according to these critics, neither (*i*) nor (*ii*) appear to be viable in the current capitalist dispensation. This dispensation is more staunchly and comprehensively inhospitable to socialist aspirations than it has ever been (thereby blocking off any socialist path to socialism). Moreover, the completely integrated character of actually existing world capitalism ostensibly makes any attempt at decoupling a sure-fire recipe for economic collapse or extinction.

In examining this claim, or set of claims, it can be suggested that a good case can be made for the retention of notions of uneven development and dependency, albeit ones framed in very different terms. After all, persistent international economic inequalities remain, and if anything are becoming even more intractable which compel on part of dependency theorists to think of the developmental state in other than purely mechanistic or monolithic terms. So there have been certain attempts at reformulation, to arrive at a reanimated version of dependency theory through the construction of an account of the impact, both economic and political, of transnational financial capital on the so-called less developed economies. Since the proponents of the theory of uneven development have tended to be viewed primarily in terms of a country's flawed or incomplete negotiation of the processes of industrialization, they could not take into consideration the impact of "global financial markets", the large-scale effects of which on economy of the Southern countries particularly of the Least Developed Countries (LDCs) have been felt only fairly recently.

With this objective in mind, the dependency or uneven development paradigm is required to involve assent to some version of each of the following related propositions which are not mutually exclusive:

1. Disparities in wealth between nations as a group are due fundamentally to asymmetries of economic and political power that are constitutive of the capitalist system of development, and indeed of world capitalism generally. This approach attempt to concentrate upon the obstacles which confront "national development": while echoing the same distrust of the free markets like in ECLA approach, this approach argued that an essentially "inward-looking" pattern of industrial development must be deliberately engineered in order to initiate the developmental process and *find* the *dynamic* (instead of the *static*) *comparative advantage* through learning-by-doing. This prescription coupled with the general notion that the main constraining factor in Countries of the North came mainly from the supply side and not from any deficiency of effective demand, and this was used as a justification for the policy bias of protecting domestic (*infant*) industries through import-licensing and capital-cheapening policies. The success stories of the East Asian countries clearly contradict the crude versions of dependency theory as the East Asian phenomenon assert in a generalized way the prime importance of internal policy, these countries pursued for a long enough time like the adequacy of market price signals as guides to resource allocation, the effective insignificance of the potential problems which might be associated with the role of foreign capital, the transfer of technology, or the generation of external economies in the course of industrialization.
2. The asymmetries of economic and political power that exist in terms of large technological gap, extensive market

instability, existence of important external economies associated with industrial development etc. between groups of nations which cannot be removed or significantly ameliorated within the structures and strategic possibilities that are integral to the prevailing system of capitalist accumulation. The essential characteristic of this approach is that it attempts to construct Frank's "theory of under-development" in which the dependent character of the peripheral economies is the *hub* on which the whole analysis of underdevelopment turn in which the surplus generated at each stage is successively drawn off towards the centre: the dependent character of these economies would trace certain processes casually linked to underdevelopment, so that it follows there is no real possibility of sustained development within the system.

Notes & References

1. Colman, David and Nixson, Frederick (1986, 2nd edition), *Economics of Change in Less Developed Countries*, Philip Allan, Oxford.
2. Durkheim, Emile, 1893/1894, *The Division of Labour in Society*, translated by W.D. Halls, London.
3. Rostow, W. W., 1960, *The Stages of Economic Growth : A Non-Communist Manifesto*, Cambridge University Press, MIT.
4. Apter, David, E. (1965), *The Politics of Modernization*, University of Chicago Press, Chicago.
5. Weber's *Gemeinschaft* in which every part fulfils for itself all needs, like a society based on organic solidarity where everyone is either is all warrior, or all hunter, or all tool-maker.
6. *Gesellschaft* Communities arranged in a manner that goods, tangible and intangible, symbolic and material are distributed. Such a distribution is always unequal and necessarily involves power. Classes, status groups and parties are phenomena of the distribution of power within a community. Status groups makes up the social order, classes the economic order, and parties the legal/political order. Each order affects and is affected by the other.
7. Lloyd, I. Rudolph and Susanne Hoeber Rudolph, 1967, *The Materials of Tradition: Political Development in India,* Chicago University Press, Chicago.
8. *Cited from* p. 56, Forbes, D. K. 1984, *The Geography of Underdevelopment: A Critical Survey*, Johns Hopkins University Press, Baltimore.
9. Bernstein (1971), *Modernization and the Structures of Society*, Princeton University Press, Princeton.

10. Eisenstadt, S. N., 1973, *Tradition, Change and Modernity*, Wiley, New York.
11. Massey, Doreen (1995, 2nd edition), *Spatial Divisions of Labour : Social Structures and the Geography of Production*, Routledge, New York.
12. Bagchi, A. K., 1982, *The Political Economy of Underdevelopment*, Cambridge University Press, Cambridge.
13. *Cited from* Forbes, D. K.,1984, *The Geography of Underdevelopment : A Critical Survey,* John Hopkins University Press, Baltimore, *ibid.* pp. 66. Marx in his article on the *Future Results of British Rule in India* in 1853, fleetingly analyzed British expansion in India as brutal, but a "necessary step towards *Socialism*". Such an expansion would have a destabilizing and disintegrating effect on Asiatic mode of production, but since it didn't possess a dialectic of internal development, so the capitalist mode of development would have a destabilizing and disintegrating effect on its societies to generate an internal dynamic leading to higher stages of development. In this respect Marx makes no distinction between endogenous capitalist development (which occurred in Western Europe) which he witnessed in North America but Marx was not aware of divergencies attributable to changes in circumstances or diagnosis, since Marx himself warned in 1877 a historical sketch of metamorphosis of genesis of capital cannot be used as one's master-key for a general geographical historiographic theory.
14. R.J. Johnston, 1993, *The Challenge for Geography—A Changing World : A Changing Discipline*, Blackwell, p. 250, London.
15. Rostow, W. W. 1960, *The Stages of Economic Growth : A Non-Communist Manifesto,* C.U.P., MIT.
16. Taylor, Peter, J. (1989), "The Error of Developmentalism in Human Geography" in, Gregory, Derek and Rex Walford, (eds. 1989), *Horizons in Human Geography*, Macmilan, New York.
17. United Nations proclaimed the decade of 1960s for the first time as a Decade of Development and kept bold targets for economic growth and social development.
18. Appadurai, Arjun, 1997, *Modernity at Large : Cultural Dimensions of Globalization,* Oxford University Press, Delhi.
19. Western Europe where capitalism first developed, went through a "classic" sequence of transformation which runs from subsistence economies through slavery, feudalism, merchant capitalism, industrial capitalism to advanced capitalist mode of production to another. But since the "logic" of capitalism requires ever-expanding markets, this set sequence of transformation was not followed elsewhere so that within the landscapes of world economy, regional variations came about by different social, cultural and political contexts which themselves are a product of particular sequences and stages of economic organization.
20. The total debt stock of all the developing countries which were $ 61 billion in 1970 rose to $ 647.30 billion in 1980 to which at the end of 1982 rose to an estimated $1,510 billion at the end of 1990 – an 132 per cent rise followed by a further 37 per cent increase to $ 2,067.70 trillion by 1995—a jump of 250 per cent since 1982, *cited in* Kappagoda, Nihal, 1997, *The Debt Crisis : Fifteen Years On*, Development Express, Canadian International Development Centre, Ottawa.

21. Incidentally, Latin American and Caribbean countries where three-fourths of population live in urban areas, are having great inequality in income and wealth—measured in terms of Gini Coefficient, the region is having the most unequal distribution in the world.
22. Development studies originated in the aftermath of World War II and Lewis, W. Arthur is known for his developmental ideas since 1954,which resulted in a bias towards seeing the secondary sector as the engine of growth and the primary sector as a source of surplus labour which he tried to balance out in his subsequent growth models for Countries of the South.
23. Rimmer, P. J. and Forbes, D. K., 1982, *Underdevelopment Theory : A Geographical Review*, Australian Geographer, 15, 4, p. 197-211.
24. *Das Capital* the legendary work in 3 volumes of Karl Marx, of which 1st volume was published in 1867 which analyzed the capitalist process of production. In *Capital*, Marx elaborated his version of the labour theory and his conception of surplus value and exploitation which would ultimately lead to a falling rate of profit in the collapse of industrial capitalism. Volumes II and III were finished during the 1860s but they were published posthumously by Engels.
25. David, Harvey, 1996, *Justice, Nature & the Geography of Difference*, Blackwell, Cambridge, USA.
26. Frank attacked the whole of the then dominant modernization theory in his *The Sociology of Development and the Underdevelopment of Sociology* in 1967.
27. Streeten, P., 1980, "Development Ideas in Historical Perspective" in, *A. D. Hirschman et al. Toward A New Strategy for Development,* Pergamon, Oxford, pp. 21-52.
28. *Note* this first stage of dependency theory involves also an obvious analogy that the relationship of exploitation between nations has its counterpart at the regional level between the regions. Known as *transfer dependency,* this regional dependency is of different type.
29. In Marxist theory, the capitalist class or "bourgeoisie" is the social group, opposite to the working class, the *proletariate.*
30. Aggressive expansionism not by establishing colonies but by aggressive capitalism is the characteristic of American imperialism.
31. The meaning of the term "Third World" has been in a continual state of evolution since the phrase was first coined in 1952. On August 14 of that year, the decolonized countries of Asia (except Japan), Africa and Latin and Caribbean America were given a pedigree given in France *Tiers Etat* (or *Monde*) or the "Third World" by a French demographer, Alfred Sauvy in an article in *L'Observateur*—the poor, powerless, marginalized people—in France prior to the 1789 French Revolution. Such a useful concept would not be confined long to the people in the history of one country. It was quickly widened to include countries that were marginalized in the international system, compared to the First World which is a fairly coherent unity consisting of the industrialized countries, including the United States and Canada of North America, the countries of Western Europe, Australia and Japan. All these are having democracies, per capita income is relatively high. Thus definitions of the "Third World" have been contested, as have

the origins of the phrase. Yet the concept of three worlds can hardly be said to convey a precise meaning. The geographical boundaries of the "Third World" vary enormously according to which statistical indicators are used and the idea has different meanings in different cultures.

32. Frank, A.G., 1969, *Latin America : Underdevelopment or Revolution,* Monthly Review Press, New York.
33. Frank, A. G., 1971, *Capitalism and Underdevelopment in Latin America,* Penguin.

4

GLOBAL SOUTH IN INTERNATIONAL SETTING OF TRADE

Dependency Revisited

Trade: An Expansion Generated Economy

Foreign trade is an integral facet of the total growth phenomena which has not been crowned with conspicuous success for the LDCs of the South. In the present context since the development process is being studied as a whole, so foreign trade has to be viewed as an integral part, facilitating perhaps the development strategy for countries of the South in most crucial aspect. Again since trade and aid takes normative aspects, hence, the role of foreign trade has to concentrate on such matters as Import Substituted Industrialization (ISI) export promotion and the economizing of foreign exchange, all heavily tinged with policy connotations and based on a minimal theoretical foundation. In short, this study often appears to border on an "art" rather than constituting any scientific analysis aimed at a better understanding of the total growth process. And scope of this analysis is also handicapped by the lack of homogeneity among the Countries of the South though commonly carrying the label of "labour surplus variety" for most of them.

Regarding trade, with respect to Countries of the South, neither with respect to one nor the other it is possible to generalize from the experience of any one. For example, South East Asian countries are in a position to export agricultural products while many others must import a considerable proportion of their consumption requirements. Venezuela exports oil, other Latin American countries import it. Such impressive instances of international comparative advantage as tin in

Bolivia, guano in Peru, rubber in Malay, may after all be viewed as historical accidents for which it is hard to find parallels, and from which it may thus be difficult to generalize. The problems implied by such heterogeneity of the economy's resource base is further compounded by the fact that growth itself very often implies a continuing revision of comparative advantage in the direction of broadening the human ingenuity and capital component and narrowing the natural resource component of exports. This constantly changing nature of the resources base means that the comparative advantage scale is dynamically changing at an unknown pace and in only generally definable directions, thus rendering precise economic analysis and prognostication difficult, if not impossible. Figure 4.1 summarizes the analysis of growth in the form of an abbreviated framework.*

Resource Allocation in Closed Economy : *The Model*

In Figure 4.1, two circles are shown, the circle on the right labelled *A* represents the relatively large agricultural subsistence sector and the circle on the left labelled *I* represents the relatively small industrial sector. In a successfully growing economy, with the labour allocation process going forward on a continuous basis, the agricultural population shrinks relative to the industrial population, this is symbolically denoted by the directions of the arrows, which point inward in the case of the larger circle and outward in the case of the smaller circle.

Arrows drawn between circles *A* and *I* represent the direction of real resources flow in such a "dualistic economy". Thus, the fact that

* It is intuitively clear that whatever a country is, it must participate in the international trading community according to its own particular comparative advantage, which is a function of its natural and human resource endowments. In talking about development, a landscape context is brought in geography which the present author also does and he comes across the term "landscape development" in the book written by two economists, John, C. H. Fei and Gustav Ranis of Yale University as early as in 1964 on a theory of economic development for labour-surplus countries. Describing the aggregate economic growth in an eclectic manner through capital accumulation, market and the sectors, they finally took trade as the engine of growth. The model they developed in normative aspects in their concluding chapter, is practically reproduced here.

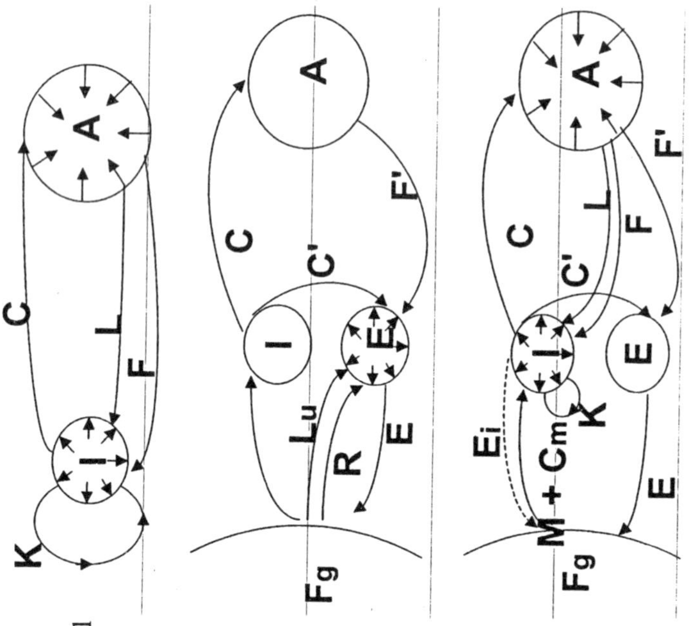

(a) Colonial Economy in Closed System : Initial

(b) Dual Economy in Closed System : Maturity

(c) Turning Point Development Oriented Open Systems

Fig. 4.1 : Growth Development Model

the agricultural sector's role in this context is to fuel the expansion of the industrial sector is indicated by the fact that Labour *L* and food and agricultural raw materials *F* must flow from *A* to *I*. On the other hand, the output of the industrial sector is seen to flow in part to the agricultural sector in the form of manufactured consumer goods, fertilizer and a limited amount of capital goods, *C*, and in part back into the industrial sector itself in the form of capital goods, *K* for purposes of further industrialization. The movements indicated here represent the bare essentials of the major resource flows, as the closed economy *moves forward on the road to turning point and economic maturity*. Flow of consumer goods within each sector (agricultural goods for consumption by urban industrial workers) are omitted from Figure 4.1. As the consequence of the growth-promoting force of the agricultural productivity expansion, agricultural workers, *L* are released by the agricultural sector and reallocated to the industrial sector. The absorption of labour by the industrial sector is accommodated by innovation activities in conjunction with the accumulation of new capital goods financed out of industrial profits *K* and the agricultural surplus delivered to the industrial sector mainly in the form of food and raw materials *F*. The expansion of agricultural productivity through time is essential, moreover, for other reasons, in that it provides a market outlet for the absorption of the increasing flow of the industrial consumer and capital goods *C*. Without these the deterioration of the industrial sector's terms of trade would impede the development effort by depriving the industrial sector of the means (*i.e.,* the investment fund) as well as the incentives for continuous investment activity.

This portrait of the resources allocation pattern underlying the Figure 4.1(a), while sketchy, nevertheless serves to focus our attention on all the crucial relevant phenomena, for example: (1) the relative size of the two sectors, (2) the overall direction of resources allocation over time, (3) the commodity characteristics of the development process, and (4) the nature of development financing. In broadening this simple framework to accommodate additional, *open economy*, dimensions of the problem we may, *heuristically* examine the new resources allocation pattern emerging in the same fashion. For example, when a foreign trade sector is appended to Figure 4.1(a), in an open economy model we may once again raise these very questions concerning the relative size of the various sectors, the anticipated direction of resource flows, the commodity characteristics of such flows, and the nature of finance. Following is the description of these basic structural characteristics which constitutes, in fact, the first step in our analysis of the foreign trade and involvement of the dimension of foreign capital in context of Countries of the South.

Resource Allocation in the "Enclave" Economy

Historically, the "opening up" of labour surplus underdeveloped economies to international trade and foreign capital has most frequently come through the establishment, by the industrialized countries, of commercial and extractive footholds abroad on what may be called a *colonial basis*. The less developed economy's export sector emerging as a consequence may be characterized as typically composed of the processing and export of one or a small number of natural resource-specific commodities, such as jute, oil, tea, cotton, etc., in which, thanks largely to an act of nature, the colonial area enjoys conspicuous *absolute*, as well as *comparative advantage*. Typically, roads, railways, and other social overhead capital, as well as the financial and commercial institutions, are also oriented toward the operation and expansion of these raw material-oriented export industries. The result is the creation of an export sector in the underdeveloped colonial economy which is really but "an extension of the industrial *mother country*" even though geographically located in the underdeveloped economy. The *spill-over* effects from such dynamically growing enclaves in the sea of stagnation in which the colonial countries were customarily embedded were likely to be of very little significance.

The resource allocation pattern of such an *enclave economy is described in* Figure 4.1(b). Here in addition to the large subsistence agricultural sector, *A*, and the small industrial domestic sector, *I*, of Figure 4.1 (a), we indicate the existence of a huge foreign sector, *Fg* and an expanding export sector, *E*. The subsistence agricultural sector must then supply food, F' and the domestic industrial sector must supply manufactured consumer goods C' to feed the labourers working in the export sector *E*. However, the output of the export sector, which is of a raw material-specific nature, is almost entirely exported. In return, the *core country* supplies the *periphery* with certain light manufactured consumer goods, mainly for the industrial urban sector households C_l, and with luxury C_c consumer goods and capital goods *C*, mainly for the expanding enclave sector. *Notice* that the resource flow arrows identifiable with the industrialization and labour reallocation process, *i.e.*, the movement of labour *L* and food *F* from *A* to *I* in Figure 4.1(a), are now deleted since they are of relatively little significance.

A close examination of the flow represented by *R*, which corresponds to the profits of the export sector, will serve to give us a firm grasp of the notion of the enclave economy. In case, the trend of foreign demand for the particular resource-specific export of the less developed economy is favourable, much of *R* will be repatriated to the core, thus advancing the course of capital accumulation and growth at the centre rather than the periphery. This source of funds for investment in the underdeveloped economy seldom finds its way into the domestic industrial sector, *I*, if only for the reason that the lack of purchasing power arising from the essentially stagnant agricultural sector, *A*, provides no incentive for such investment. In fact, whatever profits can be generated in the domestic industrial sector (likely to be small relative to *R*) also tend to flow into the export sector.

Under these circumstances the colonial economy's welfare depends almost entirely on the strength of the demand from abroad for the raw material-specific exports of the enclave sector. The underdeveloped economy thus participates in the international trading community on the principle of an international division of labour which relegates it to the task of exporting goods with a heavy natural resources component and very little value added in terms of human ingenuity or capital. In fact, if under favourable foreign demand the enclave sector spurts forward, economic life in the subsistence agricultural sector—is hardly touched by the growth-promoting forces emanating from the export-oriented activities of the enclave sector. The underdeveloped economy undoubtedly enjoys certain advantages from its enclave status as far as its participation in the international economy is concerned. For one thing, it receives an automatic flow of capital and technical assistance from the "core" and the assistance of a protected market for its export products—all part of the quasi-economic, quasi-political package called *colonialism*. There remains nevertheless the overwhelming disadvantage that the growth-promoting forces in the export sector, *E* do not spill over to the rest of the economy. The historical experience of all major *enclave economies,* such as India, Myanmar, China etc., indicates that particular growing periods in the export sectors of such economies may spurt ahead from time to time or even grow steadily with steady demand expansion abroad in their core countries, but the rest of the economy remains relatively untouched. The drive of any economy as a whole toward economic maturity must originate mainly from the overwhelmingly large domestic sector in the typical *dualistic economy*; what happens to the 5 or 10 per cent of the population in the enclave sector is not likely to be instrumental in deciding the success or failure of the labour surplus economy's drive toward "turning point and economic maturity".

Resource Allocation in the Open Development-Oriented Economy

By the end of the World War II, the political system underlying this colonial set of relations between the less developed and the mature economies had become substantially eroded, with more and more countries at the "periphery" achieving political independence. The desire to achieve a measure of economic independence by means of a more or less full-blown economic development effort has become an indelible sign of our times. Under these circumstances the triangular pattern of resource allocation as it prevailed under the enclave system threatens to break down. Accepting the goals of economic growth as their overriding national objective, most of the governments of the less developed ex-colonial countries have tried more or less consciously to assess their total actual and potential resources and deploy them in a more or less well-articulated national plan or programme. In this context the resource allocation pattern of the underdeveloped economy ideally undergoes considerable change—*i.e.*, it becomes development-policy-oriented, with foreign trade and foreign capital assigned important roles in the overall development strategy of the government. In short, the dynamics of growth in the enclave economy have given way to a growth pattern dominated by a concern for the development of the heretofore neglected domestic economy.

Schematically, the resources allocation pattern under these

conditions is depicted in Figure 4.1(c). Figure 4.1(c), it will be noted, contains the same sectors as Figure 4.1(b), since it depicts an open economy, but it also contains the same sectors as Figure 4.1(a), since it depicts a "development-oriented economy". As far as *E*, the economy's export sector, is concerned, the commodities exported retain their overwhelmingly natural resource-specific nature, an economic fact of life inherited from the colonial past which cannot be changed overnight. The most dramatic change which does occur concurrently with the abandonment of the enclave type of growth is the altered trading pattern resulting from a change in the disposition of these export earnings. As seen in Figure 4.1(c), export earnings previously deployed for purposes of reinvestment in the enclave sector, to finance the luxury consumption of the foreigners and their associated engaged in these activities —or siphoned off for repatriation to the "core"—are now redirected. This new flow is to the industrial sector, to finance a particular type of imported good, *i.e.*, capital equipment and raw materials, for the growing industries. Such relatively drastic structural change can be accomplished as the underdeveloped economy's government assumes the power previously in the hands of the foreign capitalists or governments to decide on the allocation of the scarce foreign exchange resources. While the colonial heritage may continue for some time to dominate the nature of the export trade, it is the growth-conscious intervention via import and foreign exchange controls which represents the prominent institutional feature of the postcolonial era.

Comparing the "closed" development-oriented economy depicted in Figure 4.1(a) with the "open" development-oriented economy of Figure 4.1(c), the significance of the export and foreign sectors (*E* and *F*) may be seen in at least four related ways. (First and most important) the export sector, in conjunction with the foreign sector, constitutes a powerful *new production function* available to the economy through which the inputs of F' and C' (the food and industrial consumer goods required by workers in the export sector) are converted into imported capital goods and raw materials *M*. Access to this more efficient foreign production function—more efficient, that is, any alternative, domestically available, transformation process—undoubtedly represents the most important benefit accruing to the underdeveloped economy as a latecomer on the international scene.

A (second) and related, advantage accruing as a direct consequence of the economy's full-fledged participation in international trade is that the whole range of technological alternatives, assembled abroad through a prolonged and costly method of trial, error and obsolescence becomes available virtually for the asking. Picking and choosing from this formidable shelf of historical technology, the underdeveloped economy is at least in a position to import the type of capital equipment most suited to its resource endowment and development strategy. Whether it does so wisely is another matter, but the opportunity is there.

(Thirdly), the foreign sector may be viewed as an important additional source of savings to finance the expansion of the domestic industrial sector. While, under the enclave colonial system, foreign private capital also did flow in, it was, severely circumscribed in direction and impact, certainly always well "paid for". On the other hand, the large-scale post-war flow of untagged grants and liberal-termed loans into the Countries of the South cannot but be said to have substantially eased the great resources tightness of their underdeveloped economy as well as the specific shortage of imported commodities—thus lightening the load of those guiding the development effort. In this context, we require to say more about the benefits —and the limitations—of the potential role of foreign capital in the development process.

(Fourthly), and perhaps least conspicuously, the role of the export sector is to provide in itself an additional source of savings available for the expansion of the domestic industrial sector. The nature of these savings arises from the fact that a portion of the total imports ($M + C_1$) takes on the nature of an economic surplus from the point of view of the export sector, in the sense that the total value of the resources ($C' + F'$) delivered to the export sector by the rest of the economy is worth less than the total value of ($M + C_1$), the resources delivered in return. The most concrete evidence that such a surplus exists is seen from the fact that in many underdeveloped countries there obtains an overvalued official foreign exchange rate at which the exporters are generally reluctant to sell and importers anxious to buy. To the extent that such overvaluation does exist, the export sector is forced to subsidize the importers of capital goods and producers in the domestic industrial sector. Where fiscal systems are highly inadequate it may not be possible to secure any such additional contributions to the economy's saving fund in an alternative fashion, *e.g.*, via direct agricultural taxation.

In the development-oriented economy of Figure 4.1(c), the export sector *E*, in fact plays a role quite similar to that of the domestic subsistence sector, *A*. Both sectors are "squeezed" for the benefit of the expanding industrial sector *I* with the agricultural surplus $(F - C)$ and the trading surplus $(M + C_1 - C' - F')$ representing the savings of sector *A* and *E*, respectively. There are, however, substantial differences in the manner in which the export sector *E* and the domestic agricultural sector *A* manage to fuel the expansion of the industrial sector, *I*. The most obvious and conspicuous differences is with respect to the commodity composition of the savings being channelized in two cases, *i.e.*, food and raw materials in one (from *A* through the wages fund) and capital goods and raw materials in the other (from *E* through imports). Because of this difference, the latter form of fueling development seems at times to receive much more attention than financing via the wages fund derived from agriculture. The nature of the goods is perhaps more glamorous and the foreign exchange bottleneck lends itself more readily as a focus of public attention than resistances to agricultural productivity change. But in view of the overwhelmingly preponderant size of the agricultural sector, on the one hand, and the obstacles abroad to rapid expansion of the relatively small export sector, on the other, it is doubtful that trade can be relied upon to serve as the "main engine of growth". Let us now turn to a further examination of this issue.

Trade as the Engines of Growth

The shortage of foreign exchange to meet the demands for imports arising as development proceeds is most frequently cited as the crucial bottleneck in the way of successful growth. In order to overcome this shortage, planners and government officials usually advocate either the promotion of exports or cutting back of imports via import substitution. This can be examined in the light of the resource allocation pattern of Figure 4.1(c).

As far as the promotion of exports is concerned, such an effort can conceivably directed either toward a further expansion of the traditional export sector *E* or toward the expansion of exports based on output changes in the rest of the now no longer stagnant economy. These basic alternatives may now be briefly explained. With respect to promoting the further expansion of the traditional exports of the *E* sector, such exports have an overwhelming raw material component and a relatively low value added in terms of human ingenuity, management and capital. The prospects for substantially expanding such exports in the light of well-recognized trends in world demand are, however, by no means favourable. To the extent that the primary product exports of the less developed world are consumer goods depending on changes in per capita income levels abroad, relatively low income elasticities of demand for such goods are often encountered. To the extent that they are raw materials directly related to the industrial output abroad, the continuous rapid evolution of mature economy technology has led, on the one hand, to well-known difficulties associated with the advent of synthetic substitutes and on the other, as the production process becomes more and more complicated and products more finely differentiated to a continuous decline of raw materials component in total output. Though the subject is not without controversy, the inductive evidence shows that the traditional exports, often a heritage from the colonial past, are unlikely to provide increasing foreign exchange earnings for the less developed economy in the future. The often heard complaint about the "secular" deterioration of the terms of trade of the primary good exporting countries refers specifically to this problem.

Turning our attention now to the possibility of promoting exports arising out of the increased output of the domestic sectors *A* and *I*, the prospects are somewhat more encouraging, especially with respect to the growing industrial sector. While a number of labour surplus less developed countries may currently be food exporters, Thailand and Myanmar who have to compete with the highly efficient agricultural sectors of the advanced economies that customarily protect their domestic markets and accumulate large surpluses, but could easily bring the *terms of trade* in such commodities as wheat and rice tumbling down.

With respect to the potentialities of expanding exports out of the growing output of the industrial sector *I*, the situation is more promising, since there is no reason why, over time, the economy should not increasingly be in a position to enhance its human capacity and ingenuity, in conjunction with a growing capital stock, to a point where it can produce and export a number of industrial products on a price and quality competitive basis. In labour surplus economies subject to severe population pressures, the tendency would be to export light labour-intensive industrial products with an increasing skill and ingenuity component over time. Part of the same effort would be to produce at home consumer goods and capital goods previously imported and thus, by so-called import substitution, ease the balance of payments pressures of the underdeveloped economy.

In the schema of Figure 4.1(c), this attempt to solve the foreign trade problem could be indicated somewhat as follows: there are, basically, two sources of export earnings, (one) a stagnant component, E representing the traditional natural resource-specific exports, and (two) a dynamic component, E_i, representing the export of products of newly created industries in the industrial sector, I. Through time, if all goes well, we would expect the weight in total export value to shift gradually from the former to the latter as the economy becomes increasingly capable of incorporating cheap labour with indigenous entrepreneurship, organizational talent, and capital. This tendency toward what we may call import substitution should provide the necessary foreign exchange to satisfy the industrial sector's need for imported capital goods, M. In this fashion there will be a gradual relative shrinkage of traditional export sector, E.

The important question which raises itself at this point is whether the open-development-oriented economy can take a short-cut in this fashion and achieve its goal by concentrating on the growth of the industrial sector I, fueled by foreign trade, while leaving the development of the agricultural subsistence sector A pretty much to one side. Such an attempt to achieve growth through trade may, in fact, seem likely the only choice if, for one reason or another, the labour surplus economy cannot, or will not increase the productivity of its subsistence agricultural sector. In this case, the logical pattern of resources flow is for the industrial sector to promote the export of industrial products in exchange for the needed food, raw materials, and capital goods to sustain the continuous expansion of the industrial sector. In case there is a simultaneous expansion of agricultural productivity there results the kind of natural pattern of coordination and interaction between the industrial and agricultural sectors, that is, the expansion of agricultural sector provides a natural market outlet for the products of the industrial sector, while it simultaneously supplies the latter with manpower, food, and raw materials.

In case the size of the agricultural sector *A* is relatively small it is conceivable that such a process of growth through trade can provide enough leverage to pry the stagnant subsistence sector loose. Small areas lacking a huge subsistence "backyard", such as Israel, Porto Rico—especially when they have a special relationship with mature economies abroad—may provide examples. In the case of the more typical labour surplus underdeveloped country, in which the centre of gravity lies overwhelmingly in the subsistence sector *A*, instead of the industrial sector, *I*, pulling along a non-responsive agricultural sector, *A*, it will be more likely that the dead weight of *A* will drag down *I*, *i.e.*, that the *dog will dominate its tail*—this process brings the risk factor that the growth process may grind to a complete halt. Under the pressure of population growth food shortages caused by a stagnant subsistence sector become so overwhelming that the country has to import food at an increasing rate, thus depriving the industrial sector of the needed foreign exchange to import capital goods. In effect, under such circumstances, the industrial sector shares some of the characteristics of the enclave economy in that neither is able to "pull" the rest of the economy along with it. Both may be seen as small sectors integrated and interacting with the foreign sector rather than with the domestic subsistence sector which constitutes the bulk of the economy.

There is undoubtedly to be said about this massive drive of ISI approach to achieve growth through industrial exports. (First of all), it at least recognizes the dynamic nature of the problems of trade and growth and the urgency of the need to find export substitutes. It is politically appealing but this approach is cheap because over 80 per cent of the people would thus be for all practical purposes be ignored which is a sophisticated form of escapism and substantial opportunities for effecting a successful transition to economic maturity consequently is foregone.

Trade as a Facilitator of Growth

Thus the notion that the impetus for growth in the underdeveloped world is likely to come primarily from contacts with the developed world is seen to fall considerably wide of the mark. Only a development strategy which encompasses and integrally involves the entire economy can hope to be successful. Trade and aid can then be viewed in proper perspective, as playing a possibly important but nevertheless facilitating or supporting role.

What this means, essentially, is that the basic structural characteristics which dominate the preponderant domestic subsistence sector in the closed economy must continue to receive the planner's prime attention in the open economy. In other words, what we believe to be the essence of the matter, the expansion of agricultural productivity to fuel the development process via a consistent and balanced pattern of resource allocation within the dualistic economy, is still at "centre stage" when the economy is opened to trade and foreign capital inflow. Only when the labour surplus economy manages to initiate growth by dint of its domestic effort can the opportunities of the open economy become a significant and welcome appendage to this formula of growth.

A further understanding or facilitating role of foreign trade in this context emerges from the fact that the unlimited supply of labour regime governing the economy is tantamount to growth under the conditions of austerity, thus "there should be an adherence to the path of natural austerity" in the sense that increases in productivity and output in both the agricultural and industrial sectors do not result in substantial increases in per capita consumption, though customary differential or "hill" between the agricultural and industrial real wage may result in increases in per capita consumption. But in the absence of an expanding market for industrial consumer goods, the private entrepreneur in the closed economy cannot be expected to be farsighted enough to engage in a large volume of investment in capital goods and economic overheads. Thus the ruling force of natural austerity, both necessary and desirable from the point of view of maximum industrial formation, raises the problem of sufficiency of investment incentives in the private sector, and consequently puts a heavier burden on the capacity of the public sector to implement investment projects.

Once the economy is "opened up", and we shift from development-oriented open economy of Figure 4.1(c), the expansion of the capital stock in the industrial sector I can be oriented toward the foreign sector F rather than the domestic economy. That is to say, the economic system is in a position to produce and export new industrial goods E_i, for which there presumably exists a market abroad, in return for the importation of capital goods *C*. The economy can, for instance, produce labour-intensive consumer goods more suited to its factor endowment, more amenable to the application of labour-using innovations, and providing more opportunity for private entrepreneurs to respond to the investment incentives presented by the open economy.

The development-oriented economy's search for export substitutes, *i.e*, the move away from the natural resource-oriented export base and toward a continuously higher human ingenuity and skill component in exports, must be viewed in this general context. In the early stages of development before the turning point there will need to be a good deal of government investment activity in social overhead and/or heavy industry construction not based on short-term private profit calculation. But at the same time, such activity, in the open economy, can be accompanied by a large-scale expansion of exports which do depend on such private profit expectations and which concentrate on goods more capable of incorporating indigenous labour, entrepreneurship, and organizational talent.

There admittedly exists a good deal of skepticism concerning the possibilities of such export substitution over time, but it should be recalled that much of this skepticism is based, implicitly or explicitly, on the view of the stagnant underdeveloped economy and does not visualize the full potentiality of change as a fundamental concomitant of domestically generated growth. In the Figure 4.1(c) what is being pointed out is the economy's ability to acquire new skills, use its indigenous ingenuity in more and more ways, and thus become better and better equipped to export increasingly sophisticated products. To deny this is tantamount to denying the likelihood of structural change as the companion of growth. As part and parcel of this increasing ability to produce and export goods with a higher human resources component, import substitution will take place, with only the more sophisticated goods or sophisticated stages of production being imported. Such import substitution, as changes in comparative advantage take place over time, should not, however, necessarily be interpreted, as it so often is, attempting to duplicate domestically the production of previously imported items virtually overnight. Sometimes a different type of good, perhaps of somewhat lower quality (*e.g.*, a less refined sugar), can be produced at home; and a domestic substitute for the latest imported machinery is likely not only to accommodate considerably more reallocated agricultural workers per unit of investment but also to represent a much reduced drain on the balance of payments in terms of raw material and service requirements over time.

It is, of course, not possible to sit in an armchair in New Haven and predict the precise commodity composition of this push toward export substitution and/or import substitution in any particular country context. But it is perhaps equally difficult for those sitting in the armchairs of the local planning commission. In fact, it is precisely these difficulties which only the talents of the private entrepreneur can successfully surmount. To fill out the precise dimension economy's changing comparative advantage position as domestically generated growth occurs requires the involvement of large groups of private entrepreneurs who perfect their profit and loss calculations in the process of participation and trial and error. Neglect of this vital function of the private sector which cannot be emulated by the burueaucrats is likely to hamper seriously the development effort.

The ability to produce (and export) things more in keeping with the economy's changing capacities over time, should accelerate the arrival of "the turning point". Trade makes possible a higher rate of labour absorption and industrialization each year. This is due to the fact that while, in the closed economy, the rural disguised unemployed labour force must be converted directly into domestically manufactured investment goods, in the open economy the same labour force can be converted first into industrial export goods closer to the capabilities of the economy and then, via, the foreign trade machine, into the appropriate goods needed for the development effort. This provides a much more efficient method for obtaining the goods required. As the consequence of this improved technical efficiency of the production process, the chances of beating the *Malthusian Trap* are greatly enhanced, the economy's agricultural reserve army depleted earlier, and *the advent of the turning point get advanced*. Once the turning point has been achieved, real wages and per capita consumption will be in a position to rise significantly for the first time, thus enhancing the strength of the domestic market and the possibility for entrepreneurial response mechanisms to turn to the domestic scene. Thus the key distinction between the two key policies for development are either by (one) inward-looking import or export substitution and the (two) outward export promotion, in the latter there is an involvement of domestic entrepreneurship and innovative capacities across the vast landscape of the underdeveloped economy and it is these elements which constitute the critical components of the development effort so that once the economy is dynamically moving ahead, the potentialities offered by foreign trade can substantially ease its path and accelerate its pace of development. In contrast, in the former, the underdeveloped economy moving along a natural austerity path does neither visualize the full potentiality of change as a fundamental concomitant of domestically generated growth nor does take into account the encouragement of domestic entrepreneurship so that denying all these becomes the "tantamount to denying the likelihood of structural change as the companion of growth and development".

Thus following the three stages of the above abbreviated model, the economic relationships between any countries in general and especially between the Countries of Global South and Global North are largely definable in terms of the patterns of international trade and exchange. In today's economically integrated world, it is indeed trade matters more than ever before.[1] The prolonged expansion associated with high employment, growing wages and welfare expenditures, high consumption, and benign business cycles that lasted in the advanced industrial countries of the North from 1945 to 1973, and which effectively ended when the quadrupling of oil prices in that year threw most of the world's economies into recession is mentioned as "Golden Age of Capitalism" by Juliet, 1990. [2] The recession what the world economy underwent was for a short period because since 1983, the pace of world trade has accelerated again, reaching an average annual rate of 5.7 per cent. This well *outpaces the expansion of world GDP, which has risen at an average of only 3.4 per cent per year.* However, the world economy, in fact has undergone the radical transformations in the mean time but whether this is the transformation from Juliet's "Golden Age" in the earlier post-World War II period to its successor since 1983 to a "Leaden Age"[3] is a matter of considerable debate which reflect various well-known titles "late capitalism", "advanced capitalism", "disorganized capitalism", "deregulatory capitalism", "actually existing world capitalism", "integrated world capitalism", "post-Fordism", "post-industrialization", "overconsumptionist economic regimes", and so forth. The deliverances of such reflection have profound implications for any understanding of the forces and structures associated with uneven development, late industrialization and "peripheralization"—all these being notions vital to the explanations and theories of systemic international inequality sponsored by the various marxist traditions.

Hence it is hardly surprising that problems of trade between them should dominate the agendas of international agencies, like the IMF and World Bank concerned with changes in the world trading system to make it more supportive of development at the world level and to maximize the developmental impact of regional trading agreements.

Certainly the patterns of trade between these two worlds, North and South yield a fairly picture of economic relationships between them, the predominant features of which still reflect a past in which, under colonial domination, the trade of the South was "subservient" to the needs of the metropolitan countries—this expansion was being used as fuel for the economy—like adding wood to a fire. In that epochal era the colonies provided a captive market for the manufactured products of the colonizing countries while in return supplying them with industrial raw materials and tropical foodstuffs. Although trade patterns between these two blocks have exhibited certain modifications in the last 50 years, topical questions still revolve around whether this stereotypical relationships of the colonial period have got changed? Does trade contribute to the growth and development of poorer nations, or does it serve to reinforce the hierarchy of rich and poor nations? Is the pattern of trade preserved by protective policy actions of the richer countries designed to maintain their economic advantage over poorer countries? Do the institutions controlling international trade hinder economic development by passing on most of the gains from trade to the richer nations? In other words, is the system of international trade inherently unfair to the poorer nations? Do prices and revenues of primary commodities exhibit greater short-run instability than those for manufactured goods, and, if so, does such instability hamper development? These are essentially long-term questions about the capacity of international trade to promote economic development.

Primary *Versus* Manufactured Exports

In addition, to the above long-term issues to be answered on, there are shorter-run issues concerning the possible disadvantages of reliance upon the export of natural resources, *i.e.*, primary products for the bulk of export earnings. For example, for many Countries of South, the export of their primary products are the "prime" (and sometimes the only significant) source of investment capital and foreign exchange. This, known as *primary specialization*, may not be a development strategy so much as a necessity for some countries. This is basically a "policy of deferred industrialization". *Fiscal linkage* occurs when raw or minimally processed agricultural products, mineral ores, fuels, or fish are used as a means of obtaining foreign exchange for investment in factories. This can be a "trap", however. For very small countries,[4] with little domestic demand, it may be difficult or impossible to expand beyond the primary-export stage. With some significant exceptions, the smaller the country by area as well as by population, the more likely it will be a "monoculture", that is, it will depend on only one or two exports, such as coffee in Columbia or bauxite in Jamaica.

The result is: it crucially constrains their ability to sustain sales cut for significant periods of time and to maintain export limits (which are needed to force up price) in international trade and internally it may be extreme concentrations of wealth in very limited enclaves, with relatively little of the spin-off effects that could lead to development. Demand for primary goods tends to be much more sluggish than demand for manufactured goods, which is one reason why *terms of trade* have declined at an average rate of about one per cent per year. Countries trapped at this stage may be quite poor and have inefficient and corrupt governments that fritter away foreign exchange rather than investing them in development. Ghana, for example, was the richest country in Sub-Saharan Africa at its independence in 1957, with a per capita income of $ 500 per year However, its reliance on cocoa as its single dominant crop, comprising 60 per cent of its export earnings, and its failure to diversify were major factors in the country's economic decline. By 1983, its per capita income had dropped almost 40 per cent. *Note* share of agriculture in GDP which has declined substantially between 196[illegible] and 1990, in Sub-Saharan Africa it is still responsible for generating a third of the output in GDP.

"The *dispute settlement system* must consider the development needs of countries (especially the most vulnerable), not just whether free trade rules have been violated. For instance, in the recent dispute over the banana trade, the WTO ruled in favour of the USA over the European Union's (EU's) traditional arrangement of preferential access for Caribbean banana exporting countries—a ruling that may have devastating economic consequences for Caribbean economies that depend solely on banana exports. If developed and developing country farmers are to compete in the same markets, then the $ 280 billion in annual subsidies that developed countries provide to their farmers should be reduced to the negligible amounts the developing countries provide. Otherwise, developing countries should be allowed to increase both their subsidies and their tariffs to protect their markets from the highly subsidized exports of the developed countries. Small farms in both developed and developing countries should be encouraged, not squeezed out—especially in developing countries, where farming is the source of livelihood for millions".[5]

Exporting commodities and resources are seen as favourable to help earn foreign exchange with which to pay off debts and keep currencies stable. However, partly due to the price war scenario mentioned above, commodity prices have also dropped. Falling (commodity) prices have meant that large increases in export volume by commodity producers have not been translated into greater export revenues, leading to severely declining terms of trade for many commodity producing countries. When the purchasing power of a country's exports declines, a country is unable to purchase imported goods and services necessary for its sustenance, as well as generating income for the implementation of sustainable development programmes. Furthermore, reliance on just a few commodities makes producing countries even more vulnerable to global market conditions and other political and economic influences. A vast majority of developing countries depend on commodities as a main source of revenue. Primary commodities account for about half of the export revenues of developing countries and many developing countries continue to rely heavily on one or two primary commodities for the bulk of their export earnings.

The *commodity crisis* is the principal challenge to the achievement of economic and social development, and ecological sustainability in the majority of developing countries today particularly of geographic determinants of their area, location and size of population. Commodity prices had plunged dramatically in the ten years preceding the Rio Summit in 1992 and have almost consistently continued to spiral downwards since then. In 2001, real commodity prices for agricultural raw materials and minerals, ores and metals have fallen to 60 per cent of their price levels in 1980, while prices for tropical beverages have declined nearly 70 per cent from 1981 to 2002. For many countries, the declining terms of trade associated with a fall in commodity prices have also led to a build-up of unsustainable debt. Only four of the 27 non-oil commodity-exporting LDCs did not have an unsustainable external debt in line with the criteria set by the enhanced Heavily Indebted Poor Countries, (HIPCs) Initiative (UNCTAD, 2002).[6]

By no means all poor countries are having monocultural economy: Thailand specializes in rice, rubber, tapioca and sugar. Although diversification is certainly preferable and gives the country more flexibility in riding out the inevitable fluctuations in export earnings, still it may be no better at supporting the development of a manufacturing base. At the other extreme from these impoverished primary-good exporting countries there are oil-producing countries that fail to industrialize because they do not have to: their wealth what they get out of royalty by exporting oil, is sufficient without the industrialization.

Balanced development would include both primary and manufactured exports. The Southern Countries, are, indeed, increasing its sales of industrial goods, but only a small group of countries like Singapore, Taiwan, South Korea, China in East Asia, Mexico and Brazil in Central and South Americas, and a few others—amount for almost all such exports. Most Southern Countries are still stuck at the primary-export level.

In recent years, the problems posed by export dependence "have been exacerbated by the need to service massive overseas debts". Many of the Southern Countries are heavy lender to the international banks, servicing of which by itself now consumes 40 to 50 per cent of their export earnings. Since 1980 the dollar market price for non-fuel minerals has fallen by 12 per cent per annum.

As far as international trade theory is concerned, the answer to the question "whether trade lead to development?" The answer to the question is undoubtedly affirmative: Commodity flow through international trade provide the "template" and the "context" for evolution of the world's economic landscapes. Between 1950 and 1990 the average annual growth rate of the value of world exports was 7.4 per cent in real terms: a figure that was well above the growth of world's production and several times greater than that of world population growth. But the developed economies dominate the export and import structure for most groups of commodities : *intra*–countries of the North is the dominant element in world trade. Only in the case of the export of metallic minerals and fuels (oil) is that status reduced, and the role of the Southern Countries becomes more evident. Thus main lines of orthodox reasoning rest on incompatible and challengeable assumptions.

The "theory of comparative advantage of trade" by David Ricardo (1772-1823), is based upon assumptions of full employment, complete factor mobility and the existence of different techniques of production in nation-state context as the unit analysis. From this latter assumption, it follows that relative production costs (that is, the price formation in a "pre-trade" or *autarkic state* and comparing this with a trading situation in which international prices have been formed) differ between countries almost exclusively on the availability and costs of labour. Following the Ricardian tradition, it is now recognized that the decision as to what goods will be produced for export is dependent on numerous factors, such as natural resources, technology, capital, and education levels of the populace which govern the productivity of labour. Each product requires a unique set of factors of production. Individual countries will possess these factors, or be able to attain them in different degrees by the dichotomy between the internal and external mobility of capital and labour. By using their most abundant factors of production most intensively, all countries, including the South, can have the relative "lowest production costs" by specializing in the production and trade of those commodities while importing goods that would require scarcer factors of production, so that they can attain the high levels achievable for increasing their consumption of goods. *Theoretically everyone benefits.* By selling what they can produce most efficiently, countries can get the foreign exchange to purchase that which they cannot make at all or that which they cannot make as cheaply as they can purchase. And all the items what a country require but does not produce them, have to come, initially at least, from outside the country through import. Even if loans from the World Bank are used, the loans themselves must be paid off through trade. Export industries also create jobs that did not exist before. It is argued that the opportunity for mutually beneficial trade derives from inter-country differences in *factor endowments.* In this context, the comparative advantage (or "factor endowment theory") arises from the different factor intensities in producing different products rather than from differences in production techniques.

It leads to the conclusion that poor countries (those with abundant labour) should specialize in producing and trading labour-intensive goods while the richer countries (with relatively scarce labour and abundant capital) should specialize in capital-intensive ones. For example, coffee is a labour intensive crop, a tropical country with much fertile mountainous land and a large cheap labour force can produce it more economically than a country which is similarly endowed in terms of physical resources but has a highly skilled labour force. The latter country might choose to assemble radios. The coffee-growing country would require an adequate road system and vehicles to transport workers and crops. These might be irrelevant to the radio-producing country, which would do almost all its work in the city but it needs a good educational system to train workers and managers and sufficient domestic wealth for the start-up investment. In a sense, both countries could grow coffee and produce radios, but in the international economy, it is to each one's advantage to focus its efforts on those products for which it is best adopted. Countries thus can gain comparative advantage by using their most abundant factors of production most intensively while importing goods that would require scarcer factors of production.

It also implies that free trade and specialization in production will tend to bring about *factor price* equalization, and thus pull the returns to labour in the South up towards the levels in the North. This theoretical outcome hinges very much upon the strong assumption that at least some factors of production are internationally perfectly mobile. In this version it is argued that trade as a "vent for surplus" will not only help reduce the income gap between the poor and rich countries but also to emphasize trade expansion as a means of promoting development.

There are, however, "serious problems". *Trade often is associated with growth but not necessarily with development.* (First), benefits are not distributed equally—the owners of production may get rich while peasants and workers, the actual producers remain poor. It is true that coffee provides jobs, but only during part of the year and at extremely low wages. Agriculture that employ tractors will inevitably displace workers. For meat exports, formerly subsistence agricultural land that supplied food for thousands might be turned over cattle ranges that require very few workers. (Second), local producers may be harmed. If soybeans are traded for cloth, a few large land owners will be benefited, but local textile workers will be put out of business. (Third), there are few spin-off effects of primary exports; that is, agricultural goods and raw mineral ores have few backward or forward linkages that promote other industries services in the society. (Fourth), declining terms of trade mean that countries get less and less for their raw materials relative to what they must pay for imported manufactured goods. (Finally), many markets are highly unstable : synthetics have significantly reduced the demand for rubber, sisal, jute, wool, cotton, and many other products. Also, there is always the chance that a country with a lower-paid workforce will undercut another country's export price, leaving the latter with expensive equipment or mines and no place to sell the product.

Again, the persistence of large international differences in per capita income has led to doubts about the relevance of the assumption of neo-classical trade theory and about the capacity of existing terms of trade to improve the relative incomes of the South. It is said that in a changing world of capital accumulation and technical progress, free trade does not tend to maximize welfare of the South nor to ensure their balance. The industrial core countries of the North have dominated international trade, while the share of world trade accounted by countries of the South has decreased (a trend that is particularly pronounced if oil-exporting countries are excluded from calculations). In general terms, there has been an intensification of the long-standing domination of an intra-Core and inter-Core trade at the expense of the Core-Periphery trade. The importance of Core-economy is further understood by the fact that innovations in transport, communications and manufacturing technology has diminished the index of "commodity concentration of exports/imports". The low value on this index has diversified export bases of most of the Core countries in trade and this include the countries with newly indistrialized economy (NIE) as well as the large countries of the South like China, India, Brazil, Argentina, Mexico and Peru. At the other extreme are the peripheral countries where the manufacturing sector is poorly developed and the balancing of national accounts and the generation of foreign exchange is dependent on the export of one or two agricultural or mineral resources like Iraq, Iran, Libya, Nigeria, Bolivia and Venezuela, for example. Meanwhile, we should note that a significant number of countries exhibit a high degree of *autarky* from the world economy. That is, they do not contribute significantly to the flows of imports and exports that constitute the geography of trade. Typically, these are peripheral countries of the South, the Least Developed Countires (LDCs) : examples include Sub-Saharan African countries like Mozambique, Malawi, Tanzania, Uganda, Samoa and Burkina Faso who barely participate in world trade in manufactures. (Trade Statistics, 2002).[7]

Hence if international trade is not assisted by consciously used transfer of factors of production, it has no inherent tendency to increase income and productivity levels. Indeed the trade mechanism works the other way round—in favour of the progressive countries and against stagnant ones. Underlying this viewpoint is the argument that, although some developed countries may well have high wage rates relative to those in underdeveloped countries, because of traditionally high rates of productivity growth and technological change the ratio of labour productivity to wage rates may still be higher in the richer countries. In this situation, capital will be attracted to these rather than the poorer nations, thus extending the technology gap and increasing the international difference in wage levels. It follows that if free trade is unlikely to increase the relative capital endowment of the South, international income disparities are only likely to be reduced if conscious decisions are taken to direct capital (both "official' and "financial") to the South in the form of foreign aid, technical assistance and preferential trade arrangements.

In the question of whether trade can cause development, if statistical analyses are undertaken on the data about exports of

Countries of the South and their respective GNP, there can be a confirmation that faster growth of exports is associated with higher GNP growth. However, since in a large cross-section of countries considered, the overall statistical result may be dominated by the behaviour of a subset of those countries. It is found that among the Countries of the South, for the poorest countries (which are mostly from the African primary producers), it will be found that since these countries remain overwhelmingly dependent upon primary commodity exports (like, Zaire for Copper etc.), exports do not appear to provide a strong stimulus for growth in per capita GNP, there may not be any virtual relationship between any per capita GNP and export. But among the Southern Countries, if the selection is done with reference to the more successful countries like NIE with higher per capita income and their rate of growth of exports as a share of GNP, there would be a positive relationship which indicates that growth is affected by export performance only once the countries achieve some minimum level of development.

For attaching importance to identify and examine different approaches by the Countries of Global South to trade policy strategy, trade itself should not be regarded as a *pacesetter* in any relevant development, it should be taken as a *derivative* so that the countries first define a strategy for attacking their problems of unemployment and mass poverty. In this context, since the links of growth in trade with industrial development policy are particularly strong, so decisions should be taken whether the industries are to be primarily oriented to domestic or to international markets, and whether they are to be developed through the imposition of import tariffs or through other forms of industrialization, all manifest themselves directly to the volume and pattern of trade. The same applies to policies of expanding agricultural production in simultaneous relationship between the physical economic attributes, *i.e.*, between trade, production, consumption and investment.

In the broadest terms, trade and production strategies pursued by the South ranged from (one) policies of near self-sufficiency (autarchy), through (two) ISI designed to limit reliance upon within classes of imports, (three) to export-led growth at the other extreme and presently to (four) open-economy policies of globalization through MNEs and foreign capital. Examples of autarchic policies are few and the clearest examples are those of Myanmar and the People's Republic of China (PRC) which until 1978 pursued a policy of very restricted trade, based on the highly critical philosophical view of capitalist economies and of the deleterious effects of consumerism. However, since 1979 deliberate steps have been taken by PRC to open up to international trade (the year when, PRC, the country of "closed economy" in their first step of liberalization opened up Shanghai) for private enterprise though the steps in this direction have been cautious, but the views of the present political leadership clearly indicate that faster development and greater economic efficiency in the PRC require more open competition internationally and internally.

The pursuit of ISI, and trade restrictions that accompany it, has been widespread in Countries of the South and has reflected a complex of objectives. In these cases policies of ISI have been influenced by notions of trade as an unusual exchange, about the exploitative nature of trade, or by sheer export optimism. In others, the ISI has reflected well-articulated policies of industrialization in which a sequence of industries is developed in turn to replace specific imports. However, in many cases, ISI policies were far from articulated, but were adopted in an *ad hoc* and indiscriminate fashion in a number of developing countries. In many cases, these have required the imposition of extreme measures of trade protection (*tariffs, import quotas, embargoes*) to overcome the limitations inherent in producing for what is often a small market. Such policies, as we shall consider elsewhere, in many cases led to reduce incentives to exporting industries, since they have to pay higher prices for some domestically-produced inputs than their overseas competitors, or to under-emphasis on the agricultural sector, or resulted in overvalued exchange rates with respect to "free trade" levels. Nevertheless, this inward-looking trade and development strategies held an important place in many of the Countries of South during the 1960s and 1970s.

In recent years the tide of opinion has moved much against ISI in favour of export-led growth so that at the risk of oversimplification, it can be said "there has been a *volte face* in terms of policy perspectives among the Countries of the South (Narula, 2002).[8] Partly this reflects the difficulties experienced with ISI, particularly in Latin American countries as early as in 1970s, but it also reflects the influence which the Newly Industrialized Countries (NICs) have had upon thinking about development, plus the statistical evidence (mentioned above) of a positive relationship between export growth and GNP growth. Certainly the main international agencies involved in this development are the World Bank and International Monetary Fund (IMF) and since 1995 by World Trade Organization (WTO) are strongly committed to opening up the South to trade by more export and reducing import barriers. But in the process since many of the countries in South (as well as few in the North) developed "debt problems", so when these countries approach the IMF for loans and other assistance in rescheduling debt problems, the support is usually conditional upon undertakings about changes in economic policy. In developed countries of the North, the requirements are typically that public expenditure (including on subsidizing the agricultural sector as well as the industries) be reduced, while in the Countries of South conditionality by IMF entail *inter alia* both cuts in public expenditure and removal of import restrictions. That is, the standard medicine of the IMF is to expand the role of private enterprise in production and trade. Justification, according to IMF, lies in the gross distortions and inefficiencies which frequently arise when political and bureaucratic controls impede the normal working of market mechanisms, examples of which are readily found in industries operating at very low capacity utilization, in ever-increasing price differentials of domestic goods over their imported equivalent, and in uncompetitive industries.

Despite widespread evidence that efforts by central bureaucracy to control and integrate the many complex processes of the economy which are often seriously deficient, there are vociferous critics of monotheistic views of international bodies which favour free enterprises. They see these open-economy policies as denying the South the unity to pursue independent and different policies of development, critically as forcing them to submit to the unwelcome embrace (or, control by) MNEs and foreign capital.

The debate about export-led as opposed to ISI-led trade strategies, hinges upon expectations about export prospects. In view of a rapidly expanding and demanding trade prospects of four countries of East Asian (South Korea, Taiwan, Hong Kong and Singapore) economy, export-led strategies advocated a dismantling or reduction of many of the measures of production for domestic industries—these four countries with only 3 per cent of the total population of Southern

Countries (excluding the People's Republic of China) accounted for more than 60 per cent of total manufactured exports in 1978. It is argued that the scope of increase of demand for primary exports of the South is currently as great as it was for the these NICs during their period of rapid industrialization lately, while the rate of increase in demand for their manufactured products is substantially greater. If this is correct, this undermines any arguments that currently underdeveloped countries are disadvantaged by markedly worse export prospects than those facing the industrialized countries at a comparable stage in their development. Of course, not all the Global Countries of South have promising export prospects, but it can be argued that this may be due to deficiencies in the supply conditions within those countries rather than to any deficiency in export demand.

The deficiencies in export conditions may arise from geographical factors —the landlocked countries such as Nepal, Rwanda, Malawi or Zambia suffer from the transport costs in trade, the absence of strong domestic and complementary markets in neighbouring countries, and weaknesses in their locational, transport, communication, and general institutional infrastructures. Such factors as these may go a long way to explaining the already observed distinction between the NICs plus "balanced exporters", which have developed relatively successfully through outward-looking trade and industrialization policies, and the primary commodity exporting countries which have had disappointing growth records. Given the progress of the first group, the performance of countries in the latter group cannot be necessarily be attributed to the failure of export-led policies of development. But at the same time it is apparent that no truly satisfactory general development and trade policy has been devised which meets their needs. To get new industry moving in these countries will demand special measures of aid and protection which will inevitably include elements of isolation policy of ISI. Infant industries generally need protection; the skill of national economic management is to know when to wean infants from the breast of state support and to switch attention to their younger infants.

Barriers in Manufacturing Export Growth: Despite the fact that a number of comparatively small developing countries—which are not among the poorest of that group—have succeeded in creating the necessary industrial supply conditions to rapidly expand their manufactured exports, the Southern Countries as a whole are experiencing difficulties and obstacles in penetrating industrial exports. The export shares which they hold of the manufactured goods market of the North are substantially smaller than those for primary products (of course, oil remains exception). However, the market share of both agricultural and mineral products which are the principal exports of the poorest Southern Countries, declined steadily over the period 1960 to 1980, so that in 1980 both stood at around 28 per cent. While exports of manufactures from the South rose rapidly during the same period and their share in imports of the North steadily increased, the degree of market penetration achieved by 1980 was still comparatively low, only 5 per cent in machinery and 15 per cent in the other manufactured category. Earlier data, from 1964 to 1972 indicated that developing countries were having difficulties in expanding and even holding onto their share of what were classified as raw material-intensive manufacturing and labour-intensive manufacturing markets. These are precisely the categories of manufactures in which the Countries of the South might be expected to have a comparative advantage since they supply a substantial proportion of the Northern Countries' raw materials and have large quantities of cheap unskilled labour.

The failure of the Southern Countries to achieve greater shares of the manufactured imports of the industrialized Northern Countries has been assessed by many to be a demand problem arising from policies by the latter countries to restrict the growth of certain classes of imports. The pattern of world trade has changed in the twentieth century, with a decline in free trade. For example, common markets, notably that of European Economic Community (EEC), now the European Union (EU) has been designed to promote agricultural self-sufficiency within the community, which goes to the detriment of exporting countries of primary agricultural producers of the South. The global recession of the 1980s and the anxiety of Countries of the North to protect their economies, trade and job, has meant that progress in promoting free trade has been slow. The creation of the North American Free Trade Area (NAFTA) in 1993 involving Canada, USA and Mexico, provides opportunities for Mexico, but probably at the expense of other Caribbean and Latin American countries' access to the large North American market. Farm subsidies have been given to the agricultural sector in developed countries to the tune of $ 365 billion annually which prevent export of farm products from

48-member Least Developed Countries (LDCs) of the South for whom farming provides more than 70 per cent of their population. The result is products of these countries continue to face tariff escalations, they remain marginalized in the world trade system, debilitating their economy. Such policies by developed countries like USA have been pursued for a variety of reasons, including the prevention of rapid rundown of their farming sector and domestic industries such as textiles but not opening their markets to the primary exports of the LDCs, is quite unfair when rich countries thmeselves ask poor countries to open their markets. More than 50 developing countries depend on three or fewer commodities for over half of their export earnings. Twenty countries are dependent on commodities for over 90 per cent of their total foreign exchange earnings, says the World Bank.

Thus policy measures employed to control imports can be broadly classified into "tariff" and "non-tariff" barriers, and it is argued that the industrial countries have used these in such a way that they discriminate most heavily against the exports of the Southern Countries. For example, it is observed that many developed countries employ a "cascading" or "escalating" system of import tariffs in which the rate of tariff increases with the degree of processing an item has undergone. This is clearly revealed in estimated of nominal and effective tariffs erected by the Northern Countries like USA, Japan and European Union (EU) which revealed that raw materials could enter them with a "zero or very low tariff". The combination of this plus progressively higher nominal tariff rates for more processed forms of the product provides an encouragement for the raw-material processing industries of the developed countries, combined with protection from competition from processors with comparable production costs in Southern Countries. Thus EEC shoe producers can import hides skins at international prices, and could charge 12 per cent more for their products than a Southern shoe producer would receive for comparable shoes from the same EEC wholesale market price—the Southern shoe producer thus receives market price less the 12 per cent tariff (UNCTAD, 2001).[9] This represents effective protection for the EEC shoe producer's production margin, whereas the price of "effective protection" is defined to be the percentage by which the value added in domestic production, of say, shoes is increased as a result of import barriers and other forms of support.

The discrimination by the North in most processed and manufactured imports from the South led to pressure orchestrated within UNCTAD for the granting of certain tariff exemption for such imports in 1962. As a result 19 developed market-economy countries and 6 socialist Eastern European countries introduced Generalized Systems of Preference (GSPs) to grant preferential access through abolition of reduction of tariffs to certain cases of imports from the stagnated Countries of South. The USSR is in fact credited with the first GSP in 1965, followed by Australia in 1966, with the majority of remaining items being started in 1971 and 1972. But many of the restrictions written into the schemes have been progressively withdrawn and their limited value of concessions have been criticized. Thus most minerals and raw materials enter the North with zero or low tariffs and, therefore, offer no scope for preferential tariffs for the South. Many agricultural products, particularly those produced by the South with high levels of subsidy or tariff protection were also excluded or admitted only with small margins of preference under GSP. It is found that in 1976, of the 19 Countries of North granting "preference" in their imports to the beneficiary Countries in South was only 6.6 per cent of their total imports. Nevertheless, the GSP schemes do have positive trade creation effects as far as the Southern Countries are concerned. The reduced tariffs have meant an erosion of the benefits of GSPs, but their own trade-creating influence would have more than outweighed this and reduced the need for GSPs. However, non-tariff barriers are of various types and include various regulations like health, marketing ones for foodstuff, import quotas, production subsidies in importing countries, and policies whereby government departments purchase necessarily domestically produced goods. These barriers discriminate against the manufacturing exports of the least industrialized countries of the South.

Thus it appears that trade restrictions by rich countries do limit the opportunities of the Southern Countries to profit by outward-looking trade strategies. At the same time, however, it seems and it is logical that the main beneficiaries of any general dismantling of tariff barriers to trade will be the better-off, more industrialized Countries of the North and that it might do little to many of the poorest countries (of the South) to increase their exports of manufactured products. This may explain why so many poor countries have ignored outward-looking policies in favour of an inward-looking policy of the ISI, for them free-trade policy is the same old order with new means of control, new means of oppression, new means of marginalization.

The Question of International Distributive Equity

Thus the debate over the role of the export trade in generating economic growth and reducing *spatial inequalities* in the distribution of income is a "classic case of the blind talking to the deaf". This debate is merely one facet of the larger clash of opinion over the role of capitalism in the development of the South (Rees, Judith, 1989).[10] On one side, *neo-classical economy* argues that the market contains equilibriating mechanisms which work to reduce spatial inequality. Left to themselves, according to them, economic forces would ensure that capital was attracted away from the North, with their high cost labour, and into the South, so generating growth, employment and higher income levels. Trade is regarded as "engine of growth and a support for political stability" which has advanced the countries that have engaged in it to ever increasing income levels. Taking this perspective, governments according to them, should confine their role to establishing a favourable trading climate and should avoid intervening in the market.

This is not the view of the international trade system held by most governments in the South, nor of the "dependency" and "uneven development" schools of theorists. They regard the system of international exchange to be inequitable in two senses: (first), it has conspicuously failed to narrow the income gap between rich and poor nations and (second), it consigns the Southern Countries to a position of economic dependence. The whole debate over unequal trade relations; involves much wider issues than the way primary product sector operates. Nevertheless, the value of crude mineral exports *vis-a-vis* manufactured goods and the stability of prices were a crucially important element in the so-called "North-South dialogue".[11]

It is common to ascribe the ills of international trade to the way the Multi-National Enterprises (MNEs), which dominate international trade, operate. (First), it can be argued that they repatriate a disproportionately high amount of their profits, rather than reinvesting them in the exporting country. A much quoted figure is that the formation of OPEC in 1960, some 45 per cent of the gross receipts of the Middle East oil operations were remitted abroad, chiefly to USA and UK, and even higher profit "export" figures have been reported in copper ore and concentrates in Chile, Zaire, Peru and Zambia. To the countries concerned this amounted to "inequitable exploitation" which deprived them of capital much needed for growth-generating investment. To the MNEs, it was just payment for their risk capital, technological know-how, equipment and marketing networks, without which the resource has no value. Moreover, high returns on working mines were needed for them to cover the losses made on exploration and development efforts which did not yield workable results. Thus Southern Countries are suspicious of the power of MNEs, which dominate "free global markets" regarding the MNEs' pressing the Southern Countries for greater concessions in terms of price and quantity for specific commodities.

(Second), it was argued that the vertically integrated multinationals manipulate the *transfer price* (paid between affiliates in the same company) of the crude materials to minimize their tax and royalty liabilities; this not only deprived host governments of income but also reduced foreign exchange earnings. (Third), the companies were blamed for using inappropriate capital intensive technologies. These not only restricted indigenous employment opportunities, but also required the import of plant, construction materials, equipment, energy and skilled labour. And, (finally), much of the spin-off developments associated with mining (refining, the growth of equipment manufacturers and service industries) were being "exported" to the developed nations of the North.

For all these reasons, together with their ideological opposition to their foreign ownership, the governments of the South rejected the notion that free trade in primary products in minerals could fulfil their economic growth and their national income objectives. A plethora of interventionist measures were, therefore, introduced. These included imposition of much increased export taxes on unprocessed minerals to encourage the indigenous development of smelting and refining; physical export controls (quotas) designed to force price increases; restrictions on imports to promote local construction, servicing and manufacturing industry; major increases in profit taxes (including the removal of tax holidays and similar concessions, plus the imposition of extra taxes on windfall profits); price fixing by producer cartels and asset nationalization or equity sharing schemes with the mining multinationals.

While such moves have undoubtedly increased the proportion of mining profits accruing to producer governments, they clearly have not brought the hoped for development in many Countries of the South, nor have they reduced global income disparities. An exporter government's ability to gain effective control of its mineral production is critically determined by several factors like (*a*) there is no reasonable cost substitute material available, (*b*) that the mineral is essential to the industrial base of the importing nations, (*c*) few opportunities exist for increasing the efficiency of mineral use or its conservation and (*d*) the need to market the products and must control the output of all potentially important export centres. Since these conditions do not hold, the market powers are still predominantly in the North and this places further constraints on their ability to sustain aggressive price or supply control measures. Thus, any attempt for them to control the market, there will be retaliatory actions, to any of which the exporting Southern Countries will be exceptionally vulnerable. Many would argue that the authoritarian regime could not afford to go against the interests of international capital.

While the Southern Countries have the role of crude mineral exporters, the system of international trade will tend to exacerbate, not reduce, spatial inequalities. The whole trend in the economic development process is towards the production of increasingly complex processed goods. This means that the prices of the basic primary commodities must fall in relative terms as a proportion of the price of final goods; thus reducing the relative incomes of primary producers. There seems little possibility, therefore, that the resource trade can act to redress global inequalities, unless, that is, strong and effective international institutions are created. This seems unlikely and certainly in the present economic climate there is little evidence to support any optimistic thesis that a global welfare state will develop with the evolution of international institutions to plan and co-ordinate economic development and to pursue social welfare objectives. *In an interdependent world, it is often contended that the dependency links are symmetrical, with the North being as dependent on the South as the South is on the North. In a resource context, however, it can be argued that whereas the North has acted to reduce its supply dependence, no such reductions have occurred in South's dependence. In other words, the dependence ties are not symmetrical but work against the Southern producers.*

Issues relating to Export Dependency on Primary Commodities: Terms of Trade

One reference is made to the "terms of trade" which means that the value (or price) of one commodity bundle in terms of another. A decline in terms of trade for a specified bundle of goods means that a given quantity can only buy a smaller quantity of some other bundle than was formerly the case. To calculate a *terms of trade series* therefore requires construction of two indices of value. At this level we may value (price) of a commodity in terms of some (or all) others to obtain a measure of the *commodity terms of trade* (CTT), or we may compare the price of a *standard bundle of exports*, P_X to that of a *standard bundle of imports*, P_M to obtain the *net barter terms of trade*, NBTT, such that : $NBTT = P_X / P_M$.

The general debate about development issues therefore is, however, a tendency to draw inferences from terms of trade measures about the welfare complications of trade for Countries of the South. Unfortunately, neither the commodity nor the net barter terms of trade are directly suitable as trade welfare measures. Hence a number of modified measures have been proposed which have a greater welfare content. For example, a decline in a country's NBTT may be caused by a rapid expansion of export supply pushing down prices; if export volume increases more rapidly than the NBTT declines, the country's export purchasing capacity will have increased, which may be loosely interpreted as an increase in that country's welfare through trade. The *net barter terms of trade* (NBTT) multiplied by a *volume index of exports*, Q_x, yields the *income terms of trade* (ITT), where $ITT = NBTT.Q_X$.

Any growth of export volume may result from productivity growth in the exporting economy and such increases in productivity would be exported typically to result in a decline in export prices. Using the so-called *single factorial terms of trade*, defined as $SFTT = NBTT.\, P_X$, where, P_X is the *labour productivity in export production*, the terms of trade may be said to improve provided that SFTT changes positively. This indicates that the import purchasing power of the exports produced by one unit of labour have increased, which again may be taken to indicate an improvement in welfare. It is appropriate to adjust the terms of trade for productivity in export production, it is equally right to adjust changes in labour productivity abroad in the production of imports, P_m. Thus the *double factorial terms of trade,* (DFTT) are defined as :

$$DFTT = NBTT.(\pi_X / \pi_m) \quad \text{......... Eq.1}$$

An increase in the DFTT would signify that a unit of labour employed in producing exports can purchase, through trade, more units of labour in the form of imports. Yet further modifications are possible to allow for the implications of reduced unemployment, and to obtain employment adjusted measures of the terms of trade.

Another reason for rejecting a simplistic welfare interpretations of Commodity Terms of Trade (CTT) indices relates to the "quality factor". A ton of wheat or bauxite remains much of the same quality in terms of productive services irrespective of years of production. The same is not true of manufactured items such as motor car, textile machines or refrigerators. These have changed markedly over time in the sense of providing improved services per unit cost. If, therefore, the Commodity Terms of Trade (CTT) between say, wheat and motor cars remained constant over time, wheat producers would become progressively better off in terms of motor car services. In this situation even declining wheat-to-car prices would not necessarily rule out a welfare gain to wheat exporters.

Trade Pessimism and Structuralist's Theory: The conflict between the *organic development* of indigenous peoples and the *imposed development* of European conquerors is a universal theme for the Countries of South and this is seldom as blatantly evident as in Latin American countries. Even though Latin American countries had become independent in the nineteenth century, the independence in Latin America did not bring immediate and rapid economic progress and structure of the continent's economies remained largely colonial for much of the country despite bursts of prosperity. In the former Spanish territories there was strife, disorder and dictatorship. South

American agriculture had by and large become dominated by big typically inefficient plantations, and virtual serfdom continued in several countries of the continent. Discovered by Spain, the *conquistadores* came to plunder gold and silver or to be paid off large land grants; the local people who survived the European-introduced epidemics came to be viewed as a labour force to produce agricultural exports or to be conscripted into mining. The colonial pattern of exporting primary goods in return for finished products deepened throughout the nineteenth century. British merchant houses took the place of those of the Spanish and Portuguese. What emerged to replace colonialism was an agrarian economy closely tied to the interests of Europe not of the newly independent Latin American countries, and a political order dominated by authoritarian *caudillos* (or, strongmen), who ruled in alliance with the agrarian elites.

The ground slowly started to shift as, late in the nineteenth century, small numbers of private industrialists began to appear in Chile and in Peru, often calling upon government to change policy direction and nurture their development. Though they made little impact over the following four decades, but their importance emerged during the changes came in the 1930 Depression followed by World War II, when the continent had the loss of European markets and supplies: "trade pessimism" followed, the revenue from exports of primary goods plummeted and the resulting lack of foreign exchange restricted importing manufactured goods and Latin America found itself confronted with the necessity of industrialization.

The world market suddenly appeared volatile, certainly not the type of horse to which one would want to hitch the cart of a national economy. Greater independence from the First World seemed now a distinct virtue. To secure this goal, Latin American governments decided to build up their industrial bases and trade more among themselves. By creating large state farms and encouraging private firms there has been a grand convert which Dependency theory saw in 1970s as a reaction to Modernization theory, but it actually had its origins in 1940s, when a group of economists from the United Nations' Economic Commission for Latin America (ECLA) began to focus on "international capitalism" as the source of underdevelopment in Latin American countries. Since the emergence of the period of "mercantilism" (merchant capitalism) in Western Europe in the sixteenth century, thoug.. ere have been major changes over time in the ebb and flow of powei, the basic characteristics of world system remained the same: this argument that the world remains characterized by persistence of divergences between regions and at the same time remain dominated by a single economy such that all peoples are integrated into the sphere of capitalist production, has been well known. Way back in 1950, Raul Prebisch (who went to Peru to direct the ECLA) in his report, questioned the idea of universal benefits from free international trade. He argued that industrialized economies were organized and interested in defending the rate of profit and higher wage level they enjoyed in their countries so that real sharing of the technical progress remained within the industrialized countries and in contrast the peripheral countries remained weak with decreasing prices of raw materials they uere paid by the core countries. Prebisch (1950)[12] in his study of "First World-Third World Trade" pointed to the *institutional dactors* in developing the concept of centre-Periphery for the world market system which favours the Centre (developed nations) with their industrial economy rather than the periphery (developing) economies, he argued that the industrialized nations "exploit" the less-developed primary producing ones—the latter being less-developed so though while possibly still they try to move forward by exporting more of their primary commodities over time, they would nevertheless fall further behind the frontrunners. The main challenge was directed at the validity of the assumption in trade theory of classically competitive factor and product markets in both the Centre and the Periphery. Prebisch called this syndrome the "declining terms of trade". Neo-classical economic theory suggests that, where competition exists, technological changes which result in increased *total factor productivity*, will lead to reduced product prices. Prebisch argued that this process operates at Periphery, competing down export prices of the periphery exports (which are mainly raw agricultural and mineral products) to the benefit of consumers at the Centre. However, at Centre, in the industrialized countries, more monopolistic forces are believed to operate, particularly in the labour market where labour union monopoly power and other pressures push up the wages which is seen as extracting the gains from increased productivity in the form of higher wages, and in the product market where firms are seen as having sufficient market power to pass on higher wage costs in the form of higher prices to consumers at the Periphery and at the

Centre. Thus it was argued by Prebisch that as a result of differences in competitive structure, prices in more technically advanced economies rose more quickly than those in the less-developed primary economy. Differences in income elasticities of demand strengthened this effect further. In particular it is noted that income elasticities of demand for primary commodities, (stock resources, like minerals and agricultural) fall as income, creating "the primary producers' revenue problem" but on other hand demand for finished goods rises, thus no matter how rich the people become, they would buy only so much coffee but demand for finished goods rise with income. It follows that any tendency for the ratio of productivity to demand growth to rise more rapidly in agriculture than in industry will cause a relative decline in agricultural product prices. Any global tendency for decline in agriculture's terms of trade has been magnified for the Southern Countries by the industrial countries' pursuance of policies of paying their domestic producers more than world prices. The output expansion effects of these policies have increased the industrial countries' exportable surpluses of the agricultural commodities and reduced their import demand. Added to this dilemma of income inelasticities for primary products, the traditional producers have to operate usually in very competitive markets so that structure of primary product markets are *atomistic* and hence their prices and profit margins for the peripheral countries are low compared to *oligopolistic structures* of industrial economy in which firms feel less competitive pressure to lower prices and their profit margins rise. Over and above, it is also noted that, as a result of technological advances, the synthetic structures are being continuously developed to compete with those agricultural commodities which are industrial raw materials, *e.g.* nylon polyester for natural fibres, and plastics for rubber. Thus the search for substitutes among industrial producers as well as the continuous productions in these synthetic substitutes brings down the real production costs, again all these, coupled with the ability to control their quality and quantity, has played a powerful constraint upon the demand and prices for natural raw materials. Then there is also the tendency for structural change say from such industries as textiles, shipbuilding and iron casting towards those like electronics and aeronautical industries, which are much less dependent upon raw materials.

Thus, there are several hypotheses to explain a decline in the terms of trade for primary commodities and consequently for those of the Countries of the South. It is, therefore, clear that any general evidence of this effect cannot be interpreted as confirming the validity of any one hypothesis, but will be consistent with all of them. In the global context, it has been found in the relationship between prices of primary and manufactured goods, there were short periods during which the Commodity Terms of Trade (CTT) moved sharply in favour of the primary products like in 1950-54 and 1973-74 but these short periods are interspersed by longer ones in which prices for manufactured goods' prices rise relative to those of primary commodities. It is on the basis of the evidence of these longer periods that most commentators tend to argue that the "primary commodity terms of trade show a secular tendency to decline, so that in 1983, the CTT for food and in agricultural non-food commodities have stood below their 1950 level (Lewellen, 1995).[13] As individuals and countries become richer, they spend a smaller proportion of their incomes on raw materials and a larger proportion on manufactured goods. To the extent that prices follow demand, and relative demand decreases for foodstuff and other raw materials and increases for factory-produced goods. People can drink only so much coffee, so the demand for the coffee product will be relatively stable, even if people have a rise in income, however, the demand for electronic goods expands as people get wealthier, so such manufactured goods increase in price.

The problem of unequal terms of trade resulting in widening gap between prices for raw materials and manufactured goods—was seen as critical. If it continued unchecked, the periphery would fall farther and farther behind. Fidel Caestro in 1983 complained that in 1959, the sale of 24 tons of sugar provided income to purchase one 60-horsepower tractor, by 1982 the same tractor required the sale of 115 tons of sugar. In such circumstances, he argued, increases in productivity are not matched by increases in wages, so the level of the masses stagnates despite arduous attempts by exporting countries to produce more.

The problem was not perceived by the Economic Commission of Latin America and Caribbean (ECLAC) country economists as capitalism *per se*, but rather as a *conflict between domestic capitalism and international capitalism.* Countries are prevented from developing local manufacturing capabilities by foreign competition and by the inequalities in the terms of trade, but they desperately need foreign capital. Throughout Latin America, the state stepped in to implement ECLAC-recommended "import-substitution" policies, prohibitively high tariffs were placed on formerly imported goods that could be produced internally. The logic of *import substitution* became so persuasive that, sooner or later, virtually every Southern Country adopted it to some extent. This policy is not necessarily anti-capitalist (though anti-ethical to classical market-driven economies); rather, the underlying assumption is that indigenous capitalism needs protection from international capitalism. If cheap imports can undercut any attempt by entrepreneurs to build even modest industries, it will be impossible to develop a domestic capitalist class, and foreign capitalists will control the economy to their own advantage. To be sure, most countries were able to export primary products—agricultural and industrial raw materials that might be locally owned and controlled—but such exports would not lead to the emergence of a manufacturing sector.

A government can stimulate investment in domestic manufacturing through quotas and high tariffs on competing foreign goods. The emergence of a domestic manufacturing sector provides jobs that, in turn, train a factory labour force. Ideally, one factory acts as a catalyst to the creation of supporting enterprises. When domestic demand is saturated, factories can, it is hoped, export their manufactures. Though industry is the focus of import substitution, domestic food crops also often are protected.

Some problems were obvious even before the import substitution policies were put into effect. At least initially, the quality of indigenously produced goods would be less than that of imports, and prices would be higher. Smuggling and corruption that goes with it, would be inevitable if a country with high tariff protection bordered on a country without import substitution. Countries with small populations have very limited demand for manufactured goods, and thus, it might not be worth it to even attempt such production. So much government control of the economy would lead to an unhealthy concentration of political power. Officials would be tempted to unrealistically overvalue their exchange rates, such a policy would, indeed protect local industries but would have the negative effect of making the country's exports too expensive for foreign buyers. The Countries of the Global North, can retaliate against import substitution by closing down their own markets.

Nevertheless, the advantages were seen to overweigh the drawbacks, and as it turned out, the early stages of import substitution industrialization (ISI) were relatively easy. In the early phases of ISI, there is little need for a well-trained managerial force or skilled labourers, because industries grow out of existing artisan workshops. Most manufacturing at this stage is small-scale, perhaps a family operation making ready-made garments, apparels, so even less-populous countries can take advantage of limited demand. As more and more people become employed in these enterprises, often moving out of subsistence agriculture and into wage labour, the market grows. Profits at this early stage are virtually assured.

Early ISI was often, though, by no means always, successful in its limited goals of creating indigenous factories, a domestic capitalist class, experienced managers, and skilled factory workers. Before being derailed by the rising prices of oil in 1973s, Brazil's "economic miracle" took place in an environment of one of the most severe ISI programmes anywhere in Countries of the South. The Brazilian government assumed a strong guiding role in the economy, combining ISI policies with an openness to foreign investors and the multinationals saw an opportunity to profitably invest as "state-protected monopolicies", especially in countries like Brazil, with large populations, the country which became the "bellwether" of ECLAC policy. Domestic production of basic commodities often evolved into more sophisticated manufactures, so that in the second phase of ISI—reached as early as in the 1950s in Latin America—even radios and automobiles fell under tariff protection. In these cases, large MNEs such as Sony or General Motors set up factories to assemble imported components into finished products. It might be though that MNEs would have been the strongest opponents of import substitution, but many of them were quite happy to invest as monopolies in protected markets and reaped handsome profits. The military dictatorship in these countries held down wages by force of law and restrained expenditures on social programmes. This, plus a national stability enforced by severe repression, created an attractive investment climate, and multinational money and industrial technology flowed in.

In large countries like Brazil, domestic markets may be big enough that protected industries can prosper indefinitely. More often, however, import substitution will reach a point of crisis when markets became saturated. Indigenous capitalism is as expansionistic as its international counterpart. When there are no new customers for products that were formerly imported, companies will find themselves with stagnant markets. Ideally, they then would be able to expand their market through export. However, a company that grew up in an environment protected from international competition may not have a product that can compete in the international marketplace. Although import substitution has its value in the initial phases of industrialization, the limits of internal demand may bring an economy to a dead halt.

Unfortunately, after a period of rapid growth in GNP, Brazil found itself with a large marginalized population (most benefits accrued to the top 20 per cent) and the highest debt in Countries of the South. Such experiences led to disenchantment with the ECLAC model. Import substitution is now routinely criticized for its anti-export bias, overvalued exchange rates, market inefficiencies, and concentration of political power. Early successes get forgotten. It may not be the ISI model itself that is deficient but the failure of the countries to make other changes—such as land reform, educational improvements, and political reform—at the same time. But the ECLAC model put equal emphasis on internal and external factors since both traditional landowners and international terms of trade were impediments to development. If land reform and import substitution policies were enacted, then investment from the centre would be a positive force for progress.

It appeared that the "terms of trade" theory of underdevelopment, although perhaps accurate, was simplistic. *There would be no magic formula for rapid development.* But, because vested interests benefit from protectionism, import substitution is often continued as the major economic model long after it has reached the limits of development potential.

It should be mentioned that there has been a cyclical nature of CTT which makes an assessment of their long-term trend difficult as well as ambiguous. After all, stocks of many non-reproducible minerals (and slow-to-reproduce timber) are being steadily depleted and may be expected to impart increasing scarcity value—this happened with oil (prior to 1986) and applies equally to some other commodities which reproduce biologically (like fishing), which are renewable if the rate of harvesting/use is at or below the rate of natural replenishment. At the higher usage levels, the resource is being "mined" to exhaustion as surely as coal or oil stock. In this context since the conditions vary greatly between one market and another, if the attention is switched from the CTT to the NBTT, the picture becomes somewhat clearer: since oil price rise in 1972, the rise in NBTT of the oil exporting countries in the South rose six times larger in a decade between 1972 and 1981 so that NBTT for all the Countries of Global South got doubled, however, for the non-oil exporting nations, there had been a remarkable decline, a significant contributory cause of which had been the increased cost of their oil import costs. During this period of 1970 to 1982, there had been a remarkable decline in the purchasing power of the exports of the least developed countries in Global South. This contraction of import purchasing power has been due to a complex of factors such as failure to increase supply of primary commodity export supply, inability to diversify exports, and general failure to improve productive efficiency in all sectors, as well as to weak export prices for some primary commodities. In this context, it was argued by Prebisch that as a result of the institutional factors of differences in competitive structure, like the prices of some primary exports from the periphery are depressed relative to those of imported manufactured items: thus the international trade acts as an "exploitative force" on behalf of the core, the Countries of Global North.

However, as Spraos, 1983[14] noted, such competitive difference in labour markets between the Periphery and Centre are neither necessary nor sufficient to cause a decline in the net barter terms of trade primary for manufactured goods. He indicated that Prebisch himself did not long adhere to this simple version of the argument, and came to accept "more empirically plausible arguments" for the declining terms of trade. Spraos identified two major sustainable arguments for this position, neither of which can be classed as *institutional.* These are : (one) primary commodities suffer from a downward demand bias, the measure of which is their lower income elasticity compared with manufactured one; (secondly), they suffer from an upward supply bias because they draw on the pool of excess labour in the less progressive sectors of the developing countries' economies. However, these are not conclusive arguments for declining primary commodity terms of trade since there are developmental factors which offset them, such as the progressive exhaustion of reserves, the move away from specialization by non-primary sector production, and the possibility of slower overall progress in non-manufacturing than in manufacturing. The acceptable argument Spraos identified is an "institutional one". Thus this theory goes with the name of Prebisch that "primary commodities of developing countries have had to contend with the constantly increasing power of labour unions in developed countries".

Short-Term Instability of Export Earnings

The widely accepted hypothesis which has exercised an important influence on international trade policy is that, because of their reliance upon primary commodities, the export-earnings of the Global South albeit a high-degree of *short-run instability*. As an ancillary proposition it is argued that such instability is damaging to the growth and development aspects of these countries. This follows because it is assumed that fluctuating export receipts will cause fluctuations in the capacity to import the capital goods on which their growth and development depend. Here it should be noted that what is being considered is fluctuations in earnings about an underlying trend. If this underlying trend is downward, that is, of course, an unambiguously detrimental factor, but it is not what is meant by a fluctuation.

With regard to variations in the value of commodity trade at the world market level it is popularly argued that, because of price-inelastic demand, short-term shifts in aggregate demand to supply of primary products lead to price changes of greater proportionate magnitude than those in the volume of trade. It should, however, be observed that a given proportionate reduction of demand will tend to have a greater adverse effect upon total export earnings for a commodity than an equivalent proportionate supply increase. From the standpoint of international trade policy this is an important distinction, since demand shifts are beyond the control of the exporters but are mainly dependent upon fluctuations in the level of economic activity in the industrial countries. It is evident that such fluctuations have been large —witness the world economic boom of 1972-74 and the subsequent recession of 1975-83— and precisely because they have not originated in the economic conditions of the underdeveloped countries there is a basis for arguing that the rich countries should compensate poorer ones for any damage caused to their economies by such fluctuations.

Of course, the export earnings of an individual country for a particular commodity need not move in line with total world export earnings from that commodity. Exports from a particular country may increase at a time of world supply shortage and high prices (and *vice versa*), so that they may fluctuate more than world export earnings of that commodity. Few countries, however, are dependent solely upon one export product, and export concentration ratios have declined indicating a steady process of export diversification by the majority of Southern Countries. It is, therefore, probable that there are offsetting movements in the earnings from different products which will damp down fluctuations in total export earnings, except at times of extreme world economic recession or boom when all commodity prices will tend to move together.

Export-earning instability tends to increase if (1) a primary commodity constitutes a higher proportion of total exports; (2) a smaller number of commodities dominate the primary export fraction; or (3) if exports are mainly destined for one geographic market, *e.g.*, the former colonizing nation. In this respect certainly some commodities such as cocoa, copper and cotton have experienced greater export-earnings stability than commodities *e.g.*, bananas, coffee and rubber. However, it does not appear that more volatile commodities are responsible for the higher levels of instability in those countries where this is greatest. Rather, exceptional export-earning instability is mainly due to special internal factors often fiscal, which influence the volume of exports.

The concern with short-term export instability is due to hypothesis that it hampers development. Since instability is defined in such a way that on average downswings equal upswings, "the hypothesis is equivalent indicating that the damaging effects of downward export fluctuations are more than the beneficial effects of upward deviations". Support for this form of the hypothesis is along the lines that unanticipated increases in foreign exchange earnings cannot be exploited readily for investment because of shortages of complementary local factors of production (particularly skilled human capacity) and tend to be dissipated on consumer goods, whereas unexpected shortages in earnings lead to cutbacks in capital equipment and delays in investment projects.

Thus the above argument is well known: It also means paying more attention to basics such as solid macroeconomic management, good public education, a thriving private sector, a legal system that protects property rights, and a society where prosperity is widely shared. These are the kind of pro-growth policies that most of East Asia employed to develop from sweatshop havens in the 1960s and 1970s to industrial powers. Countries that have failed to get these basics right often benefit the least from globalization. Whereas almost everyone in South Korea, Taiwan, Singapore, and Hong Kong received a high school education when their countries' export drives began decades ago, in much of Central America half of the population never makes it to the fifth grade. The record is worse in Sub-Saharan Africa. Since the 1960s, the gap between rich and poor has narrowed steadily in most East Asian countries, a sign that the benefits of growth were shared widely. In the rest of the developing world, however, income gaps have yawned. Such factors go a long way to explain why big increases in trade and foreign investment in countries such as Guatemala, Nigeria, and the Dominican Republic haven't translated into major reductions in poverty. Weak government also partly explains the appalling working conditions still found in many developing nations. China, for example, has good labour laws. But they often are not enforced. Even in China, the issue of labour oppression is complex: Talk to the workers and they'll take issue with many Western stereotypes of exploitation.

Largely due to $ 328 billion in foreign investment, urban per capita incomes in China have grown more than tenfold, to $ 705, since reform began in 1978. The volume of imports into Asia as a whole grew at an annual rate of 10 per cent between 1990 and 1995. In 1995, the imports of the ten leading Asian developing economies amounted to almost $ 750 billion. Asia's import growth of 13 per cent in 1995 exceeded the world average for the fourth consecutive year. 1995 was also the fourth consecutive year in which the volume of imports into Asia expanded more rapidly than Asia's exports. In the next five years, the increase in the volume of merchandise imports into the ten leading Asian developing countries could well be as large as that into the European Union (EU) and the USA combined. More than a third of the real increase in global imports could go to just these ten markets. *Chinese industry pumped out $ 195 billion in exports on 2002. Much of the income from the thousands of shoe, toy, garment, and electronics factories on China's east coast has spread far into the poverty-ravaged interior. China also shows that, from humble beginnings, industry can advance rapidly, given the right conditions and a determined government. With two decades of experience with simple assembly plants under its belt, Chinese industry is now making great strides in electronics and machinery. Millions of engineers and scientists have helped the country becoming a big exporter of computer equipment, auto parts, and telecom gear. Homegrown manufacturers such as Legend Computers and appliance maker Haier Group not only dominate the local market but have begun exporting to the rest of Asia and the USA. Other countries are just now learning to imitate this pattern of public and private investment.* Mexico isn't all the way there, either. While exports in Mexico have soared five-fold since 1987, to $ 137 billion, 47 million people still live in poverty—about half of the population. And only in 2002 did real wages catch up with 1994 levels, when the *peso* crashed. One problem is that half of exports come from assembly operations that have done little to create local industry because they import most parts and materials. Another: Only 50 per cent of Mexicans have completed sixth grade—and just 20 per cent enter high school.

Following Adam Smith and Ricardo, theory of specialization according to *the theory of comparative advantage*—that is the countries specialize in exporting those things they can produce at competitive prices (at lowest relative cost), trade is beneficial, which may be interpreted as suggesting that trade leads to development. However, these theories leave open the issue of the distribution of the gains from trade between trading partners, which opens the gates for the argument that because of unequal exchanges (*i.e.*, trade in an unequal marketplace): the Southern Countries are not receiving a fair share of the expansion of the world trade, and that they are not being paid adequately for the share that they do have, and the main benefits of trade accrue to the North. However, while it is clear that there is no theoretical basis for stating how any welfare gains that arise from trade will be distributed between countries, this is not equivalent to asserting that changes are unequal. Thus, the theory of comparative advantage is "half descriptive and half prescriptive. The descriptive part is simply that this is the way the system works : this is very much what countries actually do and have to do in order to play in the international game. There do not seem to be many other alternatives. The prescriptive part, which is more desirable, is that this is morally good, because it leads to the benefit of all. *But since the export-earning instability is a problem for Southern Countries, it is not surprising that in international trade negotiations there is* a priori *view about the harmful effects of export variability upon development and of the need for stabilization policies to be implemented. This pressure has led to the introduction of a variety of policies to help offset export-earning instability for some Countries of the South by stabilization measures in terms of smoothing out of payments over time; since this measure involves smoothing out of payments over time without any need for continuous long-term transfers from one group of countries to another, it is much easier to reach international agreements on stabilization policies than for policies to counteract any long-run tendencies for decline in the terms of particular nations, because these latter require a net secular transfer of funds to disadvantaged nations.*[15]

Increasing awareness of these above problems that the Southern Countries are not receiving a fair share of the expansion of world trade and they are not being adequately paid for the share that they do have, amount to the concern that their export earnings are not rising fast enough. Added to this is the concern that too high a proportion of their export revenues which do accrue to them are subsequently exported as profits by MNEs. In addition, there is wide acceptance of the view that heavy reliance upon primary commodities leads to fluctuations in export earnings which disrupt the development plans of poorer countries.

All these led to the resolution of a Programme of Action in UN in 1974 on the establishment of a New International Economic Order, (NIEO). This was a six-point plan covering buffer stock arrangements, an "international commodity agreements (ICA)" for creation of common funds towards a common financial facilities for the management of commodity markets like bananas, bauxite, iron ore, rice, wheat, wool, cocoa, copper, cotton, rubber, sisal, sugar, tea, wool and tropical timber, a network of supply and purchase agreements, transfer of primary processing activities from rich to poor countries, index linking of prices of primary commodity exports from poor countries to the prices of the imports, and finally as a residual measure, a system of compensatory financing to stabilize and guarantee poor countries' export earnings from commodity exports. These six proposals were shaped into an "Integrated Programme for Commodities (IPC)" but the full grandeur of this vision was never realized. Only the most faltering of steps taken was on "to concentrate on developing co-operation on research, marketing, technical information etc., rather than on the adoption of the IPC. The failure to make the IPC realized stems from two principal causes. The (first) of these, is the intrinsic difficulty of managing markets through ICAs. The (second) has been the fundamental difference between the Countries of the North and those of the South about the objectives of the IPC. The Southern Countries were interested not only in stabilizing the commodity prices but to increase them, and in this way help to create a new international economic order. But the North, while content in principle to countenance self-financing price stabilization, were not willing to accept general measures to raise commodity prices or to shoulder the burden of the financial transfers to commodity prices which such measures would entail.

A key requirement for successful market management by an ICA is a strong political commitment by major signatories to bear any costs which may be involved. Such an agreement exists, as with the International Wheat Agreement (IWA) between 1949 and 1971, when international market control could be achieved without buffer stocks or the restrictive application of export quotas. During the 1950s and 1960s, the domestic wheat policy of USA, the world's main wheat producer, supported by Canada, was based on accepting wheat subsidies for surplus wheat which stabilized wheat prices within the agreed international price band. Thus the USA bore most of the cost of managing the international wheat market. When in 1971, the USA cut support to wheat farmers and sharply reduced wheat and other grain stocks, a more commercial approach to wheat exports was adopted and willingness to underwrite the IWA disappeared. Since that date it has not been possible to achieve agreement between wheat importers and exporters on price stabilization, and the IWA got withered. The fact that whenever ICA is effective in the above sense, either importers are paying higher prices or exporters are receiving lower prices than they otherwise would. Thus the costs involved are high, pressure builds up to break or negotiate the agreement. Tension is, therefore, an automatic consequence of effective action. Thus the examples of successful actions to push up prices by international supply restrictions have been few. Favourable conditions for successful agreement to restrict supply exist when there are no close substitutes for the commodity so that the price elasticity of demand is low even in the medium and long run.

There are few producing countries, so that potential conflicts of interest in reaching agreement are reduced, or there is a small group of countries dominating export sales which can undermine the position of other exporters if they fail to comply with any agreement, for example, Brazil has sufficient power to manipulate the international coffee market independently,

1. The production and distribution systems within countries are *oligopolistic* (as in metals) rather than *atomistic* (as in agriculture). Where atomistic structure exist, it is very difficult to control non-quota sales.
2. The commodity is storable so that the sale of surplus product can be deferred.
3. It is also helpful if storage is cheap, as with oil and phosphate rock which can be stored in the ground at "zero financial cost".

There are very few commodities which satisfy even most of these favourable conditions for the implementation of output restriction and it is doubtful whether any such scheme would emerge to assist the level of stability of the commodity export earnings in the Countries of South.

Thus it is questionable whether a commodity-by-commodity approach is appropriate to the trade problems of the Southern Countries. A more fundamental approach would be to try and increase the level and stability in earnings from total commodity and product exports. One proposal which recognizes this calls for the index linking of commodity export indices to the import prices paid by developing countries. It is, however, unlikely that much progress will be made in this direction in view of the formidable technical problems of producing the indices and of the potential for conflicts of interest even among developing countries. For example, the interests of the oil and non-oil producing countries have suffered more than any others from the changes in relative commodity and product prices since the first oil price hike in 1973.

Since 1981, the IMF operates a Compensatory Finance Facility (CFF) for Southern Countries—this is used to compensate for solving the countries' overall balance payments: borrowing allowances under this CFF as a whole have been up to 125 per cent of their respective IMF quota. As a stabilizing device the generality of the IMF scheme makes it far more attractive than the individual commodity agreements discussed above, since in contrast to them, there is no need to negotiate trade quotas, to agree upon price ranges for specific qualities of produce or to consider a wide range of other technical details. However, the repayment of IMF loans, in spite of the provision of allowable repayment periods of 3 to 5 years, the crippling burden of indebtedness has become difficult for Southern Countries taken as a whole and a major issue to the individual poorer countries. In 1990, the Southern Countries as a whole paid about 15 per cent of their exports of goods and services in debt repayments, a level that is significantly higher than the average of 10 per cent in the mid-1970s, but this level varies enormously, rising to over 30 per cent for certain countries.

And moreover, the CFF scheme is not a substitute for the net transfer of resources to the Southern Countries as envisaged in the United Nations General Assembly Resolution of 1974 which proposed a New International Economic Order (NIEO). Thus it gets clear that no spectacular progress has been achieved by the Southern Countries in the international organizations where major decisions are taken affecting international trade. However, the 1974 UN Resolution calling for NIEO was an important flexing of political muscle by the "so-called group of 77 Southern Countries". It signified the increasingly cohesive voice with which the Southern Countries approach negotiations in United Nations Conference on Trade and Development (UNCTAD) and General Agreement on Tariffs and Trade (GATT), and their dealings with the World Bank and IMF. This togetherness initiative brought the Most-Favoured Nation (MFN) tariff reductions from the Northern Countries for their exports. Similarly limited progress has been achieved in increasing the flexibility in IMF lending and certain changes in World Bank policy. The Southern Countries have persistently voiced their dissatisfactions at the continued domination of the policies of these two organizations by the North and they claimed to have a larger say in shaping the policies of these two organizations. This pressure has, nevertheless, been largely resisted by the North which take the line that since they are the main suppliers of funds alongwith some of OPEC countries, they should be able to control the IMF and World Bank policy.

Conscious of the resistance of the North to their demands for international trade reform and dissatisfied with conditions of exchange of their currencies, the Southern Countries on occasions have sought to join forces in "regional customs unions" with the hope the creation of larger regional market units would enable local industries to take advantage of economies of size in production which would lead to increased industrial activity within each community but the outcome of these expectations has fallen short of expectations because the neighbouring Countries in the South possess similar resource endowments, structured economies and industrialization plans. Also their transport and communications infrastructure is often better developed for trade with countries outside the customs union. Thus their main opportunities for trade are not with neighbouring countries but with countries with completely different resource endowments.

Hence, the primary responsibility lies with the development policies of the Southern Countries themselves. It is argued that the policies of industrialization they are pursuing (which should create implications for trade) have reduced their potential benefits from trade and international specialization—in this respect, the LDCs in the South experience great difficulty in devising policies which could greatly increase their rates of development and export growth. Again it is argued that the share of the Southern Countries has been declining as the value of the world trade expands, though of course, the rapidly increasing component of exports by developed countries is accounted for by reciprocal trade in similar manufactured products, *e.g.*, motor cars changed for motor cars. This failure of the Southern Countries to reap greater benefits from expanding world trade is substantially due to deficiencies in their economic policies and the internal structure of these countries, and it cannot be attributed to deficient demand of their products. There is widespread acceptance of the view that the *international trade has not worked to the advantage of the Southern Countries in the manner anticipated by neo-classical economic theory.*

The efforts of the Southern Countries to achieve more rapid economic growth and trade since the 1960s have been overshadowed by the global economic recession which started with the Oil Price Increase in 1973. While there is widespread recognition of the mutual interdependence between economies and the acceleration of growth depends upon acceleration of the industrial economies, the dominant concern to contain inflation and the policies to achieve this, have both restrained economic (and hence trade) growth and affected the developed countries' attitude to the North-South negotiations. The dramatic economic boom which started with the 1950s gave rise to high levels of public spending for welfare measures plus a substantial expansion of lending to the Southern Countries to finance trade expansion. Reaction to the inflation in the 1973s in industrial countries have generally been to cut back public expenditure, to reduce protection for declining industries, to look less favourably upon subsidies and transfers to productive activities, and to rely more on "private and less on public enterprise". These attitudes have spilled over into the arena of international negotiations. They have (1) hardened the North resistance to the incorporation of transfers of funds into the Integrated Programmes for Commodities (IPC) and International Compensatory Finance (ICF), (2) militated against using International Commodity Agreements (ICAs) to raise commodity prices, (3) led to no attempt to enact the New International Economic Order (NIEO), rather attempts by national governments, or intergovern-mental agencies, like UNCTAD, NIEO are regarded as counter-productive, fostering inefficient production arrangements and reducing global economic growth rates, (4) led to tougher terms for credit and loans to South, coupled with tighter attitudes of repayment of outstanding debt, (5) led the USA to reduce its contributions to the World Bank, thus restricting the Bank's ability to increase its aid programmes, (6) led to greater insistence by the North that the Southern Countries reduce their own barriers to trade, and (7) led to emphasis of the virtues of trade rather than aid. This switch of emphasis in the North's policy from willingness to use public funds for priming investment bumps (both national and international) to increased reliance upon unsubsidized free enterprise, has undoubtedly conflicted with the aspirations and expectations of the South. Thus the end of the Cold War in 1989 should have given the Southern Countries more freedom of action, but, in reality, the developmental options of the countries most in debt were reduced to that prescribed by the IMF.

Hence the more integrated global economy based on the ideology of comparative advantage is similar in many ways to the colonial period of the 1930s, when certain countries earned their keep through exporting raw materials and importing finished goods. A difference now is that colonizers are no longer "mother countries" but multinational enterprises (MNEs) and financial institutions. Those least developed Southern Countries as well as the middle range ones, which lack the depth of entrepreneurial talent or the general infrastructure needed for successful broadly-based commercial competition have found this increasingly globalized regime exceptionally abrasive and many have recorded negative growth rates. For them, at least, specifically tailored programmes involving special trade concessions and aid are essential to their progress, adherence to the slogan "trade not aid" is seen by them as unsatisfactory. The IMF's report, World Economic Outlook, 2000 describes the plight of a fifth of the world's population: "too many countries, and nearly one-fifth of the world population, have regressed in relative and sometimes even absolute terms. This is arguably one of the greatest economic failures of the twentieth century."[16]

Whereas the IMF measured inequality between countries, World Bank economists are also measuring inequality within countries. As a result, they paint a bleaker picture. Conversely, various recent studies also point to wealth concentration. The 1999 UNDP Human Development Report reports that whereas the income gap between the 20 per cent of the world's richest people in the richest countries and the 20 per cent of the poorest was 30:1 in 1960, this gap has jumped to 74:1 in 1997. Furthermore, half the world's population— nearly three billion people — is subsisting on $ 2 per day or less. On the other side, the combined wealth of the top 3 billionaires exceeds the combined income of the 600 million people in all the LDCs. This concentration of wealth reflects in the concentration of global trade, investment and financial transactions in the hands of a few countries and a few transnational corporations. Worse still, is the fact that the failure is expected to deepen. There are many analyses showing poverty deepening as a consequence of globalization. Thus globalization has not affected all countries and regions to the same extent, especially in the Global South. While a small handful Pacific-rim countries of Asia have prospered, a vast majority of the South as well as of the world have experienced a corresponding divergence of their income level and consumption patterns away from their counterparts in the industrialized world.

Global Economic Trap

The commodity crisis encapsulates many of the controversies surrounding the debate on sustainable development and highlights in particular the disjunction between the interests of the North and the needs of the South. While the brunt of declining commodity prices is shouldered by commodity producers in the South, the main beneficiaries of these low prices consumers and companies are located in the North. The bulk of commodity trading still moves through the main trading centres in the West and most of the processing of raw agricultural consumables remain almost exclusively in the North. Global economic structures have undermined or effectively prevented efforts by commodity-dependent countries to shift out of commodity production into other areas of industry or, for countries dependent on one or two major commodities as the source of export earnings, to diversify into other types of crop cultivation.

The history of commodity price stabilization efforts and the collapse of commodity agreements is indicative of the vested interest of northern industrialized countries (under pressure from domestic corporate lobbies) to keep commodity prices at a consistently low prices. It is also illustrative of the legacy of colonial economic relations between the North and South that continues today under the guise of structural adjustment programmes and multilateral trade rules favouring developed countries. The marginalization of commodity-dependent countries, particularly the LDCs, from international trade and financial systems have weakened their control over their terms of trade and severely undermined their capacity to set remunerative income for their commodity producers, particularly small farmers.

Inappropriate policy advice from international financial institutions, notably the World Bank and the IMF, and enforced through lending conditionalities have exacerbated the situation of overproduction. Some policy prescriptions, motivated by ideology, have taken away the policy space and dismantled institutional structures, such as state marketing boards, enabling the governments of Southern Countries to intervene in the production of commodities and to ensure a steady income for its commodity producers. Depressed commodity prices have meant that for many countries dependent on primary exports have been forced to increase production in attempts to compensate for falling prices. This has led to a situation of more land resources being used for export crop production, reducing land available for domestic food cultivation, and increasing deforestation of non-agricultural land.

Rural livelihoods are also threatened by an overemphasis on cash crop production. The increasing use of mechanized and large scale agricultural farming has led to the demise of many forms of subsistence farming and smallholdings. Many small-scale farmers have been forced to abandon their farms as a result of low commodity prices, and forced into the wage economy as labour in plantations or urban sectors. Similarly, the liberalization of mineral sectors under the Structural Adjustment Programmes (SAP) have also led to the displacement of small-scale prospectors from mining in many parts of the developing world.

Declining prices for primary commodities also prevent primary commodity producers from investing in safe and environmentally sound methods of production. The lack of transfer of technology prevents commodity producers from developing and least developed countries, particularly those involved in the production of agricultural commodities and resource extraction, such as mining and minerals to invest in ecologically sustainable modes of production. Efforts preventing diversification of the economies confine many developing countries to extract more and more out of the natural resource base and creating greater environmental degradation and social displacement.

Ways Forward and WSSD Outcomes

Action to reverse the above trend in falling commodity prices is essential to any initiative undertaken at the international level to facilitate sustainable development, poverty reduction and debt relief. Without Price Stabilization Mechanisms (WPSM) to ensure remunerative prices to commodity producers and commodity producing countries, sustainable development programme are unlikely to achieve long-term sustainability. Any attempt to ensure remunerative prices for commodity producers and commodity producing countries must address the root of the problem and the issue of overproduction and oversupply. The proposal calling for the establishment of an international mechanism to stabilize commodity prices in the World Summit on Sustainable Development (WSSD) Draft Plan of Implementation must be strongly supported and firm commitments must be made by both producing and consuming countries to adhere to an international supply management programme. This is the key to resolving the commodity crisis.

A strong case can still be made out for reviving the Integrated Programme for Commodities (IPC) framework as an approach for an international commodity policy aimed at negotiating a series of price stabilizing agreements for commodities suffering from high price volatility and/or structurally depressed prices. Such an initiative would also involve reviving the functions of the First Window of the Common Fund for Commodities to provide financing for the purchase and maintenance of buffer stocks for this purpose.

Proposals calling for the provision of financial support and technical assistance for diversification activities for commodity-dependent countries must also be supported. And as efforts by Countries of the South to shift away from commodity dependence have also been thwarted by restrictive trade rules bias against agricultural products from developing countries, diversification plans must be supported by corresponding commitments by industrialized countries to eliminate tariffs and non-tariffs on commodity-based products from developing countries.

Tariff reduction has been one of the key functions of the multilateral trading system. Calling for greater market access for developing country products and the elimination of tariff and non-tariff barriers on products of export interest to countries of the South or as appropriate, eliminate tariffs and non-tariff barriers on products of export interest to primary producing countries, (particularly agriculture, textiles and clothing) must be supported. Without the requisite reduction or elimination of high import tariffs on processed and semi-processed commodities and export subsidies for domestic agricultural products from developed countries, commodity-dependent countries will be unable to diversify into higher stages of the commodity value chain as there is little guaranteed market access for these products in the North.

The linkages between the declining prices for primary commodities and issues of sustainable development are clear. Depressed commodity prices lead to declining terms if trade for commodity-dependent countries, leading to balance-of-payments problems and resulting in debt. The lack of financial resources available within the country as a consequence of debt servicing and the decline in the purchasing power *vis-a-vis* imports leads to diminished funds. This will, of course, lead to impoverishment, social conflict, a decline in social welfare provisions and environmental degradation. Lack of funds and pressure to generate more revenue from finite resources will severely hamper efforts to implement sustainable management of natural resources. Unless international action is taken to stabilize commodity prices and ensure a remunerative income for commodity producers and commodity producing countries, many small and land-locked countries of the South will not emerge from the poverty trap, exacerbating their dependence on foreign aid and blocking efforts to invest in social and economic capital for sustainable development.

"A fall in commodity prices have also led to a build-up of unsustainable debt." The lack of greater revenues from exports has knock-on effects, as described above. The irony is that structural adjustment programmes were prescribed by the IMF and the World Bank due to debt repayment concerns in the first place.

Adam Smith, in his 1776 classic, *The Wealth of Nations*[17] regarded as a bible of capitalism also provided some insights. That is, he was highly critical of *mercantilist practices* of the wealthy nations, while he recognized the value of local industry and the impacts of imported manufactured products to local industries. Though the encouragement of exportation and the discouragement of importation are the two great engines by which the mercantile system proposes to enrich every country, yet with regard to some particular commodities it seems to follow an opposite plan, to discourage exportation and to encourage importation. Its ultimate object, however, it pretends, is always the same, to enrich the country by the advantageous balance of trade. He discouraged .... "the exportation of the materials of manufacture, and of the instruments of trade, in order to give our own workmen an advantage, and to enable them to undersell those of other nations in all foreign markets; and by restraining, in this manner, the exportation of a few commodities of no great price, it proposes to occasion a much greater and more valuable exportation of others. It encourages the importation of the materials of manufacture in order that our own people may be enabled to work them up more cheaply, and thereby prevent a greater and more valuable importation of the manufactured commodities".

Reading the above, we can say that *structural adjustment policies are also mercantilist*. We are constantly told that we live in a world of global capitalism, and yet we see that while free markets are preached (in Adam Smith's name), mercantilism is still practiced! Of course, today it is a bit more complicated too. We do have, for example, products being exported from the poorer countries (albeit some facing high barriers in the rich nations). But exporting rather than first creating and developing local industry and economy, means the "developing" country loses out in the long run, (hardly "developing") because there is little multiplier effect of money circulating within the country, as mentioned above.

Furthermore, with labour being paid less than their fair wages in the poorer nations, wealth is still accumulated by and concentrated in the richer nations of North. While in trade terms more free trade for all might therefore seem to be the direction and the claim, the wealth that is accumulated by the richer countries clouds over the fact that this is still the age-old mercantilism or "monopoly capitalism" (as opposed to free market capitalism) processes being played over again. It is a system that Adam Smith himself criticized so much. When looked at in this light, poverty is more than simple economic issues; it is also an ideological construct.

What is the IMF/World Bank Prescription?

The IMF encourages the aggressive opening up of countries for trade, but is coupled with too much privatization and too much deregulation with fewer safety nets able to be put in place by the governments (which would be there for the benefit of the people of that nation in the long run). That some of the poor nations may not be as aggressive as the IMF would like, in privatization and other conditionalities, is also one of the reasons in the continual delay in theirs' receiving the debt relief. This inevitably means that the poor suffer, while the rich get richer.

The World Bank's do the *same* :—

1. Privatisation : state's selling-off the public sector enterprises.
2. Capital market liberalization : by dismantling trade barriers.
3. Free trade : by "opening markets". Europeans and Americans today are kicking down barriers to sales in Asia, Latin American and Africa while barricading their own markets against the Third World's agriculture." The IMF's structural 'assistance' led to Africa's income dropping by 23 per cent (Stiglitz, 2002)[18].

What is Structural Adjustment?

This standard policy package imposed by the World Bank and IMF was termed "structural adjustment." This referred to the purpose of correcting trade imbalances and government deficits. It involved cutting back the role of the state and promoting the role of the private sector. The ideology behind these policies is often labelled "neo-liberalism," "free market fundamentalism,"or the "Washington Consensus".[19] From the 1970s on, this orientation became the dominant economic paradigm for rich country governments and for the international financial institutions.

The basic assumption behind structural adjustment was that an increased role for the market would bring benefits to both poor and rich. In the Darwinian world of international markets, the strongest would win out. This would encourage others to follow their example. The development of a market economy with a greater role for the private sector was, therefore, seen as the key to stimulating economic growth.

Structural adjustment refers to a package of economic policy changes designed to fix imbalances in trade and government budgets. In trade, the objective is to improve a country's balance of payments, by increasing exports and reducing imports. For budgets, the objective is to increase government income and to reduce expenses. In theory, achieving these goals will enable a country to recover macroeconomic stability in the short-term. It will also set the stage for long-term growth and development. The structural adjustment programmes of the early 1980s were meant to provide temporary financing to borrowing countries to stabilize their economies. These loans were intended to enable governments to repay their debts, reduce deficits in spending, and close the gap between imports and exports. Gradually, these loans evolved into a core set of economic policy changes required by the World Bank and IMF. They were designed to further integrate African countries into the global economy, to strengthen the role of the international private sector, and to encourage growth through trade.

Typical components of adjustment programmes included cutbacks in government spending, privatization of government-held enterprises and services, and reduced protection for domestic industry. Other types of adjustment involved currency devaluation, increased interest rates, and the elimination of food subsidies. The underlying intention was to minimize the role of the state.

World Bank and IMF adjustment programmes differ according to the role of each institution. In general, IMF loan conditions focus on monetary and fiscal issues. They emphasize programmes to address inflation and balance of payments problems, often requiring specific levels of cutbacks in total government spending. The adjustment programmes of the World Bank are wider in scope, with a more long-term development focus. They highlight market liberalization and public sector reforms, seen as promoting growth through expanding exports, particularly of cash crops. Despite these differences, World Bank and IMF adjustment programmes reinforce each other. One way is called "cross-conditionality". This means that a government generally must first be approved by the IMF, before qualifying for an adjustment loan from the World Bank. Their agendas also overlap in the financial sector in particular. Both work to impose fiscal austerity and to eliminate subsidies for workers, for example. The market-oriented perspective of both institutions makes their policy prescriptions complementary.

Adjustment lending constitutes 100 per cent of IMF loans. In 2001, approximately 27 per cent of World Bank lending to African countries was for "adjustment." In the World Bank's total loan portfolio, adjustment lending generally accounts for between one-third and one-half. The remainder of World Bank loans are disbursed for development projects and programmes. The project portfolio of the Bank covers such areas as infrastructure, agricultural and environmental development, and human resource development. In some cases, the projects supported by World Bank loans do make useful contributions to development. But these occasional successes must be weighed against the negative effects of increasing debt, imposed economic policies and their consequences.

IMF and World Bank Reform?

Economics is often driven by politics. As a result of policies by the IMF, World Bank and various powerful nations, basic human rights have been severely undermined in many countries. By insisting that national leaders place the interests of international financial investors above the needs of their own citizens, the IMF and the World Bank have short circuited the accountability at the heart of self-governance, thereby corrupting the democratic process. The subordination of social needs to the concerns of financial markets has, in turn, made it more difficult for national governments to ensure that their people receive food, health care, and education — basic human rights as defined by the Universal Declaration of Human Rights of United Nations General Assembly in 1948 (This Declaration is built on the principle that human rights come from the "inherent dignity" of every person). This dignity, and the rights to freedom and equality which derive therefrom, are also inalienable on the guarantee of economic rights. More broadly, the Declaration asserts: "Everyone has the right to a standard of living adequate for the health and well-being of himself and of his family, including food, clothing, housing and medical care and necessary social services, and the right to security in the event of unemployment, sickness, disability, widowhood, old age or other lack of livelihood in circumstances beyond his control." The Bank's and the Fund's erosion of basic human rights and their perversion of the democratic process have made the institutions a clear threat to the well-being of hundreds of millions of people worldwide.

In compelling countries to adopt such policies, the IMF and the World Bank have not only threatened communities' right to social security — they have also undermined countries' democratic systems. Democracy rests on the principle that government officials and elected representatives are servants of the citizenry at large. Elected leaders' ultimate accountability is to the people they serve. The Bank and the Fund have severed that chain of accountability by making national leaders more concerned with the interests of international investors than with the needs of their own people. As soon as government officials begin worrying more about what Wall Street will think than what their own people think, democracy has been perverted. Since 1976, at least 100 protests against IMF and the World Bank policies have occurred in dozens of countries around the world. Clearly, ordinary citizens are outraged with the institutions' policies. The continued adoption of those policies reveals the democracy disconnect fostered by the IMF and the World Bank. The IMF's and World Bank's history of socially irresponsible and environmentally unsustainable policies are too long to present in a brief statement. It is necessary, then, to focus on just a few examples of how these institutions' policies have threatened human rights and democracy throughout the world.

For more than 50 years, thus the IMF and the World Bank have advanced a form of economic "development" that prioritizes the concerns of wealthy lenders and multinational corporations in the industrialized North while neglecting the needs of the world's poor majority. The institutions work as a kind of international loan shark, exerting enormous influence over the economies of more than 60 countries. In order to get loans, international assistance, and debt relief, countries must agree to conditions set by the IMF and World Bank. Under the guise of promoting "free trade," market liberalization, and financial stability, these two institutions have forced cuts in health care, education and other social services for millions of people across the planet, thereby deepening poverty and increasing inequality. By elevating concerns about macroeconomic financial stability above all other competing values, the institutions have created a human rights catastrophe.

In dozens of countries around the world, the IMF and the World Bank have violated that "right to social security." These institutions have forced debtor countries to cut social spending on health, education, and other public services. They have pressured poor nations to charge their own citizens for the use of public schools and public hospitals. And they have demanded that countries keep their wage levels low, a policy which harms ordinary citizens but benefits multinational corporations.

The impact of this model of development, whereby the North (or the developed Nations) impose their conditions on the South (the developing Nations) through IMF, World Bank and WTO has come under observation in a wide number of countries, the few of which is described below.

In Haiti and Mexico, the Fund and the Bank have actively worked to keep wages low, making it more difficult for ordinary citizens to support themselves. Throughout Africa, the IMF's and World Bank's imposition of "user fees" for health services and the institutions' resistance to meaningful debt relief have worsened the AIDS crisis ravaging that continent. In Colombia, the IMF has demanded social service cuts even as the country suffers from recession and civil war. In Brazil, as in many other countries, the World Bank has meddled in domestic politics by inserting itself into one side of a heated social debate. And in all of these places, as in many more, the institutions have directed the most political of decisions — the allocation of national resources — thereby undermining national democracies.

The IMF and World Bank say their trade liberation policies are designed to succeed in the "long run." But after more than 20 years of managing dozens of economies, the institutions have created more inequality, more environmental destruction, and no real security. It is long past time for the USA and the other wealthy nations that enjoy *de facto* control over the institutions to call for the abolition of the IMF and the World Bank and to begin work to create multilateral financial institutions that are truly committed to human rights and democracy and which can effectively respond to the new realities of the twenty first century. Policy makers everywhere must recognize that another world is possible.

Haiti: Suffering from "Competitiveness" : For the last two decades, the Caribbean nation of Haiti has dutifully followed the prescriptions of the IMF and the World Bank. And yet today Haiti remains the poorest country in the Western Hemisphere. Approximately 70 per cent of Haitians are unemployed. Only four out of ten Haitians enjoy access to clean drinking water, and the country suffers from one of the worst life expectancy rates in the world; Haitian men can expect to live 47 years, Haitian women 51 years. Most Haitians cannot read. Fifty per cent of Haitian children younger than 5 suffer from malnutrition. Per capita income is just above $ 1 a day.

The IMF's policy mandates for successive Haitian governments— whether military or civilian— have focussed on a single goal: increasing exports. Haiti's exports have, in fact, increased dramatically under the IMF's tutelage. But that "success" has come at a high price. In an effort to attract foreign investment to Haiti at any cost, the IMF has forced Haiti to keep wages at poverty levels. While this has been a boon for multinational corporations, it has not led to prosperity for ordinary Haitians. Even coveted jobs in Haiti's garment factories provide just barely enough money for people to survive. At the same time, the IMF and World Bank's insistence on cutting assistance to small farmers has ruined the country's agricultural sector, making Haitians dependent on imported food. The IMF's and World Bank's legacy in Haitian is largely one of lower wages than ever before and widespread food insecurity.

Agricultural "Development" : The international agencies' involvement in Haiti began with a disaster. In 1982, a programme was begun to replace the ubiquitous Creole pig—owned by about 85 per cent of rural households — with North American pig breeds. The Creole pig used to serve as a sort of savings account for many Haitian farmers, with families selling their pigs on special occasions — weddings, funerals — and emergencies. But foreign agriculture "experts" determined that the pigs were chronically ill, and fearing that the pig illness would spread north, virtually every Creole pig in the country was killed in a period of 13 months. The new, imported pigs from Iowa were a poor replacement. They couldn't tolerate the tropical environment and needed special roofed pigpens. Unless the pigs had clean drinking water — which even most Haitian people don't have — they became sick. They required imported feed, which ordinary farmers couldn't afford. The re-population was a complete failure. According to one observer, Haitian peasants lost approximately $ 600 million in capital. Protein consumption in rural Haiti dropped dramatically.

The pig programme was just one part of a larger World Bank plan to overhaul Haiti's agricultural system. In an effort to transform Haitian agriculture from a concentration on local consumption to an emphasis on foreign export, the Bank encouraged Haitians to plant cacao and coffee. Farmers were supposed to grow cash crops for export and then use their earnings to buy food. But the prices of coffee and cacao fluctuate widely, and as worldwide commodity prices declined Haitian farmers found themselves unable to make a living.

The transformation from an agricultural system geared toward domestic consumption to an export-oriented system soon led to a decrease in food production. The fall in local production was exacerbated in 1986, when the IMF demanded that the country lower import tariffs on rice and end support for rice farmers. Rice from the USA—generously subsidized by American taxpayers—soon flooded the country, driving down prices and pushing farmers to the edge of bankruptcy. By 1988, many rice farmers had simply stopped working on their lands. The experience of the Haitian rice farmers reveals a striking hypocrisy in the IMF mandates. While the Haitian government — strapped by debt and strangled by poverty—was forced to end its subsidies to rice farmers and lower its tariffs, the USA was not compelled to take the same steps. Consequently, USA rice soon dominated the Haitian market. Today, Haitians are in the unenviable position of being dependent on foreign exports for one of their main food staples. A single USA corporation, early Rice, sells almost 50 per cent of the rice sold in Haiti. As one Haitian farmer has said: "The introduction of American rice hurt us terribly. But if it wasn't for this rice, Haitians would die of hunger."

Lowering Wages to Increase "Opportunity" : While acknowledging that its policies may lead to social disruptions in the short run, the IMF argues that its policies will benefit ordinary people in the long run. Poor countries, the IMF says, need to strive to specialize in a few key export-oriented industries and maximize their comparative advantages. Haiti's comparative advantage, the IMF says, is in the area of labour — and the institution has done all it can to keep that labour cheap.

The 1980s saw a severe drop in Haitians' earnings. By 1989 a full time minimum wage in Port-au-Prince provided less than 60 per cent of a family's basic needs — food, shelter, clothing, etc. In the 1990s, the IMF sought to keep wages steady to make Haiti competitive and attract foreign investment to the country. In 1991 and 1994, the IMF conditioned loan assistance on wage restraint. Then, in 1996, the IMF gave the Haitian government technical assistance to revise its labour code. Haitian labour law had required an increase in the minimum wage if inflation were to exceed 10 per cent. The IMF had that law changed, even as inflation hit 30 per cent. The IMF also mandated a three-year freeze on government wages. If the government wanted to raise civil servants' earnings, it would have to offset those increases by reducing the staff strength. The move injected new instability into the already-cut-throat labour market.

Even as the IMF demanded sacrifices from Haitian workers, it pressured the country's government to give bonuses to foreign investors. At the urging of the Fund and the World Bank, Haiti reduced telephone, electricity and customs fees for foreign companies. The new corporate incentives, combined with the attraction of rock bottom wages, have in fact attracted thousands of manufacturers to Haiti. Yet, unfortunately for average Haitians, the foreign investors contribute very little to the country's economic stability. This is due to the fact that goods to be assembled are shipped from the United States, pieced together, and shipped out again, and few, if any, Haitian goods are used in the process. In addition, much of the tax-free profits made from the assembly sector are repatriated by U.S. investors, not reinvested in Haiti.

As the IMF and World Bank demanded that Haiti offer generous benefits to multinational corporations, it forced the government to cut its budget. In 1997 IMF demanded that the Haitian government reduce its spending by 50 per cent, despite the fact that Haiti already had just one-eighth the number of public employees as its neighbour of comparable population, the Dominican Republic. The cuts in public spending made it even more difficult for the government to provide essential services. In fact, during the 1990s, Haiti has spent more money servicing its debt than it has on health or education.

The IMF and World Bank programmes in Haiti have been a boon for multinational corporations. USA agri-business has reaped new profits from Haiti's lowering of its tariffs and the elimination of its farmer subsidies. The apparel industry has taken advantage of the low wages and manufacturing incentives to exploit Haiti's desperate workforce. The Haitian people, meanwhile, struggle to survive.

Mexico: A "model student"? : For nearly 20 years, Mexico has followed almost every economic policy mandate from the IMF and the World Bank. Mexico's compliance with IMF dictates has been so reliable, in fact, that in 1994 the IMF and the World Bank lauded the country as a "model student" that other Latin American countries should emulate. Since Mexico first adopted the IMF prescriptions of trade "liberalization," privatization and deregulation, real wages have fallen, poverty and inequality have increased, and the country's massive debt burden has grown. Mexico's experience with IMF policies offers a clear example of how the Fund sacrifices the well-being of ordinary people to suit the interests of international investors. In an effort to head off short-term financial crises—and calm investors' fears—the IMF has undermined Mexico's chances of creating a stable, well-balanced economy.

The IMF arrives in Mexico: Many US citizens assume that Mexico's entry into the new global economy occurred when the country signed the North American Free Trade Agreement (NAFTA) in 1993. But Mexico's economy was opened up to the forces of corporate-led globalization long before NAFTA went into effect. Before anyone had heard of NAFTA, the IMF was already setting Mexico on the structural adjustment path.

In 1982, a fall in international oil prices combined with a rise in interest rates in international financial markets forced Mexico to announce that it was on the verge of defaulting on its foreign debt. The "debt crisis" of 1982 also impacted countries throughout Latin America, Africa and Asia, giving the IMF the chance to determine the fiscal and monetary policies of countries around the world. Mexico asked the IMF for assistance, and the Fund obliged with a $ 3.9 billion credit package. The IMF's 1982 assistance package was strictly *quid-pro-quo*. To receive the new loan — and the Fund's all-important seal of approval, which opens the door to other public and private credit — Mexico would have to embark on a series of market reforms. Public spending would have to be cut, government enterprises would need to be privatized, industry would have to be deregulated, and the country would have to open up to more foreign trade and investment. Subsequent agreements with the IMF in 1986 and 1989 further cemented this policy path.

In an effort to boost foreign investment in Mexico and decrease the country's imports — which would hopefully lower the country's trade (or current account) deficit — the IMF in the late 1980s sought to contract economic activity and stabilize wages. The IMF worked with the Mexican government and the country's businesses and government-controlled labour unions to establish a set of social pacts, or "Pactos", to keep wages in check. Workers' earnings were indexed to "expected" levels of inflation. Unfortunately for Mexican workers, inflation rose more than expected. Between the implementation of the first Pacto in December 1987 and May 1994, the minimum wage increased by 136 per cent, while the cost of a basic basket of consumer goods rose by 371 per cent. That sort of decrease in real earnings would lead Mexicans to call the 1980s the "lost decade". During the 1980s, real wages (adjusted for inflation) declined by more than 75 per cent, and between 1981 and 1990 workers' share of national income fell from 49 to 29 per cent. Government investments in education, research and development, and infrastructure were reduced. The IMF policies, which supposedly were intended to make Mexico more internationally competitive, were actually doing the opposite by limiting investment in the very areas needed to make a country more productive. It was a development "strategy" doomed to fail.

The Cost of Privatization in Mexico

While ordinary wage earners saw their incomes plummet, they were confronted with increasing consumer prices. Since 1983, Mexico has sold off nearly 1,000 public enterprises. These privatizations were intended to inject some cash into government coffers. While the sell-offs did, in the short-term, earn the government new money, the privatizations hurt ordinary consumers. The privatization of Telmex in 1990, the country's phone system, is just one example of how the IMF privatization agenda hurt average citizens more than it helped them. The winners were multinational communication corporations Southwestern Bell and France Telecom. Mexican phone users were the losers. In the months before the sale—in an attempt to make the

Telmex a more attractive buying prospect—the government increased rates on local calls from 16 pesos per minute to 115 pesos per minute.

In a 1992 report, the World Bank admitted that "the privatization of Telmex, along with its attendant price-tax regulatory regime, has the result of 'taxing' consumers—a rather diffuse, unorganized group — and then distributing the gains among more well-defined groups, shareholders, employees, and the government". The Bank predicted that rates would decrease in the long run. But the long run still hasn't arrived. Although rates on international calls have dropped, that decrease has been offset by a rise in the cost of long distance calls within Mexico.

Other privatizations have been equally distressing for average Mexicans. The World Bank has concluded that privatization contributed to a "worsening of the already skewed and concentrated pattern of ownership distribution in the economy." A glaring example of this is corn, Mexico's traditional staple crop. For years, small scale farmers have seen government support evaporate as the IMF demanded an end to tariffs and import quotas on the grain and an elimination of government assistance with marketing and distribution of locally grown products. In recent years, a flood of subsidized USA corn has caused a 45 per cent decrease in the prices corn farmers receive for their commodity. This has pushed millions of farmers off their lands and into the urban ghettoes or toward the USA. And yet — because of the monopolistic control of the Mexican corn processing industry — consumer prices have not gone down. In fact, tortilla prices have actually increased in the last 15 years.

A "Success" Built on Sand: By the early 1990s, after the "lost decade," the Mexican economy started to show signs of improvement. Inflation came down, and some economic growth occurred. But it was a success built on sand. The Mexican economy was relying more and more on foreign capital flows, most of which were short-term portfolio investments. The increasingly volatile capital flows boosted the Mexican *peso*, a benefit for foreign investors, but it worsened the country's trade deficit by making Mexican exports more expensive. By mid-1994, foreign investors — assuming the Mexican government would devalue the *peso* to make exports more competitive — began pulling their money out of Mexico. In December 1994, Mexican officials were forced to devalue the *peso*. In January 1995, the government again asked the IMF for assistance. The IMF lent the country $ 7.75 billion, and the USA Treasury loaned another $ 18 billion. A new round of privatizations were prescribed to raise quick cash and pay off the loans. Under the 1995 agreement with the IMF, transportation, banking and finance, railways and the petrochemical industries were all to be sold off. Once again, the IMF was offering policies that responded to the short-term financial concerns of international investors rather than to the basic needs of or ordinary Mexicans. The1995 *peso* devaluation — combined with an IMF-mandated rise in interest rates designed to lure investors back to Mexico—sparked the worse depression in Mexico in 60 years. Unemployment doubled. More than 12,000 Mexican businesses filed for bankruptcy. And millions of families dropped below the poverty line. As the Mexican economy came virtually to a standstill, the IMF blamed the Mexicans, complaining that human error and mismanagement of financial variables had led to the crisis. The IMF's attempt to shift blame for the economic collapse seemed to many Mexicans an insult. The country had followed all of the Fund's prescriptions. The *peso* devaluation, after all, was due in large part to the short-term capital flooding the country, and that flood was precipitated by the deregulation of the country's financial markets which the Fund had demanded.

Trading away the Future: Even as its economy was shrinking, Mexico was implementing the other pillar of the IMF's structural adjustment package—trade liberalization. In 1994, NAFTA went into effect, creating a giant "free trade" block among Mexico, the USA and Canada. The agreement contained many "reforms" the IMF had long been asking for, among them a provision allowing 100 per cent direct foreign ownership of Mexican companies. NAFTA also modified Article 23 of the Mexican Constitution, which had protected communal property, known as *ejidos*, from being broken up. The article was another long-time object of IMF and World Bank ire. The constitutional change removed many people's guarantee that they would have access to land, thereby putting more people in financial jeopardy.

In the years following the 1995 crash, Mexico did enjoy some steady macroeconomic growth. But the benefits were not evenly spread. The number of Mexicans living in "severe" poverty (surviving on less than $ 2 a day) has grown by four million since NAFTA began. The MNEs of USA relocated much of the lower value adding activities to Mexico in automobiles, electronics and apparel but it is argued persuasively that these activities have resulted in creating export platforms for the MNEs, with limited benefits for the host countries. Wages, instead of increasing, have declined. Mexican manufacturing workers are today earning almost 10 per cent less than they did before NAFTA. The 1990s, just like the ten years before, were a "lost decade" (ECLAC, 2001).[20]

Today, after two decades of following IMF prescriptions and ten years of the NAFTA experiment, Mexico remains a country racked by poverty and inequality. A majority of Mexicans live below the poverty line. Even the World Bank concedes that 15 years of trade liberalization in Mexico have not succeeded in closing the gap between rich and poor. In almost every social sector — health, nutrition, housing, education — virtually all of the key indicators show serious deterioration over the last 15 years. Today one can speak not just of a lost decade, but of a lost generation.

The Same Old Policies : And what about the debt burden that started Mexico down the structural adjustment path? It has gotten even worse. In 1982, Mexico's total foreign public debt was $ 57 billion. In 1993 it amounted to $ 80 billion. By 1997, the country owed $ 99 billion in public debt. This stranglehold of debt continues to force Mexico to attract foreign investment by any means necessary, including trading away the basic rights of its workers and failing to enforce environmental regulations. Despite the failures of this "model student," the World Bank and IMF are still prescribing the same old policies for Mexico. In May 2001, the World Bank offered specific recommendations on the country's labour policies. The Bank said that if Mexico wanted to attract more foreign investment, it would need to increase the "flexibility" of Mexican labour. The Bank suggested that Mexico eliminate its regulations regarding mandatory severance pay, collective bargaining, obligatory benefits for workers, company-sponsored training programmes, and company payments to social security and housing plans. The Mexican President said the recommendations were "very much in line with what we have contemplated". But it surprised the leading Mexican industrialists, according to whom, "Some of these proposals of the World Bank are not made even to the most developed nations. Why are they then being recommended for the emerging countries?"

The World Bank's Land "Reform" in Brazil : Unconstitutional Meddling

According to conventional macroeconomic measures, Brazil is a wealthy country. In 1997, the World Bank began a pilot programme to address land distribution in Brazil. Called the "Land Bank," Cêdula da Terra in Portuguese, the project is intended to help landless families gain parcels through access to credit and technical assistance. What may appear a well-meaning effort, however, is wrought with troubles. In developing the Cêdula, the Bank never consulted with the Brazilian citizens' groups leading the country's land reform movement. In failing to conduct such consultations, the Bank placed its enormous weight behind a single social actor—landowners—in a hotly contested political issue. To add injury to insult, it now appears that the Bank's project may even worsen rural inequalities in Brazil. The Land Bank experience is just one example of how the Bank inappropriately inserts itself in domestic politics, in the process short circuiting the democratic system. As an analyst at the Environmental Defense Fund has said, "This project is possibly the most cynical imitation of a development project that I've ever seen, in a very competitive field."

A Long History of Failures : The Cêdula project is not the first time the IMF and the World Bank have interfered in Brazil. In the 1970s, the institutions lent billions of dollars to the military dictatorship then controlling the country. The loans were used to pay for massive infrastructure projects such as dams, many of which failed to meet even the goals set by the institutions themselves. By the 1980s, the country was borrowing new money to pay off the old loans. Brazil's "lost decade" saw a massive transfer of wealth outside the country as the government paid $ 148 billion—$ 90 billion just in interest, the rest in principal—between 1980 and 1989 to foreign lenders. Much of this money went to recover the costs of Bank-initiated projects that were unsuccessful. In 1991, the Bank itself estimated that 55 per cent of its Brazilian projects had failed.

In the 1990s, under a succession of civilian governments, the IMF and the World Bank compelled Brazil to institute their structural adjustment policies. Industries were deregulated, government enterprises sold off, and government spending reduced and taxes raised. In the late 1990s, President Cardoso greatly accelerated the neo-liberal reforms demanded by the IMF. The profitable government phone system and a profitable publicly owned mining company were sold to multinational corporations. Government spending on social services was reduced even further.

In November 1999, Cardoso's administration put Brazil even more at the mercy of the IMF when he agreed to a $ 41.5 billion loan package. At that time a financial crisis was sweeping Russia, and the IMF, concerned that the "contagion" would spread to Brazil and destabilize the entire global financial structure, was putting pressure on Brazil not to devalue its currency, the *real*. The IMF forced the Brazilian treasury to dramatically increase interest rates — which at one point hit 41 per cent — to keep the *real* strong. In the end, it didn't work. The high interest rates sent the Brazilian economy into a recession, and Cardoso eventually had to devalue the currency anyway. Essentially, the IMF forced Brazil to spend billions of dollars to prevent the inevitable.

The IMF had predicted that if Brazil devalued the real hyperinflation would occur. That never happened. But sending interest to stratospheric levels did greatly burden ordinary Brazilians. The higher lending rates forced thousands of rural foreclosures. According to the Brazilian government, between 1995 and 1999 an estimated 4 million Brazilians left the countryside for the cities. The Cardoso administration's and IMF's insistence on ensuring investor confidence worsened the maldistribution of land.

Taking Political Sides: Recognizing that excessive land concentration not only contributes to poverty but also hinders sustainable development—the Brazilian Constitution of 1988, explicitly states that land must be used for the benefit of society. Other laws that establish mechanisms for progressive land distribution allow the government to expropriate unused or over-large landholdings and then pay back the owners through the issuance of government bonds. Since the mid-1980s, several citizen groups in Brazil—led by the Landless Workers Movement (MST is the Portuguese acronym), the Confederation of Agricultural Workers, and the Catholic Church—have used such laws to push for land redistribution.

In just 15 years, these groups, the MST in particular, have succeeded in securing land titles for 250,000 families through the non-violent occupation of unused land. That grassroots energy has also moved the Brazilian government to take stronger action. The current administration has distributed more than 28 million acres of land, and Cardoso administration officials concede this wouldn't have happened without pressure from the MST and its allies.

In 1997, the World Bank decided it wanted to become involved in land reform in Brazil. But instead of taking a neutral posture that supported land reform as such, Bank officials supported the government and the landowners against the MST and the civil society organizations aligned with it. By doing so, it corrupted the democratic process. Land reform is one of the most emotional political issues in Brazil today. Heated rhetorical sparing between the MST and the national government is common. In the countryside, the political struggle has turned physical. State governments, usually allied with local landed elites, routinely repress MST activities. Police beatings of MST activists are not unusual, and some people have been killed. The MST and the government, thus became political adversaries.

From the outset of its involvement in Brazilian land reform, the Bank has supported the interests of one side in this political struggle. The Bank planned its land reform project without consulting or informing any of the national citizens' organizations working on land issues. The Bank and the Brazilian government even refused to negotiate with rural workers when they made specific proposals. The land reform mechanism pushed by the MST and reluctantly pursued by the Cardoso administration relied on the idea that land can be confiscated even against the wishes of the owner if it is not being used. In contrast, the World Bank's system relies on free-market principles. Under the World Bank's plan, land reform is privatized: the landless apply for loans from the Brazilian government with which they can purchase land from landowners at market prices. However, no landowners will be forced to sell, no matter how much of their property is lying unused.

The Cêdula is just now reaching the point where a complete analysis of its success can occur. But already preliminary evidence suggests that the programme is doomed to failure. First, the majority of borrowing farmers don't even fully understand that they have taken out a loan — and even fewer know the terms of their loans. The tough terms of the loans — interest rates on the loans can be as high as 18 per cent — make it likely that many poor farmers will lose the land again within a few years. Also, under the Land Bank programme there is only a three-year grace period (an earlier government programme had a five-year grace period), and little way to get credit for supplies or seeds. Landless advocates also fear that the Land Bank, which is planning on distributing $ 2 billion of loans over ten years, could contribute to an inflation in land prices.

Most disturbing for landless advocates, many Brazilians perceive that the Bank's programme is not simply a complementary alternative to the constitutionally guaranteed land reform; rather, many people believe the Bank's market-based approach to land reform is a substitute for other kinds of land distribution. That perception meets neatly with the political agenda of landowners and some government officials who want to eliminate the existing means of land reform. It also, of course, goes against the interests of the MST and its allies, who want to pursue land reform via the constitutionally sanctioned means of expropriation. Essentially, then, the Bank has sided with the political agenda of a particular group. By failing to even consult with those at the forefront of Brazilian land reform, the Bank inserted itself into the one of the country's most heated political debates.

Colombia and the IMF: Policies that Worsen Problems

Colombia's relationship with the IMF is similar to the relationship between a poor sick patient and a costly doctor. The patient, badly in need of help, goes to the doctor for treatment. The doctor in return prescribes the patient a very expensive, controversial medication. The patient agrees, knowing that he cannot afford it, but, feeling desperate, consents to the treatment. Eventually, the patient develops terrible side effects to the treatment, and also discovers that the original problem has worsened. The patient—poorer, sicker and more desperate —needs more "medicine" than ever before. Two years of financial "infusions" from the IMF affects Colombia in the same way. The South American country's experience with IMF policies illustrates how the IMF caters to the interests of international investors and the Colombian elite, while harming the well-being of ordinary

Colombians. Ignoring massive public protests against the IMF-mandated reforms, the lending institution's policies have systematically cut health-care, education and other social services for millions of Colombians, laying off tens of thousands in the process. The IMF's relationship with Colombia has served to deepen poverty and inequality in the South American country, while claiming to do the opposite — all in the name of "free" trade, market liberalization and financial stability. Equally as damaging to Colombian society is the way that its government has imposed the IMF-mandated economic strategy. Characterized by undemocratic and authoritarian practices, and supported by billions of dollars of mostly-military USA aid, the government silences grassroots opposition, undermining the legitimacy of the Colombian state.

The 1990s —from Optimism to Emigration: Colombia in the early 1990s was a land of relative optimism and hope. The 1980s — referred to by many as a "lost decade" in Latin America because of region-wide economic woes — passed with Colombia riding somewhat high on one of the hemisphere's most stable and strong economies. A new constitution, not without its critics, had been written that included more of Colombian society than ever before. Some of the nation's guerrilla groups, weary of war, decided to involve themselves in the political process, forming political parties. Social peace, for so long an abstract, seemed tangible.

Considering this, perhaps not even the most hopeless of pessimists could have imagined the troubles that Colombians have lived through during the last decade. The economy has fallen into its worst recession of the century, with official unemployment hovering around 20 per cent. The country's bloody civil war claimed over 40,000 more lives. Today over 1.5 million Colombians are refugees within their own country. As a result, Colombians began to lose the hope that had buoyed them through the early 1990s. Between 1996 and 2000, 1,072,000 Colombians, sensing the worst was yet to come, left their homeland behind.

The IMF and Colombian Neo-liberalism —A Future for Whom? : Not coincidentally, it was in the late 1980s and early 1990s that Colombia began to experiment with neo-liberal economic policies. Beginning in 1986 till the present, the neo-liberal economic model was wholeheartedly imposed by the Colombian elite. Gaviria, blamed by 25 per cent of Colombians for the economic depression in a recent poll, proudly proclaimed to his countrymen in 1991, "Welcome to the future," referring to the new constitution and the neo-liberal economic opening. Ten years later, during a recent development forum, the same man who boasted with such optimistic pride, quizzically answered questions regarding Colombia's economic downturn. "I have to confess that the evolution of the country after my government has left me perplexed, like many other Colombians." Yet, while Gaviria scratches his head trying to figure out what happened, most Colombians — those who cannot afford the same luxury of economic theorizing — are jobless and hungry.

The debate surrounding who and what is to blame for Colombia's economic downturn points to various causes. Some blame the constitution of 1991, whose new laws turned out to be costly to create and enforce. Some say neo-liberalism and the uncontrolled opening-up of the Colombian economy to foreign investment are at fault. The other usual suspects are the reevaluation of the Colombian *peso*, and bad management of public spending by the state, among others. Whatever the reasons, Colombia's economy slid into dire straits. And the Colombian elite, needing a bailout, went knocking on the IMF's door, hat in hand.

The IMF and Plan Colombia : After the scandal-ridden Samper presidency, Andrés Pastrana Arango was elected in 1998 with a mandate of peace. Recognizing the nation's increasingly desperate social, economic and political situation, the Pastrana administration developed *Plan Colombia*. The Colombian and US governments ostensibly describe Plan Colombia as a $ 7.5 billion integrated strategy to meet the most pressing challenges confronting Colombia today — promoting the peace process, combating the narcotics industry, reviving the Colombian economy, and strengthening the democratic pillars of Colombian society.

Billing its support of Plan Colombia as part of the questionable "War on Drugs," the USA government has thrown its weight behind the plan with billions of dollars of mostly-military aid, with unconvincing effectiveness. The USA emphatically characterizes the aid as counternarcotic, instead of counterinsurgent — a *semantic tonguetwister* that most Colombians know to be untrue.

Although not directly related to Plan Colombia, the IMF's Colombia loan fits in to Plan Colombia as part of the larger strategy to revive the Colombian economy. In 1999, Colombia appealed to the IMF for a loan for the first time in its history. The IMF obliged with a three-year credit worth $ 2.7 billion and its customary mandate — a structural reform agenda, or austere fiscal policies that includes measures to streamline the revenue sharing system, reform the public pension systems, and downsize the public sector.

Failing to implement an economic alternative that protects ordinary Colombians, the Colombian state tightened its fiscal belt by cutting public spending — a move that hit the poorer segments of society the hardest. In the words of one Colombian economist, "What's the alternative? Increase the taxes on your elite? Do it if you can." Most damaging to the average Colombian are the severe cutbacks in public investment in basic social services — health care, education and social security. It is a painful practice that one Colombian congress person likened to "conducting a complex surgery without the use of anesthesia."

Colombian Grassroots Opposition to IMF Economic Reforms: To implement the IMF's demands, the Colombian government needed to pass special legislation, including changes to the constitution. Many Colombians, not keen on the idea of having their well-being on the fiscal operating table, have organized in nationwide protests against the various legislative changes.

The first IMF protests in Colombia occurred while negotiations about the loan were still underway. On August 31, 1999, a general strike was organized by a grassroots coalition called the Colombian Social Movement. Consisting of indigenous, Afro-Colombian, farmworker, labour, women, student, youth and other civil society groups, the movement presented an alternative programme of social and economic development that emphasized the necessity of investing in programmes that advance the well-being of ordinary Colombians.

Months later, in March 2001, hundreds of thousands of Colombian state workers went on a 24-hour strike to demonstrate against economic policies that resulted in mass layoffs and reduced social benefits. The layoffs are expected to total around 100,000 government employees. Starting May 15, 2001, over 125,000 health-care workers and 300,000 public school teachers went on a month-long nationwide strike to protest against the "ley de transferencias", or transfers law, a bill that would cap the federal money to departments [states] and municipalities for education, health-care and other social needs. Critics of the law warned that its implementation would result in the loss of 1,416,000 places for students in the public school system, adding to the estimated 3 million Colombian youths that are already outside the system. They also claimed that some 2.5 million more Colombians would be excluded from public health-care.

Upon the bill's passage on June 20, 2001, Colombia's largest teachers' and health-care workers' unions, FECODE and ANTHOC, publicly declared, "Since the time of Pastrana's development plan, no other initiative has created such a movement of opposition. But, it appears that the requirements of the IMF and financial capital supersede the interests of the nation and of the Colombian people." Next up for surgery is the pension system. The controversial reform is another Colombian commitment to the IMF and is expected to galvanize the Colombian public in even more anti-Fund protests.

The Authoritarianism of the Colombian State : As the Colombian state adopts neo-liberal economic strategy, it has become more authoritarian. Increasingly gaining more power to fight the USA-sponsored "drug war" and 40-year-old guerrilla insurgency, the government has used those powers to continually repress democratic liberties. Laws like the recently passed "ley de guerra", or war law, have granted the Colombian military more room to unilaterally crack down on "guerrilla terrorism." Colombian and international NGOs, basing their concerns on past experiences, worry that such increased military powers will also be used to silence any voices of dissent that challenge the will of the Colombian government.

By heavily relying upon the force of its USA-sponsored military, Colombia risks exacerbating the level of violence and state authoritarianism. This is a misguided approach that is likely to repress the sectors of the population that are suffering the most from Colombia's neo-liberal economic policies.

By silencing dissent like the IMF protests, and closing in on the political space that civilians are allowed to participate in, another result of Colombia's authoritarianism may be to further undermine the democratic process, and worsen Colombia's social, economic and political woes.

The World Bank and IMF in Africa

The IMF and World Bank's policies have been heavily criticized for many years and are seen as unhelpful and sometimes, unaccountable, as they have resulted in an increased dependency by the developing countries upon the richer nations, as also mentioned at the top of this page. At the same time, the different cultures are not respected when it comes to prescribing structural adjustment principles, either. In Africa, the effects of policies such as SAPs have been felt sharply. As an example of how political interests affect these institutions, *Africa Action* describes the policies of the IMF and World Bank, but also hints at the influences behind them too.

The World Bank and IMF became increasingly powerful in Africa with the economic crisis of the early 1980s. In the late 1970s, rising oil prices, rising interest rates, and falling prices for other primary commodities left many poor African countries unable to repay mounting foreign debts. In the early 1980s, Africa's debt crisis worsened. The ratio of its foreign debt to its export income grew to 500 per cent. African countries needed increasing amounts of "hard currency" to repay their external debts (*i.e.* convertible foreign currencies such as dollars and deutschmarks). But their share of world trade was decreasing and their export earnings dropped as global prices for primary commodities fell. The reliance of many African countries on imports of manufactured goods, which they themselves did not produce, left them importing more while they exported less. Their balance of payments problems worsened and their foreign debt burdens became unsustainable.

African governments needed new loans to pay their outstanding debts and to meet critical domestic needs. The World Bank and IMF[21] became key providers of loans to countries that were unable to borrow elsewhere. They took over from wealthy governments and private banks as the main source of loans for poor countries. These institutions provided "hard currency" loans to African countries to insure repayment of their external debts and to restore economic stability.

The crisis experienced by African countries in the early 1980s did expose the need for economic adjustments. With declining incomes and rising expenses, African economies were becoming badly distorted. Corrective reforms became increasingly necessary. The key issue with adjustments of this kind, however, is whether they build the capacity to recover and whether they promote long-term development. The adjustments dictated by the World Bank and IMF did neither.

African countries require essential investments in health, education and infrastructure before they can compete internationally. The World Bank and IMF instead required countries to reduce state support and protection for social and economic sectors. They insisted on pushing weak African economies into markets where they were unable to compete with the might of the international private sector. These policies further undermined the economic development of African countries.

The past two decades of World Bank and IMF structural adjustment in Africa have led to greater social and economic deprivation, and an increased dependence of African countries on external loans. The failure of structural adjustment has been so dramatic that some critics of the World Bank and IMF argue that the policies imposed on African countries were never intended to promote development. On the contrary, they claim that their intention was to keep these countries economically weak and dependent.

The most industrialized countries in the world have actually developed under conditions opposite to those imposed by the World Bank and IMF on African governments. The USA and the countries of Western Europe accorded a central role to the state in economic activity, and practiced strong protectionism, with subsidies for domestic industries. Under World Bank and IMF programmes, African countries have been forced to cut back or abandon the very provisions which helped rich countries to grow and prosper in the past. Even more significantly, the policies of the World Bank and IMF have impeded Africa's development by undermining Africa's health. Their free market perspective has failed to consider health an integral component of an economic growth and human development strategy. Instead, the policies of these institutions have caused a deterioration in health and in health care services across the African continent.

Poverty and Health Care Cuts : Health status is influenced by socio-economic factors as well as by the state of health care delivery systems. The policies prescribed by the World Bank and IMF have increased poverty in African countries and mandated cutbacks in the health sector. Combined, this has caused a massive deterioration in the continent's health status. The health care systems inherited by most African states after the colonial era were unevenly weighted toward privileged elites and urban centres. In the 1960s and 1970s, substantial progress was made in improving the reach of health care services in many African countries. Most African governments increased spending on the health sector during this period. They endeavoured to extend primary health care and to emphasize the development of a public health system to redress the inequalities of the colonial era. The World Health Organization (WHO) emphasized the importance of primary health care at the historic Alma Ata Conference in 1978. The Declaration of Alma Ata focussed on a community-based approach to health care and resolved that *comprehensive health care was a basic right and a responsibility of government.*

These efforts undertaken by African governments after independence were quite successful. There were increases in the numbers of health professionals employed in the public sector, and improvements in health care infrastructure in many countries. There was also some success in extending care to formerly unserved areas and populations. Across the continent, there were improvements in key health care indicators, such as infant mortality rates and life expectancy. In Zambia, the post-independence government expanded public health care services throughout the country. The number of doctors and nurses was also significantly increased during this time. Infant mortality was reduced from 123 per 1,000 live births in 1965, to 85 in 1984. In Tanzania, during the first two decades of independence, the government succeeded in expanding access to health care nationwide. By 1977, more than three-quarters of Tanzania's population lived within 5 km of a health care facility. With the economic crisis of the 1980s, much of Africa's economic and social progress over the previous two decades began to come undone. As African governments became clients of the World Bank and IMF, they forfeited control over their domestic spending priorities. The loan conditions of these institutions forced contraction in government spending on health and other social services.

Poverty and Health : The relationship between poverty and ill-health is well established. The economic austerity policies attached to World Bank and IMF loans led to intensified poverty in many African countries in the 1980s and 1990s. This increased the vulnerability of African populations to the spread of diseases and to other health problems. More than 17 million Africans have died of

HIV/AIDS. It is currently estimated that more than 28 million of the 40 million people living with the disease worldwide are in Sub-Saharan Africa. There are more than 12 million African orphans who have lost their mothers or both parents to AIDS. The social and economic effects of the AIDS crisis are reversing post-independence progress and exacerbating conditions of underdevelopment. The policies imposed by the World Bank and IMF have fueled the spread of the disease and continue to hinder the response to this health emergency.

Declining Health Care Systems: As poverty has increased Africa's health problems, World Bank and IMF intervention has also led to a breakdown of healthcare delivery systems. The role of the state in the provision of healthcare services has been reduced substantially. Cutbacks in the health sector have severely undermined existing services. And Africa's debt repayment obligations to foreign creditors have diverted money directly from spending on health. Cutbacks in government spending represent a major component of World Bank and IMF adjustment programmes. In the 1980s, real disbursements per person dropped in government expenditure in many African countries. In Madagascar, between 1977 and 1985, government expenditures declined by 24 per cent. Cutbacks in government budgets led to major cuts in the health sector. In the 42 poorest countries in Africa, spending on health care fell by 50 per cent during the 1980s. In Nigeria, per capita expenditure on health fell by 75 per cent between 1980 and 1987. The dramatic drop in health expenditure in the 1980s and 1990s resulted in the closure of hundreds of clinics, hospitals and medical facilities across the continent. Those that remained open were generally left under-staffed and lacking in essential medical supplies. Many were unable to afford even basic vaccines. In 14 Sub-Saharan African countries between 1990-1992, the level of polio vaccination dropped by more than 10 per cent as a result of cutbacks in health care services. Thousands of health-care professionals throughout the continent lost their jobs as a result of public sector cuts. In Ghana, between 1982 and 1992, the number of doctors in the government health care system dropped from 1,700 to 665. In Senegal, between 1980 and 1993, the number of people per nurse increased six-fold.

Even as government spending on health was cutback, the amounts being paid by African governments to foreign creditors continued to increase. By the 1990s, most African countries were spending more repaying foreign debts than on health or education for their people. Health care services in African countries disintegrated, while desperately needed resources were siphoned off by foreign creditors. It was estimated in 1997 that Sub-Saharan African governments were transferring to Northern creditors four times what they were spending on the health of their people. In 1998, Senegal spent five times as much repaying foreign debts as on health. Across Africa, debt repayments compete directly with spending on Africa's health care services. The erosion of Africa's health care infrastructure has left many countries unable to cope with the impact of HIV/AIDS and other diseases. Efforts to address the health crisis have been undermined by the lack of available resources and the breakdown in health care delivery systems. The privatization of basic health care has further impeded the response to the health crisis.

Privatization and User Fees

Privatization forms a centrepiece of the World Bank and IMF agenda. Reducing the size and scope of government, and privatizing state-owned enterprises and services, is a major element of World Bank and IMF programmes. Under World Bank and IMF direction, control of health care services has increasingly been transferred from African governments to the private sector. The rationale is that health care services are better financed and more efficiently delivered privately.

The World Bank has recommended several forms of privatization in the health sector. These include: the introduction of "user fees" for health services previously provided free of charge; the promotion of health insurance schemes; increased investments in private care in order to attract patients to private facilities. Through these measures, private services are made the primary focus of health care.

Throughout Africa, the privatization of health care has reduced access to necessary services. The introduction of market principles into health care delivery has transformed health care from a public service to a private commodity. The outcome has been the denial of access to the poor, who cannot afford to pay for private care. For example, the introduction of user fees for health care services has been shown to sharply reduce access for those unable to pay. The World Bank and IMF have promoted "user fees" as a means of generating revenue for the health sector. They argue that these fees will tax the rich and that a system of exemptions will protect the poor from the costs. However, the evidence shows that user fees have actually succeeded in driving the poor away from health care services. Ghana, Swaziland and the former Zaire were among the first African countries where user fees replaced free, or almost free, services. In each of these, studies showed that the introduction of fees led to reduced utilization of these services. Studies in Côte d'Ivoire have shown that those with income above the median make more use of medical services when a user fee is charged. But those with below median incomes reduce their use. Across Africa, reports indicate that attendance at hospitals and clinics drops significantly after the introduction of user fees.

Schemes intended to exempt the poor from user fees have been shown to be ineffective. A comprehensive UNICEF study discovered that such schemes are rare. The study claims that poor people are generally unaware of such exemptions, and that there are often complex administrative barriers involved. The report concludes that the implementation of exemption schemes is infrequent and is applied in *ad hoc* ways. It, therefore, does not appear that user fees have been managed so as to collect revenues only from the rich. Instead, they have had the effect of closing off access to those who cannot pay. As a recent paper concluded, user fees "increase the barriers disproportionately faced by the poor when seeking health care". Their failure to generate revenue has also undermined their economic rationale, as propounded by the World Bank and IMF.[22]

An internal Bank report has concluded that the poor were better off without structural adjustment. The report, finds that World Bank and IMF adjustment lending "means that economic expansions benefit the poor less under structural adjustment. The paper does not fully resolve why the poor benefit less from adjustment but speculates that they may be ill-placed to take advantage of new opportunities created by structural adjustment reforms because as the Bretton Woods Project[23] insinuate, they have neither the skills of human and technological infrastructure nor financial resources to benefit from high-technology jobs and cheaper imports.

The Southern debt activists stressed that :

- the debt of the LDCs is illegitimate and must be totally cancelled without conditions;

- structural adjustment programmes of the IMF and World Bank creditors should not place conditionalities on the debt cancellation process;
- ecological aspects of debt should be considered.

Foreign private capital flows as "an essential element of a strategy to speed up sustainable development and poverty reduction," although only a small proportion of these flows are invested in activities which benefit the poor. "It is essential therefore that reforms to attract financial flows be complemented by other reforms" and "road maps" should be produced to guide countries on the speed of capital account liberalization and associated reforms. The government has committed to link its bilateral Country Strategy Papers to the Poverty Reduction Strategy Papers (PRSPs) and "will continue to encourage the World Bank and IMF to make the necessary changes to their own structures and working methods in a way consistent with their commitment to the PRSP process."

PRSPs replace SAPs but still SAP the Poor

The IMF is not particularly interested in hearing the thoughts of its 'client countries' on such topics as development strategy or financial austerity. All too often, IMF's approach to the developing countries has had the feel of a colonial ruler. To those in the developing countries, the picture raised a very disturbing question: Had things really changed since the 'official' ending of colonialism a half century ago?

The IMF in 1999 replaced Structural Adjustment to Poverty Reduction Strategy Papers (PRSPs) as the new preconditions for loan and debt relief. However, the effect is still the same, as the preceding disastrous structural adjustment policies, as the World Development Movement reported. Many civil society organizations are increasing their critique of the PRSPs.[24]

Loans and Flight Capital

Liberia's current annual budget is just over $ 60 million. This is expected to pay for law enforcement, a judicial system, a civil service infrastructure as well as education, vaccination programmes, clear water and sanitation. But this budget is only a fifth of the $ 333 million owed to USA.

The largest single debt of the Philippines is the Bataan nuclear power station. Completed in the mid-1980s at a cost of $ 2.3 billion, it was built on an earthquake fault at the foot of a volcano, and has never been used. "Filipinos have not benefited from a single watt of electricity," To this day the Philippines pays $ 170,000 per day for the power station. The debt will not be repaid until 2018.

The nuclear power station was built by US multinational Westinghouse, in spite of a much lower bid from General Electric. The then president Ferdinand Marcos over-ruled the choice of General Electric; Westinghouse admitted it paid a commission to a Marcos associate, and the *New York Times* estimates Marcos was given $ 80 million by Westinghouse. The International Atomic Energy Authority noted that attempting to build a nuclear power station in an area of such high seismic activity was "unique in the atomic industry". It was good that the plant was never put into operation and never became an Asian Chernobyl. But who was responsible for the $ 2.3 billion white elephant?

Marcos was not alone. In the 1970s and 1980s, banks were giving loans to countries and then helping their leaders steal the money. For example, in the late 1970s, when Britain was supporting the Argentine military dictatorship (before it attacked the Falklands/Malvinas), British banks made multi-million dollar loans to Argentine parastatal companies—knowing that the money never went to Argentina, but remained in accounts in London. In other words, they knew the loans were fraudulent. In some cases, banks lent money to poor country governments and literally sent men with suitcases to bring back dollars from the leaders of those countries. More than a decade ago, a New York banker is often quoted who said "there is no debate in [my] bank over the dual role of lending money to a country and accommodating flight capital." It is said that banks stop sending "guys with suitcases" to poor countries, and instead press some of their private clients to use some of the flight capital to pay their country's debts.

Nigeria estimates that former military governments have stolen billions of dollars, and deposited these in British, US and Swiss banks. Nigeria's foreign debt is now $ 34 billion. The US-based Citibank has a special "private banking division" which was set up to deal with just this sort of money. On 8 November 1999, it gave the US Senate Governmental Affairs Committee details of accounts of four wealthy developing country leaders that had been frozen by Swiss courts due to illegal payments: Raul Salinias, brother of a former president of Mexico, $ 80-$100 million; former Pakistan Prime Minister Benazir Bhutto, $ 40 million; Gabon President Omar Bongo, $ 130 million; and the sons of Nigeria's former dictator Sani Abacha, $ 110 million.

The problem is there is no real impetus for change because the people who are paying the price for this reckless lending and borrowing are the ordinary, poor citizens who are often unable to raise their voices in protest. The lending institutions can keep rescheduling the loan and adding interest; the contractor gets paid, the borrowing government gets its cut and will probably be long gone when the people find out what has really been going on. The system works very well for elites in both countries, and unless a new open, transparent system is established, it is likely to continue.

Failed projects: The World Bank's own evaluation of project performance in the 1990s showed that in the poorest countries, and in South Asia and Africa, between 60 per cent and 70 per cent of projects failed. In Nigeria, at least 61 development projects financed by more than $ 5 billion in foreign loans have either failed or never opened, according to a government commission. Because exporters and poor countries are keen to gain the lucrative contracts, bribery is unnecessary, and inappropriate projects are pushed through by exporters. Tanzania owes the World Bank more than $ 575 million for 26 failed agricultural projects, according to the Tanzania Coalition on Debt and Development. The World Bank's own Operations and Evaluations Department admitted in 1998 that it failed to involve Tanzanians in appraising the projects before they went ahead, and that "pilot projects went to the Board as proven undertakings; early warnings of failings by World Bank technical staff and co-financiers were ignored; and pricing issues were neglected".

In Bolivia a lead and platinum refinery was built using untested technology and requiring more ore than could be produced by local mines. The project doubled in cost to $ 250 million and was built with $ 100 million in Belgian and German export credits plus German bank loans. The project has never functioned but the loan is still being repaid. Peru also is still paying a $ 70 million loan for boats worth $ 18 million and which never worked properly, and for aircraft engines which were never delivered.

The list of loans for bad or foolish or unwise projects seems endless, and often results from "loan pushing"—exporters and export credit agencies that want contracts and banks (including the World Bank) that want to encourage borrowing. But who is responsible when these projects fail?

Dams, Politics and Corruption

The conservative and cold war politics of the 1970s and 1980s was often directly linked to corruption. We will look at three cases: big dams, Indonesia and Zaire. In the mid-1980s as the anti-apartheid sanctions campaign was gaining global force, conservative forces were looking for ways to support apartheid South Africa. The European Community [now European Union (EU)] and World Bank agreed to back a $ 2 billion dam in Lesotho (totally surrounded by South Africa) to supply water to South Africa. In 1999, there was a court case in which 12 of the construction companies building the dam were charged with bribing the head of the project. Six of the companies—from Sweden, Switzerland, France, Italy, and Germany—were linked to previous dam corruption scandals, yet they still won contracts for the World Bank-funded project, according to the International Rivers Network. Two of the Lesotho firms were involved in a Kenyan dam in the 1980s, which cost twice what it should but was approved because of bribes to the president and energy minister.

Three firms were involved in the $ 3 billion Yacyreta dam on the Paraguay-Argentina border, which had $ 460 million in World Bank loans. In 1990 Argentine president Carlos Menem called the dam a "monument to corruption". But it was built in the era of northern support for the military dictatorships of Latin America, and the World Bank turned a blind eye. Some of the money went into foreign bank accounts, while some went to the Argentine military. Some World Bank staff call Yacyreta "the dam that financed the Falklands war".

Indonesia: In 1965 General Suharto overthrew Indonesian President Sukarno and massacred up to 750,000 alleged communists.

The next year western creditors rewarded Suharto with an independent assessment of his country's debts. The arbitration was undertaken by Herman Josef Abs, a prominent German banker, who had in 1953 negotiated massive debt cancellation for his country, Germany has recommended massive debt cancellation for Indonesia. The country was granted a four-year debt service moratorium and in 1970 Suharto's creditors agreed to a major reduction in debt service payments. Suharto's appalling human rights record continued, including the occupation of East Timor in 1975. He was rewarded with $ 120 billion in loans—making Indonesia the third most indebted developing country, after Brazil and Mexico. One of Suharto's first projects was to move millions of Javanese people to populate the other islands. Many of the militia-members who fought against East Timor independence were migrants sent there as part of this programme. This "transmigration" programme was funded, in part, by nearly $ 1 billion in World Bank loans—despite complaints about the human rights and environmental problems caused. Many of the migrants were put on marginal land and could not survive, so the World Bank had to put in more money to try to support them.

Corruption was well known from the start, but acceptable because the West was anxious to support Suharto. Corruption inflated the cost of World Bank projects by 10-15 per cent in 1977, but this had risen to 30 per cent by the early 1990s. The Bank Bali case in 1999 exposed further corruption in World Bank projects, and also exposed World Bank efforts to conceal the fact that they knew about the corruption. "Donor institutions like the World Bank have turned a blind eye to the corruption because they were afraid to offend Indonesia". Suharto's family is listed in Forbes magazine's dozen richest families in the world, with an estimated $ 30 billion in stolen wealth. As the crisis grew in Indonesia, the IMF stepped in with $ 43 billion, the largest bail-out in Asia, which continued on political grounds until, it was commented in *The Guardian* (22.9.1999), "the sensibilities of the western democracies were offended by the brutality of the militias in East Timor." And as the economic crisis increased, Britain was forced to reschedule $ 200 million in export credits for 40 Hawk military aircraft which had been used to attack independence fighters in East Timor. The circle that brings together political support, bad lending, and corruption is nowhere clearer than in Indonesia.

'Kleptocracy' and Zaire : When the former Zaire's ruler, Mobutu Sese Seko, died in 1998, he was of the world's richest men, with a personal fortune estimated at $ 4 to $ 10 billion and palaces in Europe and Zaire. Mobutu's government was the one for which the word "kleptocracy" was first coined. The *Financial Times* called it a story "about personal enrichment and the plundering of the nation, and the international complicity that made it possible".

The West backed Mobutu as a loyal ally in the Cold War. As early as 1974 there were warnings of loans being stolen. Karin Lissakers reported that in 1978 the IMF appointed its own man, Erwin Blumenthal, to a key post in the central bank. He resigned in less than a year saying that "the corruptive system in Zaire with all its wicked and ugly manifestations" was so serious that there is "no prospect for Zaire's creditors to get their money back". After Blumenthal's report to the IMF, it tripled its lending to Zaire.

Though almost none of Zaire's foreign debt arises from anything that much benefited the Zairian people who are being slowly starved to pay for it but in 1987, when Mobutu made his territory available for USA covert action against neighbouring Angola, the USA pushed through yet another IMF loan to Zaire—this time over the objections of some IMF officials. Initially, Zaire's debt was $ 5 billion, when Mobutu was overthrown and died in 1998, the debt was over $ 13 billion. Today, Mobutu's legacy is a country mired in civil war, as economic degradation leads to social and political disintegration.

The Strange Logic of Aid Officials and Loan Pushers

Rich country politicians and bank officials argue that because dictators like Marcos, Suharto, and Mobutu were kept in power with western arms and were given loans to squander on ill-judged and repressive schemes, that the people of those countries—who often fought valiantly against those dictators—cannot be trusted not to waste the money released by debt cancellation. This may seem confusing to people not familiar with the logic of the IMF and World Bank. In summary :

- Creditors colluded with, and gave loans to dictators they knew were corrupt and who would squander the money.

- Creditors gave military and political aid to those dictators—knowing arms might be used to suppress popular opposition .
- Therefore, successor democratic governments and their supporters, who may have been victims of corruption and oppression, cannot be trusted.

To many people in the South, this seems irrational and illogical—the logic of blaming the victim. It is the logic of power rather than of integrity, and is used to benefit the rich rather than the poor in Southern countries. A similar logic argues that if the World Bank and government export credit agencies promoted inappropriate and unprofitable projects, then Southern Governments proved their inability to control money because they accepted the ill-advised projects in the first place. Thus, if money is released by debt cancellation, it must be controlled by agencies which promoted those failed projects. This is the logic that says if people were stupid enough to believe cigarette advertising, then they are too stupid to take care of themselves and the "reformed" cigarette companies should be put in charge of their health-care.

The same institutions who made the corrupt loans to Zaire and lent for projects in Africa that failed repeatedly are still in charge, but their role has been enhanced because of their success in pushing loans. Preventing new wasted loans and new debt crises, and ensuring that there is not another debt crisis, means that the people who pushed the loans and caused this crisis cannot be left in charge. The creditors or loan pushers cannot be left in charge, no matter how heartfelt their protestations that they have changed. Pushers and addicts need to work together, to bring to an end the entire reckless and corrupt lending and borrowing habit.

Resurgence of Financial Capital

Another fundamental set of criticisms of free trade as a mechanism for promoting development is directed at standard theory's implicit assumptions that there are no basic institutional differences between nations in the way the production and trade are managed. One very influential criticism is that in many countries of the South, main exporting industries have been and continue to be dominated by foreign ownership. In the colonial heyday of the eighteenth and nineteenth centuries, export production in mining and plantation agriculture of the South was initiated and controlled by European-based companies, whose main function was to serve the markets of metropolitan countries and to provide cheap, secure supplies of raw materials to their processing industries (for example, in textiles, rubber products, copper and sugar refining). This, it can be argued, being assisted the competitive advantage of many industries in the wealthy countries so far it is now difficult for poorer countries with weak domestic markets (particularly those in Sub-Saharan Africa) to establish viable industries since so many industries exhibit economies of size and that way they require the assistance of international finance and business services which can only be provided by the capital flows of the MNEs. High-income industrialized countries, particularly the Triad[25] are both the most important source and destination for foreign direct investment (FDI). In industrial North there has been an average growth of 10.5 per cent between 1988 and 1997 against the growth of 2.6 per cent in trade, this evidence indicates growth in North Atlantic economic activity has been "in investment, and not trade", that is, the industrialized world Multinational Enterprises (MNEs) production is replacing national production, at least in the sense of supplying goods between countries.

Until the early 1970s, USA based transnational corporations accounted for about two-thirds of the total outflows of FDI, and about "four-fifths was directed towards Canada and the more advanced industrial nations of Western Europe. Meanwhile, another source of FDI had begun to show up : the MNEs based in the NICs .[26]

Foreign Investments: Looking back over past centuries, the capacity of the market economy system to provide (1) work opportunities and sufficient income to the maximum number of people, and (2) to create and liberate enough resources for public institutions to fulfil their collective responsibilities and to promote the common good, are reasonably clear and difficult to challenge. However, the second aspect is not a natural value inherent in the market economic system. To finance the social security and ageing of population in a developed economy of the North, most of these governments have borrowed so that public debt has sometimes reached 75 per cent of their GNPs, as it did in Western European countries in 1995. This gives a feeling that welfare state and welfare society are in crisis—the argument is that taxes have reached a maximum permissible level in the North. The only solution is to ask less from the states and more from markets and private sources of financing. This is also taken to mean that less ought to be spent on adjusting incomes and other forms of inequality. On an international level there is a dwindling element of development assistance in form of aid and loan to Countries of the South, which is being increasingly questioned and criticized. The proposed solution to this problem of the Southern Countries is private foreign investment. There are no international taxes except to provide the "user fee" to core administrative budgets of IMF and World Bank. At this juncture in history, it becomes the logic of the market to be less adverse to the logic of lending, to proclaim "URBI ET ORBI that all human being share a common humanity.[27]

By definition, Foreign Direct Investment (FDI) and Foreign Portfolio Investment (FPI) are distinguished by the degree to which foreign investors exercise management control in a company. Portfolio investors tend to limit their involvement to financial engagements, while direct investors have a significant and long-lasting management interest in the company targeted for investment. *However, the pace of world trade has been slower than the pace of FDI, which has risen at 14 per cent per year since 1985.* Although portfolio investments can make an important contribution to the financing of equity capital for local companies, concerns have been expressed by host countries particularly regarding the volatility of these flows and their effect on exchange and interest rates. Furthermore, portfolio investors can pull out of markets very quickly, and are prone to do so if short-term performance targets are not met or if other economic indicators are thought to portend weakness (such as the high levels of non-performing loans in East Asian banking systems that are said to have been instrumental in bringing about the current crisis).

Accompanying the FDI boom, foreign portfolio equity investment has also accelerated globally. Cross-border transactions in bonds and equities have soared in major industrialized countries, from less than 10 per cent of average annual growth rate of GDP in 1980 to 150-250 per cent in 1995 and 230 per cent in 1990-2000. The USA alone accounted for more than 35 per cent of outward portfolio investments in 1992-1994, with Japan accounting for 15 per cent and the United

Kingdom 11 per cent. United States investment flows into foreign securities accounted for $ 74 billion in 1982 and for $ 1.27 trillion in 1996. Foreign portfolio investment flows in the United States (defined as investments in US securities other than United States Treasury securities) accounted for $ 93 billion in 1982 and $ 1.23 trillion in 1996. In Japan, outward foreign portfolio investment outweighs outward foreign direct investment. Japanese portfolio investments abroad in 1996 amounted to 108.7 trillion yen, compared to outward FDI of 30 trillion. Inward portfolio investments amounted to 64.8 trillion yen in 1996.

It is difficult to do full justice here to the many facets of the epochal transformations that have taken place in extra territorial financial markets and institutions since the early 1980s. But in noting some of the more distinctive features of these shifts, it can be seen that in addition to a tripling in the overall volume of transnational capital flows between 1980 and 1992, there has been an equally marked change in the composition of the flows themselves, especially to the Countries of the South nearly doubled in the period from the early 1980s to the early 1990s, and in 1995 the East Asian countries received an estimated $ 53.7 billion in FDI, up 24.7 per cent on 1994. However, portfolio capital investment to these countries developed even more spectacularly, with a ten-fold increase between 1990 and 1993, with East Asia being the largest recipient. It should be noted, though, that flows of portfolio investment declined from $ 84 billion in 1993 to $ 56 billion in 1995, with a particularly sharp fall taking place in equity flows—these dropped from $ 46 billion in 1993 to $ 22 billion in 1995. This was due mainly to the rise in USA's interest rates in early 1994, the Mexican crisis a year later, the USA stock market surge in 1995 (a surge that has maintained itself pretty consistently until 1998), and concerns that the East Asian economies were overheating in late 1995, to cause crisis in the region (the Thai banking-system collapse that precipitated the crisis occurred in July 1997). It is important to get a sense of the kinds of portfolio investment flow that have been moving in and out of East Asia in these alternating phases of financial expansion and contraction, regarding what this may portend for a *new and modified version of the uneven development or dependency paradigm.*

Capital from bonds or fixed income securities amounted to $ 800 billion in developing countries in 1995, about two-thirds of their equity capitalization. Where the East Asian countries are concerned, the large direct flow of capital provided by FDI noted above has been exceeded in terms of growth rate by equity investment, which rose from $ 2.6 billion in 1989 to approximately $ 12.2 billion in 1995. This represented about 55 per cent of all portfolio equity flows to Southern Countries in 1995 (Latin America came next with 28 per cent). In 1994 foreigners made net purchases in East Asian stock markets that amounted to $ 2.5 billion in South Korea, $ 1.3 billion in Malaysia, and $ 2.1 billion in Indonesia.

This very substantial total private investment in the East Asian countries has been made possible by important structural changes in the financial markets themselves, and by the creation of several new instruments of international finance in the last two decades or so—largely because of financial liberalization and also because of the emergence of completely new international markets for securities, futures, options, swaps, international mutual funds, international bonds (these markets were opened to developing countries in the 1990s), and American and global depository receipts which gave American companies access to the stock markets of industrialized and industrializing countries. Part of this development was a new interest shown by investors in Asian stock markets that led quickly to a boom in the East Asian bourses: in 1993 alone the share indexes in Hong Kong grew by 116 per cent, in Jakarta by 115 per cent, and in Manila (the best Asian performer) by 154 per cent. The scale of this very rapid growth can be indicated by a comparison, between the relative times it took the United States and the emerging countries to reach roughly the same capitalization ratios.

The speed of development of Southern Countries' bourses (stock markets) in the recent period may be judged from the fact that it took eighty-five years (1810-1895) for the USA's capitalization ratio (market capitalization as a proportion of GDP) to rise from 7 per cent to 71 per cent. In contrast, the corresponding Taiwanese ratio jumped from 11 per cent to 74 per cent in just 10 years between 1981 and 1991. Similarly, between 1983 and 1993 the Chilean ratio rose from 13.2 per cent to 78 per cent; the Korean from 5.4 per cent to 36.2 per cent and the Thai from 3.8 per cent to 55.8 per cent.

In the 1970s, most governments viewed MNEs with a certain

amount of hostility. That time has got changed. At the same time the growing reliance of developing countries on portfolio investment has also been instrumental in creating the structural conditions responsible for the current collapse of the East Asian financial markets. To begin with, there is the sheer disparity of scale between the combined resources of the funds run by financial institutions in the most advanced industrial countries and the market capital of middle-income countries very new to this form of capitalism. The combined pool of funds managed by financial institutions in the high-income countries runs to around $ 10-15 trillion, whereas the total market capitalization of all lower-income countries (referred to by World Bank officials as 'emerging market' countries) is in the order of $ 1 trillion. *The effects of an imbalance of this kind during a rapid movement episode are potentially catastrophic for a lower-income country, whose stock exchange is likely to be dominated by foreign-owned portfolios* (as has already been noted Manila's, for example, is 44 per cent foreign-owned). This is especially true of the USA's mutual funds, which are inclined to jettison their holdings if quarterly performance standards are not reached or if there is the expectation of a falling market. Another feature of the new kinds of market capital that promotes instabilities of the kind now being seen in South East Asia has to do with the disposition of short-term investment capital not to reflect underlying economic 'fundamentals' such as output or employment. A rise in US or European interest rates, say, with no change whatsoever in the macroeconomic conditions of the lower-income countries involved, can nonetheless induce a change of perception on the part of foreign portfolio investors with holdings in the lower-income countries. The result can be a swift reversal of investment flows as funds are channelled elsewhere in a stock market stampede, with possibly devastating consequences for prices in the equity markets of the poorer country thus affected. The behaviour of this new form of short-term portfolio capital is quite different from that of financial and industrial capital as characterized by Marx, since the virtual autonomy it enjoys in relation to actual economic activity (this being the primary source of its volatility) makes it correspond more to what he calls 'fictitious capital' in Volume 3 of *Capital*, where it is used to designate a form of capital that creates money in ways completely detached from the productive process and the exploitation of labour.

Interestingly enough, the only advice the World Bank and IMF in this case can give to lower-income countries faced with a potentially damaging reversal of portfolio investment flow is the now familiar refrain: "strive for consistency in the implementation of strong macro-economic and structural policies, and ensure that borrowed resources are appropriately invested".

A country facing significant withdrawals of foreign portfolio capital will face the 'usual' problems resulting from pressure on its exchange rates and its balance of payments (with a resulting drain on foreign-exchange reserves), and from the almost inevitable price falls in domestic financial markets. This in turn can result in a weakening of the lower-income countries financial system, which is more likely if banks and financial houses are closely integrated with the securities sector, and have borrowers with a high level of investment in the domestic market: these borrowers will default on their repayments and leave the banks and financial houses facing shortfalls in meeting their obligations, which they may then try to discharge by borrowing abroad. In the longer term, however, the lower-income country relying on relatively large inflows of short-term capital provided by foreign portfolio investment (as opposed to foreign direct investment which is harder to withdraw quickly), and thus facing the possibility of overnight capital flight when financial markets become volatile, confronts a serious problem of macroeconomic management encompassing but also extending beyond the phenomena—runs on its currency, interest-rate hikes, a balance of payments and foreign-reserves squeeze, and so forth—more immediately visible when capital-market volatility starts to become more generally disruptive. For this instability makes it difficult for lower-income countries to pursue independent fiscal and monetary policies and to have a coherent strategy for managing exchange rates. This is especially so when the lower-income countries involved have only very short periods of time in which to find workable realignments in vast and unstable markets (in 1995 the portfolio investment market's world-wide transactions amounted to \$ 1.3 trillion daily or \$ 312 trillion in a year of 240 business days. Moreover, transnational capital markets are now large enough, and institutional speculators possess sufficient resources that are swiftly inter-convertible (banking, securities and insurance businesses now blend into each other, as typified by the proposed

$ 83 billion merger between the second largest USA commercial bank, Citicorp, and the financial conglomerate Travellers Group (which is the USA's third largest brokerage firm), as well as the comprehensive transformation in the UK of what a decade ago used to be 'building societies'), to neutralize the coordinated efforts of even the American and Western European central bank during a financial crisis. Lower-income central banks, such as those in South East Asia today, with resources that are a fraction of their G-7 counterparts, have virtually no chance of succeeding where much bigger central banks have failed (and will continue to do so in the absence of more permanent institutional arrangements to control such instabilities).

Lower-income countries face a constant and seemingly irremovable dilemma in this immense and fluid financial environment. To raise standards in education, health-service provision, and social welfare they may have to pursue an independent fiscal and monetary policy, but this will almost certainly result in exchange-rate instability, and setting independent exchange-rate levels will likewise cause the country in question to have less control over its domestic macroeconomic and monetary arrangements. The three situations—capital mobility, fiscal and monetary policy autonomy and stable exchange-rates— seem, therefore, to be mutually incompatible, leaving countries, and especially lower-income countries, with little or no room for manoeuvre. In addition, and the recent Asian economic crisis bears this out, portfolio funds, both foreign and local, tend not to go into those sectors, manufacturing and agriculture primarily, that take longer to produce significant yields, but instead find their way into domains where options favouring 'quick' money are more readily available, namely, the stock market and real estate. The ideal situation for such threatened lower-income countries would be for there to be a multilateral strategy that is coordinated accordingly but which still allows a country to use a range of capital controls that give it a degree of macroeconomic and monetary-policy autonomy in the face of market integration. But the coordination of multilateral efforts to stem capital market instability is difficult to sustain over the longer haul, when the crisis that prompted the initial search for multilateral coordination has dissipated, and national self-interest asserts itself again in the absence of institutional forces and principles strong enough (politically and not just economically) to counter such fissiparous tendencies. And so the structural dilemma remains: market integration places a high premium on the coordination of policies between nations, but this coordination is harder to maintain in the longer term because there seems to be no way of obviating tendencies to fall back on national self-interest when it becomes more difficult to support coordinated macroeconomic and fiscal measures for more than a just a relatively short span of time. Here of course the wealthier countries have a huge advantage because they have a better chance of using their economic and political resources to manage such crises in ways not available to their poorer counterparts: they can use these resources to allow their fiscal and monetary policies to operate independently in the short-term, and the policies they implement in this connection are not necessarily to the benefit of the less-wealthy nations (who in any case enjoy no such freedom to decouple fiscal from monetary policy). For example, the 'structural adjustment' package that South Korea had to accept as a condition of getting its recent IMF bail-out will cause a significant rise in unemployment because the government will have to lower or eliminate its subsidies to state-owned enterprises and set high real interest rates to reduce Korea's current account deficit and keep the *won* (name of Korean currency like rupee) stable, but this does not seem to figure among the IMF's primary concerns, which, as in everything else that comes under its purview, are to safeguard 'confidence' in the international banking and trading systems and to preserve and promote access to markets world-wide as an end in itself. Or, to mention another recent case, the Mexican economic collapse in December 1994 was precipitated in part by high USA interest rates imposed by the Federal Reserve to prevent the USA economy from 'overheating' and to help reduce the size of the USA's chronic current account imbalances—with high USA interest rates however Mexico's debt situation became unmanageable and the *peso* went into a free-fall.

The Trans-nationalization of Financial Capital: Some Theoretical Considerations

The absence of a viable system for ensuring adequate macroeconomic management in the face of financial market volatility is a problem that is particularly pressing for lower-income countries. It is, however, the outcome of a more profound failure that results from the convergence or intersection of two other crises: the crisis that accompanied the ending of the 'Golden Age', when the social and economic costs (inflation in particular) of Keynesian and New Deal/ Great Society policies intended to ensure that consumption was somehow always in line with production were found to be 'unacceptable' by the Reagan and Thatcher administrations; and the succeeding crisis, when the neo-liberal switch of emphasis from fiscal to monetary policy (augmented by an array of supply-side instruments) turned out to be too deflationary, so that monetarism was abandoned in the early 1980s by its leading exponent, the British Conservative government. Since then there has been no adequate system of national macroeconomic management capable of producing sustained non-inflationary growth, whether in the USA or elsewhere, and the powerlessness of the less-wealthy nations in the face of world financial market instabilities must necessarily be seen in this broader and more encompassing context. The currently prevailing neo-liberal ideology— promulgated in the USA but also espoused by multilateral institutions such as the World Bank and IMF, and which accords priority to keeping inflation low, to avoiding price 'distortions', to monetary discipline (though not quite full-blown monetarism after the 1980s), and to deregulation and liberalization— has no solutions for capital market volatility beyond the pious injunction that the affected countries should try always to have 'sound macroeconomic fundamentals' and embrace open markets. The components of this ideology were seen by the World Bank and others as the enabling basis of East Asia's 'miraculous' economic success, and the collapse of the economies of Thailand, Indonesia, Malaysia, and South Korea (as well as Japan's lingering recession) has exposed in the most abrupt and dramatic fashion this ideology's economic and political limitations. It is in this context that a new and modified version of dependency theory has its place, along with the notion that we may be on the threshold of the emergence of a new and somewhat different regime of accumulation (though as yet it is of course too early to make any predictions).

Whatever happens the claim, made by Giovanni Arrighi and others, that Japan and the other East Asian economies constitute a formation that has begun to supersede an American economic hegemony that has lasted for most of this century, is one that can no longer be upheld in this bald and unqualified fashion, if at all. For the immense proliferation of pension, insurance, and mutual funds in the last decade or so has changed radically the circuits of realization and accumulation that are at the disposal of the capitalist system, and the crisis of the East Asian economies is arguably an outcome of this new state of affairs. The impact of these new circuits of realization and accumulation on the economies of the less-wealthy nations has to be taken into consideration, especially since these nations are, confronted unrelentingly by an ideology: the 'Washington Consensus' (identified by John Williamson) which insists that all economic advancement, including progress beyond the most abject poverty and immiseration, has necessarily to be market-driven. And, moreover, the question which lies at the heart of theories of dependency and uneven development— *viz.* the possibility of implementing a post-capitalist system of production and accumulation— can be answered only by an analysis of these new and very recent capitalist circuits of realization and accumulation.

Portfolio Capital in Practice

As stated earlier, the sheer disproportion in the relative sizes of the stock market capitalizations of the advanced industrial and the so-called emerging countries affords the former group of nations more flexibility in dealing with international capital market mobility. However, the scope for economy policy-making in the less wealthy nations is constrained in another way, namely, by the particular concentrations of accumulated assets in pension and mutual funds in the advanced industrial countries, and especially by the business practices of the managers of these funds, who are concerned solely with short-term dividends and not at all with the economic and social well-being of the economies in which their funds are invested. And these assets are huge. In 1994 two American pension funds alone— had assets of $ 240 billion, and the largest pension fund in the UK, the Post Office and British Telcom Fund, had holdings of $ 35 billion. The totall figure for world-wide pension fund assets in 1994 was $ 10 trillion. In the same year, 1994, the stock market capitalization of all emerging Countries totalled a mere $ 1.9 trillion, and the GDPs for all of the world's Southern Countries totalled $ 1.2 trillion (only $ 166.2

billion when the 48 LDCs are taken). The disproportion between these pension funds and the less-wealthy nations, therefore, is palpable if not staggering: pension funds in a handful of countries, with no account taken here of mutual and insurance funds (insurance fund assets amounted to $ 2.6 trillion by the end of 1994), dwarf the total 'wealth' of these less prosperous countries.

Moreover, the forms of capital associated with pension, mutual, and insurance funds have done away with many of the traditional disciplines exercised by ownership at the point of production: many of the financial instruments created as part of the burgeoning of such funds are less amenable to established disciplines and forms of regulation, and this because the instruments in question are not positioned within relatively stable and organized markets in the ways typical of more conventional forms of financial capital. These instruments are managed by relatively few and largely anonymous fund managers, based overwhelmingly in the USA, the UK and a few other countries, and their complexity is often such that they tend to be understood by only a few specialists. The inevitable outcome is (*i*) that the creators of these novel, often hybrid instruments stand most to benefit from their deployment simply because they are in a position to be maximally cognizant of their intricate workings; (*ii*) that since many of these instruments are consciously designed to pass financial risks, transaction costs, etc., from the lender to the borrower. The financial institutions of the Southern Countries who borrow through them inevitably take higher risks than their lender banks in the North; and (*iii*) that, as with any financial-market innovation, the first users of the mechanism or channel in question tend invariably to profit disproportionally from it— as these come to be more widely used the rates of their attendant gains are inclined generally to fall. The power of those who control such funds and the instruments and channels through which they are deployed is, therefore, immense, and these highly mobile and relatively unconventional forms of capital are very much the engine of the newest regime of capitalist accumulation.

But this expansion in its present forms of transnational portfolio capital *does not augur well* for the development of the less-wealthy countries. There is ample empirical evidence of direct correlations between portfolio capital equity inflow and exchange rate instabilities;

as there is of the destabilizing 'income-effects' generated by stock-exchange volatility, of the failure of portfolio inflows in developing economies to be matched by increases in aggregate saving and investment in those countries, and of the propensity of stock markets to favour the survival of large though relatively unprofitable firms at the expense of their smaller but more efficient counterparts. Other difficulties exist: the proneness of financial markets to failure (information deficits, unenforceable contracts, and so on); the inability of the developing country relying on foreign portfolio capital to use this short-term and speculative capital as part of a long-term macroeconomic strategy; and the susceptibility of such capital to exogenous pressures (shifting interest rates of USA, the paramountcy of the needs of advanced country investors, and so forth). But the most important consideration here is that the transnational financial markets have done nothing so far (and neither do they give any indication of doing so in future) to deal with what is perhaps the single most important causal factor in the economic declines experienced by many Latin American and Sub-Saharan African countries in the 1980s, declines that in many cases are continuing into the 1990s (of the 48 low-income countries in the 1996 *World Development Report*[28] for whom such information is available, 23 experienced a fall in the average annual growth of GNP per capita during 1985-94, as did 22 of the 50 middle-income countries). This is the inability of the governments of the countries in question to maintain levels of real investment: according to the 1991 *World Development Report*, gross domestic investment grew negatively in the Latin American and Sub-Saharan countries as a group in the 1980s, and both groups experienced the largest declines in growth rate at the same time.[29] Why did real investment fall so significantly in these less-wealthy countries? And will the exponential growth in transnational financial markets do anything to alleviate the chronically low levels of investment (and output) in these nations? The answer to the latter question, in the light of the evidence available so far, is 'no'. In fact, the need of emerging countries to keep real interest rates high in the hope of attracting investor funds from abroad by ensuring higher returns for foreign capital, and to fall in line with the deflationary intent that is standard to all IMF/World Bank SAP will, all else being equal, lead to lower real wages. Given continuing high levels of unemployment, this policy will inevitably have negative effects on that country's income distribution. Given also the high existing levels of poverty, a growing population, depressed wages (for the reason just indicated), uneven economic performance, and other possible factors such as a decline in the quality of land stock (and the almost certain depletion of environmental assets), there is certainly no way that the poorer nations will be able to generate enough savings and investment endogenously to drive any kind of real growth, and this even if they try to heed the World Bank's injunction to seek their 'comparative advantage' (though many non-industrialized nations have no evident 'comparative advantage' to benefit from) and to maintain open trade and investment arrangements (as the optimal way in its eyes to ensure such growth). Moreover, significant amounts of foreign portfolio capital is not likely to flow in the direction of such countries, even if they sought to implement open trade and market arrangements. The plight of such countries is dire, and given the almost complete absence of institutions and mechanisms designed at both the national and international level to promote long-term investment and financial stability, continued stagnation is virtually inevitable for most of these countries, many of them have already been consigned, in Samir Amin's (1992)[30] words, to a 'Fourth World' that has no significant prospect of advancing even to the threshold of industrialization and of benefiting in any way from current and future expansions of international trade. The fundamental insight of dependency theory—that all claims to the contrary notwithstanding capitalism is not inherently 'progressive'—appears to be more probative than the cozening neo-liberalism of those who insist that "open" markets and the pursuit of "comparative advantage" are somehow going to be the salvation of the world's low-income countries.

Dependency Theory Revisited via a Detour through Arrighi's Analysis of the Current World System

Dependency theory has, of course, aimed to provide an explanation for the systematic underdevelopment of the non-industrialized or semi-industrialized nations (thereby filling a lacuna in Marxist theories of imperialism, which have tended to focus more on the contradictory nature of capitalist production and its property relations and less on the politically instituted relations of exchange between nations). Dependency theory has done more than this, however, because it has as one of its corollaries the proposition that as long as actually existing capitalism prevails, the low-income countries as a group are in principle not going to be the beneficiaries of any structural changes in the international system. But if it is the case that the less-wealthy countries are not going to benefit from *any* structural disposition of the international system as long as actually existing capitalism continues to exist, then it behoves the proponent of dependency theory to provide an account, however stylized, of the current disposition of this system. In this context, the most historically detailed and analytically rigorous analysis of the current regime of accumulation, its relationship to its predecessor regimes, and the conditions being set down now for its possible successor, is the one furnished by Giovanni Arrighi (1994)[31] in his *The Long Twentieth Century*. It seems that any advocate of whatever version of dependency theory has some point to engage with argument of this exemplary text.

One of Arrighi's aims is to account for the decline in profitability and economic growth associated with the ending of the "Golden Age"[32] of the Western economies from the 1970s onwards. His primary analytical notion is that of a "systemic cycle of accumulation". Adapting Marx's formula for the basic circuit of capital, *money commodities more money* (MCM), Arrighi (1994) argues that each systemic cycle consists of two phases. One is a period of "material expansion", in which profits are derived largely from the production and traffic (trade) in commodities, and which Arrighi takes to be the equivalent of the *material capitalism* (MC) component of Marx's formula. The other is a period of "financial expansion" (FC), in which *profits come primarily from financial enterprises and not from the extension or intensification of the production and traffic in commodities*, and which Arrighi takes to be the equivalent of the *capitalist money* (CM) segment of Marx's formula. Arrighi describes these expansions thus: Material expansions occur because of the emergence of a particular bloc of governmental and business agencies which are capable of leading the system towards wider or deeper divisions of labour. These divisions of labour, in turn, increase returns in capital invested in trade and production. Under these conditions, profits tend to be ploughed back into further expansion of trade and in production more or less routinely, and knowingly or unknowingly, the system's main centres cooperate in sustaining one another's expansion. Over time, however, the investment of an ever-growing mass of profits in the further expansion of trade and production inevitably leads to an accumulation of capital over and above what can be reinvested in the purchase and sale of commodities without drastically reducing profit margins. Decreasing returns set in; competitive pressures on the system's governmental and business agencies intensify and the *change of phase* is set from material to financial expansion.

Three major systemic "Cycles of Accumulation", or "Long Centuries", in Western capitalism are identified: the Dutch, the British, and the United States. It may be possible that Japan/East Asia is on the verge of constituting itself as a successor (fourth) major systemic cycle. According to Arrighi (1994), there are three different sources of financial profit that have a particular role in the processes of financial expansion which are as follows :

1. a source in which 'cut-throat inter-capitalist competition' creates excessive liquidity that finds an outlet in financial transactions.
2. a source in which the significant redistribution of income to capitalists creates conditions for the profitability of such financial transactions.
3. a source in which the liquidity generated by this profitability moves from centres that are no longer capable of sustaining material expansion to those that are developing this capability, thereby creating the conditions for the supersession of the previous phase of financial expansion.

Arrighi says explicitly that 'all financial expansions were eventually superseded by a new phase of material expansion'. This may be true of all systemic cycles of accumulation up to the present one. But unless this characterization of succeeding phases and cycles is to be elevated into a teleology that runs a staunchly dialectical course (a "logic of world economic history" in the manner of Hegel), the possibility must be left open that there may be no recognizable successor systemic cycle of accumulation to the present one, and that the primacy of the current systemic cycle may thus extend indefinitely into the (so far anticipatable) future. Or for there to be a series of transformations, however prolonged, of the current phase of financial expansion, so that by the time this series of transitions has progressed far enough, there may no longer be a form of capitalism that resembles the presently "Regnant and American-sponsored systemic cycle", or indeed for there to be anything like a systemic cycle of accumulation in the way characterized by Arrighi. If these scenarios were to obtain, then Marx's insight that the capitalist has no intrinsic interest in the value of commodities it buys or it finances so long as the commodities in question can be sold for a profit will be borne out, and the blissfully 'utopic' situation for the capitalist, whereby the capitalist moves directly from the initial *sum of money to yet more money* (the circuit MM), bypassing altogether *the production and exchange of commodities* (the circuit MCM), will have materialized. If this new regime of accumulation should be implemented, then, this will be something that occurred without the accompanying impetus to generate conditions for a new and succeeding phase of material expansion: there would simply be an indefinitely ongoing process of financial expansion. Arrighi in his *The Long Twentieth Century* outlined the two circuits or phases of capital which are in a relationship of oscillation: to put it somewhat schematically, according to him when capitalists can't make enough profits from commodity production or have over-accumulated as a result of commodity production, *they switch to financial transactions*, and return to commodity production (only) when the speculative bubbles generated by excessive financial expansion have burst. In other words, Arrighi conceives of the relationship between phases as one of succession or alternation, *i.e.* as an essentially temporal relationship. And that has become possible as a significant body of work on recent international trade and financial regimes would seem to indicate that the relationship between the financial and productive capital is now *primarily spatial*, so that what we have in world capitalism today approximates more to a complex amalgam of two simultaneously existing subregimes, one that is purely financial, and that the other represented by commodity production as well as its attendant forms of productive capital. The huge disparity between the sizes of these subregimes needs to be emphasized again: not just the scale of the pension, insurance, mutual and hedge funds noted above, but also the sheer volume and value of trading on foreign exchange markets (the world's largest financial market), which in 1994 was nearly 40 times the daily value of cross-border trade.

Thus the international system of financial markets has undergone a series of structural transformations since 1972, several of which are still taking place. These include changes in the sources of international credit and a new capital recycling mechanism, and together they began a revolutionary transformation in the 1980s that has effects which are still continuing. The changes in the sources of international credit are well-known, but, the more important changes have taken place in the capital recycling mechanism, *i.e.* "the form of credit made available to the world economy, in the networks of monetary agents which control access to this credit, and in the relationship between public monetary agents and private monetary agents within the global financial system". As a result, a new era in international finance has emerged, one that can appropriately be called *"decentralized globalization"*, and which is to be associated with the enfranchisement of a whole range of new and not always disciplined systemic creditors, like the rise generally of unstable institutional arrangements, the diminished authority and effectiveness of state and public monetary institutions (except when it came to leading the way in deregulating financial markets), the complementary growth in the authority and effectiveness of private monetary institutions (towards whom the balance of power has now gravitated), and the changing of the criteria used to govern access to flows of mobile capital (these have moved in favour of the interests of private agents). While private institutions have grown in importance, the state continues to have a role since capital mobility is not perfect, in at least two respects. (One) is that the state still possesses a degree of macroeconomic policy autonomy, though this room for manoeuvre is nonetheless circumscribed by the global integration of financial markets and by the propensity of states in this situation to allow private agents use of their policy instruments in ways which effectively make these agents proxies for state and public authority. And (the second) is the pre-eminence enjoyed by the American, German, and Japanese governments and their policies in determining the course and constitution of global financial markets, and allied with this is the systemic centrality enjoyed by the financial markets in New York, Tokyo and London. But this state capability notwithstanding, today no single state or public authority has effective control of the international financial system, which was not the case in Arrighi's two most recent systemic cycles of accumulation, the USA and England, which gave markets in New York and London a clear and decisive primacy in those systems of accumulation. The result is a growing regionalization of interest rates and the declining importance of reserve requirements as financial institutions become more hybridized and their resources more interchangeable as a result.

There is no unitary, comprehensive model of how this global financial system (which can appropriately be viewed as an MM sub-regime) relates to the production and exchange of commodities (which constitutes a CM sub-regime). As it involves the economic system of a particular country (or group of countries) or region, the dynamics of the relation between these two sub-regimes is inflected by path dependency, so that the economic activities that form this particular dynamic necessarily occur within, and have outcomes that are determined by, an always specific structure of political and social relations, a structure which *that* country or group of countries may or may not share with other countries or groups of countries. Some countries—Hong Kong and Singapore come readily in this context because their economies hinge crucially and overwhelmingly on the provision of financial services and the undertaking of a variety of *entrepot functions*—they thus display a very substantial embeddedness in the financial sub-regime or circuit of capital MM, whereas other countries—Taiwan or Brazil are good examples—which are much more dependent on manufacturing and are thus more profoundly embedded in the commodity-production sub-regime or circuit of capital CM, and still others—the United States (especially) and Japan—evince a powerful embeddedness in both of these sub-regimes. The forms of embeddedness in these subregimes are inevitably somewhat stylized in this account, since there is no such thing as a pure or perfect regime of financial enterprises nor one of commodity-production. And since there is in principle a potential multiplicity of forms of path dependency and embeddedness, different countries can relate in very different ways to one of both of these subregimes. But what of the claim, made by Arrighi that (in the latter's words) "the MM circuit of pure financial deals operates successfully only because this operation always pre-supposes a newly successful MCM circuit"? The Financial Capital Sub-Regime (FCSR) can stand in a variety of relationships to the Productive Capital Sub-Regime (PCSR), and each such relationship will have its own particular dynamism, though it will certainly be possible to categorize the different kinds of relationship and their accompanying dynamism, and to formulate principles that govern the relationships and dynamisms under consideration. Thus a great deal will depend on whether the FCSR happens to be in a relation of subordination or superordination to the PCSR. In a capitalist order, the survival or continued viability of this or that FCSR or PCSR is wholly contingent on the capacity of that particular sub-regime to generate continued surpluses. Failure to do this would amount to a crisis for that sub-regime and for the capitalist system which it embodies. This is the only absolute necessity incumbent upon any system or form of accumulation as long as capitalism prevails. To continue to extract surpluses a regime has to enable its agents or instruments to find outlets or markets into which these surpluses can be channelled. Determining the subordination or superordination of an FCSR in relation to a PCSR will therefore require a determination to be made of their respective (and always path dependent capacities) to generate surpluses.

Judged on the basis of the respective scale of the surpluses they have generated, there is compelling evidence that since the end of the 'Golden Age' (and especially more recently) FCSR and not PCSR has been the primary capitalist sub-regime. "From 1982, the beginning of the great bull market, until 1989, the end of the "leader era", the market value of non-financial corporations' stock rose to $ 1.83 trillion (from $ 1.38 trillion to $ 3.21 trillion), and the value of corporate bonds outstanding rose to $ 519.1 billion (from $ 407.0 billion to $ 926.1 billion). But while stocks rose 133 per cent, and bonds, 128 per cent (and of course most of the gain in bonds was the result of new issues, not capital gains on old ones), business sector productivity rose 12.6 per cent, and the broadest productivity measure of all, real GDP per hour worked throughout the entire economy (including government and non-profit sectors), rose to 6.3 per cent— "quite weak by historical standards".

The primacy of FCSR over PCSR in the last two decades or so is palpable, and so is the fact, that surpluses are being recycled into financial expansion and not (so much) into commodity-production. With generally weak levels of productive output, the threat of an immense over-abundance of commodities being left unsold is thereby reduced—and it is this threat which lies at the heart of the insistence that financial capital can continue to have life only if the system of accumulation it subserves maintains or increases productive output (*i.e.* incorporates a successful MCM "circuit"). Of course commodities "*have*" to be produced, but in the current system of accumulation this is being left to the low labour-cost economies who function therefore as suppliers for the markets of the high-income countries. Arrighi himself notes this when he says that "the main structural feature of the emergent East Asian regime remains the provisioning of wealthy markets with products that embody the cheap labour of poor countries". But nevertheless, the "informality" and "flexibility" of the Japanese multi-layered subcontracting system, combined with the abundance of parsimonious and industrious labour in the East Asian region, endow Japanese and East Asian capital with a distinctive advantage in the escalating global race to cut labour costs. It is precisely in this sense that the emerging East Asian regime of accumulation is "a negation of the old USA regime" (Arrighi, 1994).

However, it is clear that there is no such "global race" (let alone one that is "escalating") to cut labour costs, since the foreign direct investment, FDI that targets such low-cost labour in the South is being aimed at very specifically at a tiny minority of countries: The World Bank's figures show that *of the world's inward FDI which goes to the South, its two-thirds come to its mere 8 countries*, and that *50 per cent of the 93 less-developed countries of South receive little or none*. There is a quest for low-cost labour, while some features of its general trajectory may conform to the one delineated by Arrighi, it is not one that is by any means global: on the contrary, the search or 'race' for cheap labour has a marked regional structure (Africa and most of Latin America are effectively disqualified from participation from the outset), and its direction within that structure is even more highly selective. The quest for cheap labour subserves the imperative of stocking the markets of the wealthy nations, granted, but the over-accumulations of the wealthy nations are not allowed to saturate commodity-producing domains located in low-income countries (*i.e.* domains which belong to the PCSR): instead, in the currently prevailing capital allocation mechanism these surpluses are ploughed-back into dealings on world-wide financial markets (the FCSR), and when surpluses do find their way to the less-wealthy countries, they do so through the intermediation of these financial markets, which means that they can be pulled out of the low-income countries at extremely short notice. Money continues to be invested in a few developing countries of course, with East Asia being the primary beneficiary until its economic collapse. But even here the traffic is almost entirely one-way: in 1994, foreign direct investment accounted for 50 per cent of East Asia's net capital flow, and portfolio investment for a further 24 per cent; foreign direct investment moved from $ 1.3 billion (10 per cent of net capital flows) in 1980 to $ 43 billion (or 50 per cent) in 1994, and portfolio investment grew from nil to $ 18 billion (or 24 per cent of capital flows) in the same period.

East Asia's current crisis can be attributed in part to this almost complete reliance on foreign capital, especially on the part of the so-called 'second-tier' East Asian newly industrialized countries (Indonesia, Malaysia and Thailand). Arrighi does quite plausibly say that the centres of the emerging regime of accumulation will receive a surge of foreign capital drawn to them by the prospect of higher rates of profit than those available at the centres of the declining system. But the heavy reliance of the East Asian countries on foreign investment possesses several troubling features that immediately qualify any judgement of them as the (collective) potential successors of the dominant but ostensibly fading USA regime. In 1991-93 Malaysia had an annual average ratio of foreign direct investment to Gross Domestic Capital Formation of 24.6 per cent. This also accounted for its high 1995 services account deficit ($ 6.7 billion), caused by foreign companies repatriating their profits. Malaysia also has a problem with low total factor productivity, which grew at an annual rate of only 2.2 per cent between 1991 and 1996. At the same time Malaysia has seen a massive credit expansion, averaging over 30 per cent a year between 1994 and 1997, and constituting around 160 per cent of its GDP, with bank loans alone approaching 57 per cent of GDP. In Thailand, every $ 1 exported contains 43 cents of imported materials, so dependent are its export-oriented manufacturing sectors on imported parts. Thailand also has a large foreign debt total of $ 90 billion. Indonesia had a foreign debt total that stood at $ 60 billion at the time of its 1997 IMF bail-out. Non-performing loans in the South East Asian banking system amount to approximately $ 73 billion, over 13 per cent of South East Asian GDP. Thus therefore hardly seem nations like Japan belonging to the regime of accumulation as seen by Arrighi and many others are the potential successor to the possibly declining United States regime.

Hence, the features that most strongly define actually existing world-capitalism point inexorably to the relative autonomy of financial (and 'fictitious') capital from productive capital, that is, the autonomy of the FCSR from the PCSR. The embeddedness of a national economy in either or both of these sub-regimes is strongly path dependent, and this may, for the economy of that particular nation, either enhance or reduce the autonomy of the FCSR in relation to the PCSR, or *vice versa*. But in reality there is no such thing as an absolute autonomy or an absolute integration of these sub-regimes: financial capital markets and regimes and productive capital markets and regimes are necessarily related to each other, and this because funds are free to move between them, and policies made in regard to the one necessarily affect the other.

At the same time the world-capitalist system that has emerged since the end of the "Golden Age" is in a condition of crisis, and this because the surpluses yielded by the FCSR enable those national economies most deeply embedded in it to escape many of the disciplines that are imposed on those economies which function by being embedded primarily in the PCSR. The most telling illustration of this is the recent 'accomplishment' of the USA in eliminating the massive fiscal deficits incurred during the Reagan and Bush administrations, thanks largely to tax revenues garnered from the current stock-market boom (stocks on the New York Stock Exchange rose in value by 33 per cent in 1997)—only a country as widely and powerfully embedded in the FCSR sub-regime as the USA is today can afford itself this option (which must make the USA now able to comport itself as the 'investor aristocracy' of this world-capitalist system — an object of envy for many a poor country struggling with its fiscal deficits but with no financial-market resources at hand to bring it this kind of miraculous relief.

World-wide economic polarization in its current manifestation is driven more by the divorce (always politically instituted and maintained) between financial and productive capital, and less by the mechanisms of unequal exchange as typically understood in dependency theory. In the typical account, unequal exchange exists because of an international division of labour which allows the industrial countries to commandeer industrial productive capacity while consigning low-income countries to the production of primary commodities (this being the phenomenon of compulsory mal-development described by Samir Amin and others in the dependency and uneven development school), because international trade on these terms can never be mutually advantageous. However, in the account being canvassed here, while there still is unequal exchange—it being undeniable that even in the era of financial capital (and its fundamental divorce from productive capital), the less-wealthy countries function essentially as producers of primary commodities and as providers of cheap labour — the main source of international economic polarization today is precisely the autonomy of finance capital from productive capital. This autonomy establishes asymmetries between the high- and low-income countries that appear to be deeply entrenched, and so the question remains whether, and if so to what degree, these asymmetries are surmountable.

Surmounting International Economic Polarization

The development-policy prescriptions—trade and price liberalization, deregulation, privatization, and closer links with the world economy— of international financial institutions such as the World Bank and the IMF, as well as those nations, primarily the United States, who underwrite the neo-liberal "Washington Consensus",[32] are largely irrelevant to the economic situation of the lower-income countries, many of whom are still struggling with varying degrees of success to make the long-term structural adjustments necessitated by the major recession and debt crisis of the 1980s. For these institutions and wealthy countries swimming with the tide of global economic integration is the only way forward for the nations of the South. The dependency and uneven development paradigm has long been associated with the proposed delinking of the economies of the South from those of the North. This delinking strategy is commended because of the conviction of those who uphold this paradigm that the situation of the lower-income countries cannot be improved structurally within the terms of the prevailing regime of accumulation. As we have seen, a judicious scrutiny of the available evidence is hardly likely to controvert this conviction. More recently, Lance Taylor (1991)[33] who has made a number of careful criticisms of the World Bank over the years but who is hardly to be regarded as a Marxist, has recommended a partial delinking on "narrowly technical grounds". Taylor has analyzed extensively the data regarding the open trade and capital market strategies of a cross-section of 48 lower-income countries (going as far back as the economically more propitious 1960s) and found few gains and some losses accruing from these exogenously-oriented policies. He proposes that these countries should dispense on a piecemeal basis with linkages to the markets of the North that bring "the least benefits" or exact "the greatest costs", concluding that a limited and selective autarky of this kind offers them a better bet for economic survival— as Taylor puts it, "the inwardly oriented resource allocation strategy seems the least risky, especially for large countries". Given that openness to global capital flows makes a lower-income country more vulnerable to external shocks and to the onset of financial crises such as the one now occurring in South East Asia, it would seem to be in the best interests of the lower-income countries to buck the wisdom of the neo-classical consensus and attempt a strategic and selective delinking along the lines recommended by Lance Taylor. This would at least place the less-wealthy countries in a better position to exert greater control over their macro-economies as they implement development policies that are endogenously oriented, as opposed to placing their faith in roller-coaster markets that have the lower-income countries at their mercy. With this kind of control, and more democratic political institutions, they can at least pursue more consistently, if not more seriously, the project of an economic liberation that hopefully will start to bring to an end the newest form of economic dependency.

Portfolio investments received by developing countries moved ahead strongly in 1996, following a pause induced by the Mexican *peso* devaluation and the rise in short-term US dollar interest rates in 1994. Portfolio investments accounted for 38 per cent of all private investment flows into developing countries in 1996, with bonds accounting for a further 19 per cent and equity 18.8 per cent. The distribution of inward portfolio investments among developing regions and countries is very similar to the distribution of inward FDIs. However, the growth of textile industry in India, the nationalization of foreign oil interests by the Middle East countries in the 1950s and 1960s, and the acquisition of Malaysian owners of former British-owned rubber plantations in Malaysia are all indicators of ways in which the colonial heritage of foreign ownership of primary commodity industries has been diminished in some of the more dynamic Countries of the South. It is again in Sub-Saharan Africa that this legacy persists more clearly than elsewhere.

Thus if the demise of the World War II, has left us with two worlds rather than three, and whatever all the above interactions seem to generate, benefits disproportionately for the richer countries so that these two worlds are growing farther and farther apart economically and technologically. This gap is not merely one of economies but is increasingly one of technology. As USA, Western Europe and Japan move into a post-modern era, no longer based mainly on industrial production but rather on information processing via computer, technologically, the Southern World falls farther and farther behind. In purely practical terms, for the Southern Countries, the New International Economic Order (NIEO) is looking not much different from the Old World Order. All the trade organization at the world level, GATT, UNCTAD, WTO are all purportedly democratic institution, stating in their preamble about conducting the world trade and economic endeavours more supportive of development, but they are actually dominated by the leading Countries of the North, so that over the years the corporate interests of Countries of the North particularly of USA have expanded, volume of the world trade has also got increased by 20 per cent since 1995 but the benefits of the increased trade are not widely shared. For example, the LDCs of the South which represent 20 per cent of the world's population, generate a mere 0.03 per cent of the trade flows. Structural adjustment policies have therefore had far-reaching consequences around the world. Yet, this is just one of the mechanisms whereby inequality and poverty has been structured into laws and institutions on a global scale. *We should turn now to the theory behind the polarized growth of world economy.*

- Governments therefore *must spend less*, reduce consumption, remove or decrease financial regulations and so on.
- Over time then : as the value of labour *decreases,* capital flows become more *volatile* and the country goes into *a spiralling race to the bottom.*
- Exchange rate stable is costly due to measures such as increased interest rates etc.
- Investors are obviously concerned about their assets and interests can then pull out very easily if things get tough.
- In worst cases *capital flight can lead to economic collapses* like seen in the Asian/global financial crisis of 1997/98/99, Mexico, Brazil and many other places — of course, the blame by mainstream media and free trade economists is laid on emerging markets and their government's restrictive or inefficient policies, crony capitalism etc., which is a cruel irony.
- Keeping the exchange rates in their favour *by IMF donors, means that the poor nations remain poor, or get even poorer.* Even the 1997/98/99 global economic financial crisis in the East Asian countries can be partly blamed on structural adjustment and overly aggressive and early deregulation for emerging economies.
- *Social unrest* is often one result.

This is one of the backbones to today's so-called "free" trade as practiced by the *North with the South.*

Unequal Exchange Rate: Need for Relating Exchange Rate to Work Time Parity

An exchange rate is the amount of the currency of one country that exchanges for the currency of another country. The money one receives for travelling from one country to another is determined by the currency exchange rate. Exchange rates also determine how much the goods of one country cost the people of another country when those goods are imported. Exchange rates are fair if they are neutral, similar to how within the same country we can exchange one amount of currency for an exactly equivalent amount, say, when we trade a $ 10 bill for 10 $ 1's. Many currency Exchange rates are unfair and one-way, or extractionalist so that they are biased in favour of one country and are therefore biased against another country, like giving up a $ 10 bill and receiving only 5 $ 1's in return. Fair exchange rates would have currencies trading for equal amounts of work time.

Today, it takes a few minutes of work in some countries to equal their exchange rates while in others it takes many hours of work. The amount of work (hour) time required to equal countries' exchange rates, *i.e.*, Hour rates were calculated by dividing Gross Domestic Product by number of people employed times 2000 where the 2000 represents the number of hours a person would work in a year at 40 hours per week for 50 weeks a year. So the divisor is an estimate of the total hours of work that produced the GDP. To get an estimate of total hours worked for countries where the number of people employed is not available, a fair approximation is half the total population times 2000.

Hour rates and exchange rates of countries correlate strongly. For 1994-1995, hour rates correlate 0.78 with exchange rates. Their logarithms correlate 0.88. These strong correlations mean that work time accounts for 78 to 88 per cent of 1994-95 currency exchange rates. Strong correlations have existed throughout the 50 year history of the International Monetary Fund. The strength of the correlations vary with length of membership in the International Monetary Fund. The correlations are strongest for the 42 countries that have been in the IMF for the longest time. The weakest correlation, $r = 0.41$, occurs for the 29 countries that entered and left the IMF at various times throughout IMF history. These correlations tell us that the work hour is the centre of gravity for exchange rates. It is probably the equilibrium point dictated by economic fundamentals that economists seek.

In spite of these strong correlations, deviations from hour-for-hour exchange rates favour some countries at the expense of others. This is clear when each country's exchange rate is expressed in minutes of work (Exchange Rate/Hour Rate times 60). If exchange rates were fair, all exchange rates would equal the same number of minutes.

Instead, Japan at the head of list has an exchange rate equal to a fraction more than 2 minutes. Down the list, work time increases to a crippling 415 minutes for Bangladesh at the end of the list. Are these differences due to differences in productivity? If that were so, we should expect the products of the most productive countries to be cheaper, not more expensive, than the products of less productive countries. Instead, exchange rates make the products of the supposedly most productive countries more expensive than those of the least productive countries. It is directly opposite to what fair price would dictate.

Countries close to each other on the list can trade with each other on fair terms. The further apart on the list, the more unfair the terms of trade between countries. For example, Switzerland and Norway can trade fairly. Their exchange rates are both equal to about 2.5 minutes. But trade between Austria and Hungary is biased in favour of Austria where about 3 minutes of work equals the exchange rate of its *schilling* compared to Hungary where 13 minutes of work are needed to equal the exchange rate of its *forint*.

What needs to be done is *to move exchange rates to work time parity*—to where they trade on an equitable basis that work time parity is where exchange rates tend naturally. Deviations are probably maintained unnaturally by the desire of the countries that control the IMF to keep exchange rates biased in their favour. But this unequal exchange is because of the wage differences between the North and the South and is argued as the main reason that cause the inequality which lead trade to be "exploitative of low-wage economies" by the Countries of the North (Liang, 1997)[34].

Maintaining Dependency and Poverty

One of the many things that the powerful nations (through the IMF and World Bank etc.,) prescribe is that the Southern nations should open up to allow more imports in and export more of their commodities. However, this is precisely what contributes to poverty and dependency. One of the effects of structural adjustment is that developing countries must increase their exports. Usually commodities and raw materials are exported. Historically this has been a partial reason for dependent economies and poor nations. To export commodities (which are cheap), to be denied industrial capital and real technology transfer and to therefore import finished products (which are more expensive due to the added labour to make a product from resources), nations that are primarily commodity exporters are likely to lose out, as there would be little multiplier effect in their own economies. This was also the role enforced upon former countries under imperial or colonial rule. This is *unequal* trade.

At first glance it may seem that the growth in development of export goods such as coffee, cotton, sugar, and lumber, would be beneficial to the exporting country, since it brings in revenue. In fact, it represents a type of exploitation called *unequal exchange*. A country that exports raw or unprocessed materials may gain currency for their sale, but they lose it if they import processed goods. The reason is that processed goods—goods that require additional labour—are more costly. Thus a country that exports lumber but does not have the capacity to process it must then re-import it in the form of finished lumber products, at a cost that is greater than the price it received for the raw product. The country that processes the materials gets the added revenue contributed by its labourers.

Notes and References

1. Over the period 1950-1973, the volume of world trade (in goods and services) increased at an annual rate of nearly 8 per cent, while world GDP rose annually by 5 per cent in volume. In the decade following the abandonment of the Bretton Woods fixed exchange rate regime and the first oil price shock, the rate of expansion of both world trade and world GDP slowed down substantially. In 1996, world merchandise exports accounted for $ 5.1 trillion and commercial services exports $ 1.2 trillion. The respective figures for imports were $ 5.2 trillion (merchandise) and $ 1.2 trillion (services). In 1996, global trade (based on exports) rose 3.3 per cent from the previous year. (World Trade Organizations (WTO), Trade Statistics, 1997).
2. Schor, Juliet, B. 1990, (eds.) *Golden Age of Capitalism : Reinterpreting the Post-war Experience*, Oxford University Press, London.
3. Pollin, Robert, 1997, *Contemporary Economic Stagnation*, Cited from Surin, 1998, *Dependency Theory Reanimation in the Era of Financial Capital*, Cultural Logic, Vol. 1, No.2.
4. In 1992, 48 out of the 88 countries classified by World Bank in the low- or middle-income category had populations of below 10 million. Markets are not only small but poor for these countries. In 1992, the GNP per capita of the low-income countries averaged $ 390 and the middle-income group $ 2,490, which were respectively 2 and 11 per cent of the average for the high-income, industrialized countries.

5. An excerpt *cited.*
6. UNCTAD Secretariat, 2002.
7. UNCTAD Secretariat, 2002, "*Least Developed Countries at a Glance*", Bangkok, 48 LDCs : Trade Statistics.
8. Narula, Rajneesh, 2002, *Switching from Import Substitution to the "New Economic Model" in Latin America : A Case of Not Learning from Asia*, Research Memoranda, Maastricht.
9. United Nations Conference on Trade and Development, UNCTAD 2001.
10. Rees, Judith, 1989, "*Natural Resources, Economy and Society*" in, Gregory, Derek and Rex Walford, eds. *Horizons in Human Geography*, Macmilan, London.
11. *North—South dialogue* in the present context refers to the way rich and poor nations negotiate terms of trade. However, it can mean much more than this, like to international relations among people.
12. Prebisch, Raul, 1950, *The Economic Development of North America and Its Principal Problem*, United Nations, New York.
13. Lewellen, Ted, 1995 2nd Edition, *Dependency and Development*, Bergin & Garvey, Westport.
14. Spraos, J., 1983, "*Inequalizing Trade? A Study of Traditional North / South Specializing in the Context of Terms of Trade Concepts*" , Oxford University Press, Cambridge, Ma.
15. An excerpt *cited.*
16. International Monetary Fund, 2001, *World Economic Outlook*, 2000, Oxford University Press, New York.
17. Adam Smith, 1776, "*An Inquiry into the Nature and Causes of the Wealth of Nations*", *Book IV, Chapter VII,* republished, Oxford Clarendon Press, 1976, *Ibid.*, 33, Chapter 1.
18. Stiglitz, Joseph, 2002, *Globalization and Its Discontents*, Penguin Books, London.
19. Washington Consensus: refers to a set of ten rigid, almost unalterable, theoretical propositions adopted in by John Williamson in 1993 as a consensus regarding how the modern-day market laws as policy recommendations for poor countries willing to reform their economies about what makes them prosperous. The general ideas derived from the Washington Consensus had a huge influence on the economic reforms of many countries. Yet, the way the countries interpreted this "consensus" varied substantially and their actual implementation even more so. Moreover, the original ten policy prescriptions of the Washington Consensus reigned unchallenged only for a short time. Changes in the international economic and political environment and new domestic realities in the reforming countries as different as Russia, Argentina, Rumania or South Korea, created problems so that in practice, the differences in the reforms of these countries often clearly exceeded their similarities so that the only thing they had in common was their designation by proponents and detractors as "Washington Consensus policies".
20. Economic Community of Latin American and Caribbean, 2001.
21. The World Bank and IMF were important instruments of Western powers during the Cold War in both economic and political terms. They performed

a political function by subordinating development objectives to geostrategic interests. They also promoted an economic agenda that sought to preserve Western dominance in the global economy. Not surprisingly, the World Bank and IMF are directed by the governments of the world's richest countries. Combined, the "Group of 7" (U.S., Britain, Canada, France, Germany, Italy and Japan) hold more than 40 per cent of the votes on the Boards of Directors of these institutions. The USA alone accounts for almost 20 per cent. It was USA policy during the Reagan Administration in the early 1980s, to expand the role of the World Bank and IMF in managing developing economies. The dependence of African countries on new loans gave the World Bank and IMF great leverage. The conditions attached to these loans required African countries to submit to economic changes that favoured "free markets."

22. *Less than 10 per cent of Africa's labour force is employed in the formal job sector.*
23. The Bretton Woods Project: Created as a networker, information-provider, media informant and watchdog, it monitors, scrutinize and influence the overall management of the Bretton Woods Institutions like World Bank and International Monetary Fund, IMF through its briefings, reports and the bimonthly digest.
24. PRSP process is simply delivering repackaged structural adjustment programmes (SAPs). It is not delivering poverty-focussed development plans. According to the fifteen African countries, PRSPs were simply window-dressing to improve the IMF and World Bank's declining legitimacy. The content of PRSPs continues to put corporate rights before social, human and environmental rights. Rather than enable local people to decide their content, PRSPs meant more IMF and World Bank control "not only over financial and economic policies but over every aspect and detail of all national policies and programmes", said the statement. In particular, the "anti-poverty programmes are expected to be consistent with the neo-liberal paradigm including privatisation, deregulation, budgetary constraints and trade and financial liberalisation. Yet these ignore the role of international/global factors and forces in creating economic crises and poverty."
25. *Triad* refers to USA, UK and Japan.
26. NICs are the Newly Industrialized Countries which include South Korea, Taiwan (and the city-states like Singapore, Hong Kong) who have gone upscale from a marginal producers of chintzy clothes and toys to high-tech manufacturing in the recent two decades, to be known as NICs since 1979 as against the traditional industrial countries of the North.
27. Mihaly Simai, 1999, *The Democratic Process and the Market: Challenges of the Transition*, UN University Press, Tokyo.
28. World Bank, *World Development Report*, 1996, Oxford University Press, New York.
29. World Bank, *World Development Report*, 1991, Oxford University Press, New York.
30. Samir Amin,1992, *Delinking towards a Polycentric World*, London.

31. Arrighi, Giovanni, 1994, *The Long Twientieth Century: Money, Power and the Origins of Our Times*, London.
32. *Ibid.*, 19.
33. Taylor, Lance,1991, Economic Openness: Problems to the Century's End in, Tariq Banuri, ed. *Economic Liberalization : No Panacea (The Experiences of Latin America and Asia)*, Oxford University Press, London.
34. Liang, Hong, 1997, *Rationales of ISI—A Strategy*, University of Denver.

5

WORLD–SYSTEM THEORY OF POLARIZED GROWTH

As economic development proceeds, an economy's spatial structure changes from the one characterized by "small, isolated and functionally undifferentiated communities" to one mode of interdependent regions with a centre-periphery structure. As a process in the 1950s and 1960s, in many countries an advantaged central region invariably emerges which drains resources from the peripheral regions. This leads to a geographically differentiated patterns of economic growth and decline between places, for example, when established regional economies were beginning to experience challenges to their competitive advantage while the "poorer regions" went on declining *e.g.* Italian South or US South were seen as lagging behind other regions in respective countries. Result of this decentralized reactions of one kind or another led to the widespread acceptance of the need to encourage regional *balance* in economic growth at a time. It is in this context that appeals for government action arise to "help" a particular depressed region or a set of regions either by "adjusting" to a new encouraging situation or providing compensating investment by means of fiscal measures such as tax breaks or realocation subsidies. It will not occur "naturally"—in the absence of intervention, writes Friedman (1963)[1], "even the advanced economies of Western Europe and USA continue to be plagued with problems of depressed areas".

The 1950s was the period of "relative prosperity in the world economy" so that affluent societies could afford to indulge in *redistributive policies*, and the inter-regional equilibrium was seen as possible for encouraging regional authorities to encourage regional economic planning and foster local industrial regeneration with the active intervening of government.

In contrast to this "decentralist reactions" of one kind or another at the regional scale within a country, in the global scale as the world gets increasingly dominated to a single economy such that all peoples are integrated into the sphere of capitalist production. They are linked by a series of metropolis-satellite chains which John Friedman in 1972 has described as Polarized Growth Theory.[2] This paradigm, Friedman (an American regional planner working in Venezuela), applied to regions [3] within a single country. He defined *heartlands* (or, core regions) as territorially advantaged central region with organized sub-systems of society possessing high wages, advanced technology and having a high capacity for generating as well as innovative change in areas possessing favourable physical qualities with good accessibility to markets : they display a diversified profile of secondary, tertiary and quaternary industries, they are characterized by a highly urbanized and concentrated population which participates in a well-integrated urban system and they are able to influence and usually control through the metropolis—economic, social and political decisions of national importance. *Hinterlands* (or, peripheral regions) are characterized *by the obverse:* they are all the regions beyond the heartlands having an emphasis on primary resource production: scattered population and weakly integrated urban systems, limited innovative capacity and restricted political prowess. They are all the regions (lying beyond the heartland) whose growth and change are determined by their dependency relationships to the heartlands.

Heartlands share in common a heavy concentration of their labour force in secondary (manufacturing) activity and advanced (quaternary) services, representing a shift from development based on natural resources to development based on human resources so that both industrial and service activities become less concerned with the processing and marketing of primary resources. As they become more sophisticated and more dependent upon the products of scientific research and development, the distinction between manufacturing-related and service-related activity becomes increasingly blurred. It is this new nexus of activity that has become the "modern engine of growth", making the term "heartland" almost completely synonymous with high per capita income and income potentials.

Looked at modernization perspective, in those days of affluence in economy in 1950s and 1960s as a *liberal model of development*, for both national as well as global scales, heartlands were looked upon as that they set the development path for the hinterlands, stimulating economic growth in the peripheries differentially according to the resource needs of heartland industries and consumers. For the conventional economists the *core-periphery* regional differences are not in the context of the core as the advantaged central region, not exploiting the periphery by draining its resources but developing the periphery. This process of regional development is often called "modernization". From the perspective of modernization theory, regional differences are only temporary and disappears with time. The mechanism for such a change through a shift/spread/trickling down from the more developed core to the periphery lies in a comparative market-place where regional inequities are corrected by the mobility of factors of production, particularly, by labour and capital. Within a nation-state, like in Canada workers are attracted to areas of high wages, thereby creating a shortage of workers in the low-wage areas. At that point, wage levels rise, reaching an equilibrium with wages in other regions. Inter-regional equilibrium within a nation-state is also possible, provided the government intervenes—but it will not occur "naturally", in the absence of intervention, as it is mentioned earlier, that "even the advanced economies of Western Europe and USA" quoting Friedman (1963), "continued to be plagued with problems of depressed areas". From modernization perspective, Friedman (1963)[3] conceived *four types of regions* at national level: core plus three varieties of peripheries, all dominated in different ways by a core region. Three peripheral regions are the (1) upward transitional region, (2) downward transitional region, and (3) a resource frontier. This division of the periphery into three sub-regions demonstrates the existence of important variations in peripheral resource bases and their consequent development.

However, the "fiscal crises" experienced by many national governments world over in the 1970s, brought rising expenditures on "entitlements" (like social security, health care) and defence which were not matched by corresponding increases in revenues. These all challenged the relevance of such conventional regional policies, the state was increasingly focussed its attention on the national rather than the regional crisis: regional policy was in large part irrelevant, and as Massey (1984)[4] pointed out, "no longer is there really a *regional problem* in the old sense, the old spatial division of labour based on *sector*, on contrasts between industries, has gone into *accelerated decline* and in its place has arisen a dominance of a *spatial division of labour* based upon a spatial division of firm activities; since it is not regions interrelate, but the social relations of production which take place over space".

Krugman (1995)[5] suggests a more complex and more formal model to account for a centre-periphery pattern of economic development. He argues that the relationship between demand and production established during an early phase of industrialization locks into place an inter-regional imbalance through the conjoint operation of "increasing returns to scale" in plant operations, transportation costs and demand. Thus, given sufficiently strong economies of scale, to minimize transportation costs, the manufacturer chooses a location with large demand and this location will be chosen by majority of manufacturers because of their *propulsive character* so that there will be a circularity by the agglomeration economies that tends to keep the manufacturing belt in existence once it is established in terms of a *growth pole*.

Complementaries in the availability of factors of production between heartland and hinterland lay the foundations for interaction. Improvements in transportation and the organization of trade increase the transferability of the "staples" from hinterland to heartland. Intervening opportunities impose spatial regularities in the timing of hinterland development and sequence the order in which unsettled areas and areas with a subsistence economy are drawn into the heartland's sphere of influence. Thus for the upward transitional region, if its economy and population grow to a critical mass it may serve to stimulate production facilities there. Then there can be a dramatic shift in regional fortunes. The "local boosterism", that is, faith in a locality's future possibilities and policies that reflect this can also prove decisive in reversing regional imbalance. "Nothing", therefore, "is forever" so that both capital and labour flow into this rapidly developing area (Krugman, 1991).[6] This is an argument for a *rapid reversal* in regional fortunes, rather than for slow regional balancing—centres become peripheries and *vice versa*. Krugman argues that increasing returns, imperfect competition, and historical accident conspire to set in motion particular patterns of industrial development over time and space and in terms of subsequent "locking in" of those patterns via self-reinforcing effects. Once a given pattern of geographical concentration is established, a high degree of persistence (or, inertia) over time, and that this inertia can operate either to foster regional growth or to retard it. But the identity of the favoured region is not set for all time. A new pattern of concentration can break the historic mould. While initial development occurred in the primary sector, there is now a greater emphasis on manufacturing and service activities. British Columbia in Canada is an example of upward transitional region.

In a *downward transitional region, the economy declines, unemployment rises and out-migration occurs.* Often this is an "old region" that was dependent on resource development for its economic growth. Now that these resources have passed their prime or have been exhausted, the regional economy stalls. With pool of migrant labour force, Atlantic Canada is an example of a downward transitional region.

Located far from the core region, few people live in the *frontier zone* and little development takes place. Resource companies are just beginning to penetrate into this remote area. As energy and mineral deposits are discovered, the prospects for economic growth are enhanced. The "Territorial North" represents Canada's *last resource frontier.*

Unlike the others (including Friedman's), Krugman's model, therefore, does not incorporate a concept of spatial equilibrium from which the "real" world of development diverges but towards which there can be convergence in the long-term in context of regional economic development. But this *long-term never arrives.* Rather, in a dynamic world driven by cumulative processes, initial advantages can persist for decades but then they can be reversed dramatically as new conditions create *new patterns of economic concentration.* Concentration, somewhere, is "the perpetual rule". So, though apparently attentive to historical change, this model is static in its assumptions about the operation of economic-locational principles. The same principles of increasing returns, imperfect competition, and agglomeration are at work *in the same way* all the time. From this point of view, geographical outcomes can change but the process driving them does not (Martin and Sunley, 1995).[7]

Friedman (1972) saw the diffusion of innovations from the core as controlling the system growth and the form of the heartland-hinterland relationship affecting economic activity and settlement patterns, socio-cultural traditions and values, and the organization of power not only in the core, but also in the periphery. The *invention* lies at the heart of internally generated advances in knowledge—it involves new ideas, new procedures, new types of organization, new products, new technologies and it includes *basic* applications of new knowledge and new components—as in the invention of the first jet engine, computer—and also elaboration or improvements in later models based on additional experimentation and experience. Innovation in the core capitalist countries thus has allowed the capitalist class to systematically increase its surplus through increases in labour productivity resulting in cheaper goods and greater output on one hand and on other hand new technology (through invention) create powerful impacts on world trade. The problem with the "centre-periphery model" is not solely the difficulty of accepting the logical possibility of spatial equilibrium. The process driving this *paradigm* is *regional exploitation* rather than *social class exploitation* (according to the left-wing view or the neo-Marxist version which substitutes *regions for social classes*). The volume of controlling decisions that emanates from the core draw towards the centre the surplus which is produced at each stage of production. In the 1960s it became clear *empirically* that "trickling down" was not at work in many countries. Indeed at that time, "backwash effects" seemed much stronger than the anticipated "spread or diffusion effects" of economic growth. Spatial inequalities at a range of scales—international, regional, and urban—had become more rather than less pronounced. Thus the result is what Andre Gundar Frank said in a didactic statement on *Urban Location and Dissipation of Entropy* (1982) that "urban and inter-urban politics are in large measure an outcome of this global structure and process. Indeed, much of what appears on closer examination is not just inter- or intra-URBAN relations but the international relations. Thus in the contemporary context, revival of *regionalism* for instance in Europe or China is importantly derived from cities and their denser regional relations, though some of these cities may also be major WORLD CITIES, like New York, Los Angeles, London, Tokyo, Hong Kong, Shanghai that form a global network of their own. They in turn connect *minor* world cities like Chicago, Toronto, Mexico, Sao Paulo, Buenos Aires, Frankfurt/Berlin, Zurich, Cairo, Istanbul, Moscow, Mumbai, Singapore, Sydney, Osaka, as well as of course, all other cities and their hinterlands. The changing position or location of these cities and the efforts of their policy makers to promote or delay these changes largely determines the benefits or costs that its inhabitants derive from their common participation in the global whole. And it is the analysis and understanding of the actual and potential place within the contemporary and near future cyclical moment in global development that can afford policy makers the intelligence on which to act. For instance, a change in global or regional economy can promote one or more cities at the expense of marginalizing other cities and exert direct effects on imperial or other political relations among these cities or between them and their respective hinterlands. Periods of global or regional economic, technological, demographic, military, political, cultural and other expansion offer opportunities for policy making success; while successive periods of contraction or crisis impose serious limits to policy makers. However, the same crisis in leading cities, economies, empires, etc. also offer some but never all intermediate economies and polities within the global network or its regional parts to improve their place and fortune within the whole. Much of the riches and other benefits of a city, region, and sector is derived less from its alleged internal, let alone pristine capacities and strengths than they are from gaining and maintaining a favoured position or location, within the whole from which the socio-political economic unit can derive benefits at the expense of those who are pushed into a position of disadvantage. This structure and function has been in place but evolving and self-transforming throughout Afro-Eurasia over several thousand years and globally for the past five hundred years. According to Frank "globalization is age old and has long been constructed through an ever changing network, especially within and among cities, which constitute the nodal knots in regional, inter-regional, and global networks of communication and other relations. The whole system of networks is greater than the sum of its urban, hinterland, and inter-urban parts, which are shaped and re-shaped by the structure and dynamic of the global system as a whole, to whose transformation the changing parts themselves also contribute".

At the same time, the growth of those in positions of advantage generates *entropy* or disorder, which compromises the continuance of such growth—unless that entropy can be and is dissipated or exported to cities, hinterlands, or other regions who are obliged to absorb this entropy and generate disorder due to their unfavourable position in the system as a whole. For instance, no large city could survive nor maintain mutually beneficial relations with others like it except for its ability to dissipate its own entropy to its immediate hinterland and/or to somewhere else half way around the globe. The economic and demographic growth in and of the industrialized cities would not have been possible without the dissipation of their entropy to other parts of the world who are obliged to absorb the ecological costs of a world *development* that has benefited and continues to do so to the few at the expense of the many. Frank cited the case of military production by industrial powers uses local and imported raw materials—and often brain drain—to produce at huge economic resource opportunity and environmental costs "goods" of no social utility whatsoever. Most of these are then exported back to the Second World countries, the suppliers of the primary raw materials. They pay for them with foreign exchange derived from their export of other still more commodities or goods of low value added, which they thereby deny to their own populations. Thus, starving Africans and Asians export foodstuffs to the rich. Entropy is thereby dissipated already through the transfer of exhaustive and polluting industries from the rich to the poor. But perhaps more serious even is the dissipation of socio-political entropy from the richer who sell their military hardware and training that aids them better to afford "democratic order" at home to the poorer abroad who import these arms and use them to kill each other in an entropy absorbing ever more chaotic "Second World". Even so, the arms producers keep enough of them for their own use to enforce, maintain and even further extend this exploitative and entropic world division of benefits for themselves at the enormous cost to everybody else. "That is called preserving human rights, freedom, democracy, civilization and most recently also combating terrorism" (Frank *et al.,* 1982)[8].

Myrdal in 1957 also reiterated the same impoverishment on a world scale in *Asian Drama* : *Rich Lands and Poor* [9] that in advanced social systems, each change propels growth and development in the direction of initial growth impulses, causing social processes to become cumulative and to gather speed at accelerating rates. Thus poverty stricken periphery—the *satellite*—remains impoverished, whilst the heartland as the centre of knowledge creation generate more knowledge: there develops a vicious cycle, the rich become richer and the poor becomes poorer.

Opposition to Modernization: The Concept of Under-development—From Regional to Class Exploitation: The Dependency Theory

In doing this, a virtual avalanche of critical writings appeared and the discussion of development is moved away from regional exploitation to "class exploitation". These writings claimed that development and prosperity of the "core" countries in the world economy (Western Europe and USA) were based on underdevelopment and squalor in the periphery. The periphery could not "follow" the previous historical experience of particular developed countries because its underdevelopment was a structural requirement for the development of the core. The development of the core, therefore, *depends* on the systematic underdevelopment of the periphery. By means of unequal trade, exploitation of labour, and profit extraction, the underdeveloped countries were becoming *increasingly* rather than decreasingly impoverished.

Three writers exemplify the alternative models of regional economic change that arose from this critique: Andre Gunder Frank, his *Dependency Theory* (1969),[10] and David Slater's *Internal Atomization* (1992),[11] and Krugman's *Theorization of Uneven Regional Development* (1995).[12] This left-wing view takes the form of A. G. Frank's *Dependency Theory* (1969), according to which, rich industrial countries or regions (the so-called *core*) extract the wealth generated by resource industries in developing countries (or, regions, the so-called *periphery*). As a form of neo-colonial domination, cities within the peripheral region serve the interests of the *core* and thereby facilitate the development process unleashed by capitalist firms. Such branch operations of foreign companies are located in these cities, the primary role of these cities is to function as a node for the exploitation of surrounding hinterland. From an ideological point of view, the functions of these centres are controlled by decisions made in the industrial core where the main headquarters of the foreign companies are located.

In Frank's dependency theory there have been proposals that each society's development problems could only be understood in relation to the rest of the world, that is, in a "world capitalist system", which originated in the sixteenth century. *Imperialism* as exploitative system of economic, social and political relations changed the colonial societies economically, socially and culturally, changed them in order that they could provide cheap inputs to production in the capitalist societies as well as markets for their products. This arrangement always worked to the advantage of the imperial power. Although the form of the monopoly power of core over periphery (metropolis over satellites) has changed (for example, with the switch from merchant to industrial capitalism in the nineteenth century and following political independence for former colonies), the system of "surplus expropriation" or transfer of wealth from satellites to metropolis has continued to fuel growth in the North at the expense of some Countries of the South. Frank argued further that "world capitalism was itself the cause of underdevelopment".

Frank draws a useful distinction between *un*development and *under*development. Undevelopment is an early stage of economic and social evolution, similar to Rostow's traditional society; underdevelopment, on the other hand, involves the systematic exploitation of the poorer countries by the richer ones. In other words, underdevelopment was not intrinsic but was *created.* Because the wealthy countries were never underdeveloped, stage theories drawn from an analysis of Western history are irrelevant.

From this perspective, many of those activities that modernization theorists propose as solutions—such as multinational investment or the Western education and training of elites of the South—are actually causes of underdevelopment. Only by delinking from world capitalism can a poor country hope to develop.

Slater (1975)[13] took up the logic of Frank's model and applied it specifically in case of the African countries of the South. In this case interest was to examine the spatial structure of underdevelopment in the African countries and how it has changed over time. Slater identifies four stages of spatial organization in African underdevelopment :

1. Pre-colonial spatial structure (pre-1800s).

2. Colonial penetration: Initial concentration and the beginning of internal expansion (1800-1914).
3. Colonial organization and continued extension (1919-60).
4. Neo-colonial concentration and limited attempts at restructuring (1960 to present).

The *first stage* is one of inter-regional trading networks rather than isolated local economies. In the *second stage* these networks are disrupted as capitalist penetration from the coasts' reoriented regional economies to long-distance external ties. The older inter-regional networks were dissolved both by the imposition of new political boundaries that bore little relationship to them and by the construction of railways to service new plantations, mining concerns and settler estates. In the *third stage,* export crop production was extended into new areas and transportation networks became clearly focussed on ports and colonial capitals. In the *final stage*, even with political independence, the *external dependence* and *internal atomization* continued.

These two above models of Frank and Slater are examples of what can be called "dependency theory". This has been a very influential approach to explaining global patterns of development and underdevelopment. It states essentially, that development and underdevelopment are reverse side of the same process. Development somewhere requires underdevelopment somewhere else. *Independent development is impossible.* Further work by Frank and Slater puts more emphasis on the nature of political and social structures created by external dependence and how these, in conjunction with the direct effects of external dependence limit "real" development. Recently Slater (1992) has explored the possibility that the forms of oppression and exploitation that appear so obvious in the periphery reveal "that which does not appear immediately visible in what seem to be more *normal* cases (the core). From this point of view the metropolis and its problems can be seen more clearly in its mirror, the satellites or periphery, than in itself. The essential spatial duality of development/ underdevelopment has not been transcended materially, it can only be inverted intellectually. *They* can teach us and *we* can learn from *them*, so to speak. But the material world is still one of a geographical of them and us.

In their structuralist models, Krugman, Frank and Slater argued that large-scale regions are more significant economic units rather than nation-states so that the regional agglomerations of economies are emphasized at the world scale, compared to their national level concentrations. This geographical revival has stressed the importance of shifts in the world economy and the rise of intra-industry and intra-corporate trade. Thus these models are based (without stating it) on an ideal of *perpetual spatial disequilibrium* between metropolis and satellites, core and peripheries. Given that there is a circular relation between production and demand, once the regions (or, the countries) get a head start in manufacturing, will attract industry and with forward and backward linkages, and economic distributions may be "locked in" through increasing returns. However, the logic and demands of world economy are in a continual state of reconstitution. The rise of the USA from satellite to metropolis, the challenge and collapse of the Soviet Union, the appearance of the newly-industrializing countries, NICs, and the increasing importance of multinational enterprises, MNEs, in the core as well as in the peripheries, are all contradictory to the "fact" of permanent spatial disequilibrium. Most important of all, the structuralist models completely subordinate them to a world economy in which the *identity* of regions of growth or exploitation, of development or underdevelopment, has been determined from the beginning and, therefore, they are a historical and static, whatever their references are to phases or stages.

Such an approach to the question of development differs in many ways from those that we have already looked at. And Countries of the South remain distinguished by *three features,* namely, (*i*) they remain primary producers and cannot control the prices of their products, (*ii*) their production remain determined by the demand for primary produce from the developed countries, and (*iii*) they do not have the concentration of industrial and technological developments of the developed countries. Taken together, all these three factors mean that "development" in the Southern cannot follow the same path as in the Northern countries.

Imperialism Dominance: Imperialism is a sophisticated mode of dominance in which the core country establishes a "bridgehead" in a peripheral nation. This is similar to the idea of *comprador elites i.e.,* a small sector of the native people who represent the interest of elites of the Global North in the Global South, so that the peripheral nation serve as a mere appendage of the core country. Inequalities of living standards will increase both within the peripheral country and between the periphery and the centre. Imperialism is not a stage, not even the highest stage, of capitalism: from the beginning, it is inherent in capitalism's expansion. The imperialist conquest of the planet by the Europeans and their North American children was carried out in two phases and is perhaps entering a third.

Imperialist exploitation takes place in three stages. (First) is a simple looting of raw materials with nothing given in return, as we see in the conquest of Latin America. (Second), something is offered in return, but the price is ridiculously low. Incipient political and economic development has brought the peripheral country's power level from zero to some low positive value, but there is still a marked inequality between the trading partners.

At the (third) level of exploitation, there may be some balance in the flow of goods between the periphery and the core, but there will be enormous differences in the *effects* of trade. Such differences result from the different levels of processing the goods involved in the exchange, and the *spin-off effects* of such processing. For example, a peripheral country may exchange oil with USA for tractors. Oil is a natural product that can be put on a ship for export with little or no processing. The extraction of oil provides relatively few jobs, requires little research, and needs no domestic auxiliary companies to support it. The spin-off effects of production are, thus minimal. Producing tractors, on the other hand, involves great amounts of labour, high levels of education for researchers and scientists, considerable job specialization, numerous auxiliary producers and suppliers, and a strong infrastructure of roads and railroads—that is, numerous spin-off effects. The production of raw materials may make good profits but is relatively self-contained, with little discrimination of benefits to other parts of the society or economy, and, thus leads to stagnation. High-tech production, however, has a great ripple effect throughout society, leading to increased development.

After this third stage of imperialism is reached, there is no need to use force to maintain it; rather, the structure of the system assures domination and inequality; though "professional imperialism" has not been perfected, and, therefore, direct violence may sometimes be required to keep the structure intact.

The Limits of Modernization

This produced something of a crisis in the field of development studies. Thus for many people in North America and Europe, when the 1960s were a period of considerable affluence, there have been the continuing failure of the Southern countries to cater for the needs of all their citizens or to achieve their own steady growth and development. Indeed, their development instead of adapting modern industrial technology successfully to their own particular ideological system, involve a problematic combination of Western form on their own non-Western form, which according to Irfan Habib, 1973[14] is a distinctly "Asiatic mode of production"—this is termed as "dual economy". The physical expression of this *dualism* is the foreign capitalistic enclave with a life of its own : the trading city, commercial agricultural estate, the mining centre, and all the modern transport and communication links that go with these. It is also manifested in the contrasting ways of life within the primate city: the Western-style central business district (CBD) as opposed to the "bazaar economy", scattered native markets and street-sellers; the modern manufacturing establishments in the port area and suburban industrial estates as compared with the cottage industries scattered throughout the indigenous residential sector.

Simultaneously in the developed world, there has been a rediscovery of poverty and an increased concern for state welfarism—the concept of "RELATIVE DEPRIVATION" was used to show how despite apparent affluence, many people remained poor and unable to participate fully in their societies—the explanations of all of these becomes inadequate by "unilinear modernization theory".

Intermediate, or appropriate technology, non-capital intensive but innovative agricultural methods, appreciation of the value and developmental possibilities of the Southern countries' medical, cultural and agricultural practices, all these areas now receive serious attention in developmental debates. The Global South has changed, is changing and continues to change. The best future that "development" may offer is some combination of "rediscovered", but appropriate "traditional" practices, ideas and technology, culture and ideas of the First World. And this can happen only if people make it happen and they can only do this if they have the social power to protect their long-term interests.

Opposition to Modernization: *World-System* (or, Diffusionist) Theory[15]

Another way of conceptualizing the interpretation of transmission of large parts of the world over the past three hundred years through a geographical historiography that is through history as well as by a geographical perspective (that is, in time as well as in space) is to be found in the writings of Immanuel Wallerstein. The modernization theorists not only insist that history of the modern world is the history of capitalist accumulation and the dynamics of development refer to the processes which maintain and transform the world-economy as a single system so that their model is relevant not only to capitalist-dominated economic structure but also that the fundamental international political issues revolve around the relations among the hegemonic economic powers of the world, acting as the "core" on which the economy of the other countries of the world depend as the destination of a large proportion of their exports (mainly by primacy products) *i.e.*, the source of their imports (capital goods), as well as a lender of their capital. Meier and Baldwin, 1957 were the earliest of the writers to conceptualize the cleavage of the economic world into this unusual camps of "core-periphery model". But in the "world-system" theory, for Immanuel Wallerstein (1975),[16] the primary focus of analysis is not on commercial relationships between the individual nations or even the relationship between nations, but the progressive incorporation of the world working into workings of a single global (Euro-America-centred) capitalist economy (in Western Europe from the sixteenth century onwards when political energy was used to secure monopoly rights outside the territorial boundary of the nation-state and the state became less the central economic enterprise than the means of ensuring certain terms of trade) with a global periphery, so that any country's development conditions and prospects are primarily shaped by economic processes and interrelationships operating at the global scale. This argument is well known, but what Wallerstein (1984)[17] stressed in his world-systems theory that though there have been major changes over time in the ebb and flow of power, the basic characteristics of this world system have remained relatively constant and "it is futile to attempt to analyze or shape development by focussing at the level of individual countries, each of which is deeply ingrained in the world system" since he believed that the relations of the hegemonic powers are only one of many issues that confront the world-systems.

The Modern World-System: An Architectural Digest

In his book, *The Modern World System: Capitalist Agriculture and the Origins of the European World Economy in the Sixteenth Century* Immanuel Wallerstein (1974) develops a theoretical_framework to understand the historical changes involved in the rise of the modern, capitalist world economy. Although he has been both prolific and wide-ranging in his intellectual labours, Wallerstein will doubtless be primarily judged for his 3 volumes of *Modern World System.*[18] It is in some ways unfair and unfortunate to appraise that work on the basis of only the first three volumes. But they provide at once the most comprehensive application of his evolving theoretical scheme, and the richest vein from which we can extract his methodological principles. The major purpose and accomplishment of Volume I is to demonstrate the cogency of the world-system perspective through a reinterpretation of the transition from feudalism to capitalism. It is thus more concerned with establishing the primacy of the whole as a viable and necessary unit of analysis than it is with the detailed reconstruction of how and why each part came to play the role it provisionally held by the end of the long sixteenth century—though to be sure there is plenty of the latter. Wallerstein thus focuses on the new kind of totality represented by this European world economy, on its axial division of labour and structure of strong and weak states, and then on their concrete historical trajectories, most of which becomes the semi-periphery, the ascending core states of the Dutch, England and France and the East European and Spanish American peripheries which he contrasts with as yet unincorporated "external arenas" of Russia and the Indian Ocean respectively. Along with the short concluding chapter seven, the second and third chapters are the most heavily theoretical, and the rest rather more empirical.

Medieval Prelude: Before the sixteenth century, when Western Europe embarked on a path of capitalist development, "feudalism" dominated West European society. Between 1150-1300, both population as well as commerce expanded within the confines of the feudal system. However, from 1300-1450, this expansion ceased, creating a severe economic crisis. According to Wallerstein, the feudal crisis was probably precipitated by the interaction of the following factors:

(*a*) Agricultural production fell or remained stagnant, agricultural productivity got decreased. This meant that the burden of peasant producers increased as the ruling class expanded, subsequently there was an increase in epidemics within the population.

(*b*) The economic cycle of the feudal economy had reached its optimum level; afterwards the economy began to shrink.

The European Economic Division of Labour : Wallerstein argues that Europe moved towards the establishment of a capitalist world economy in order to ensure continued economic growth. However, this entailed the expansion of the geographical size of the world in question, the development of different modes of labour control and the creation of relatively strong state machineries in the states of Western Europe. In response to the feudal crisis, by the late fifteenth and early sixteenth centuries, the *world economic system* emerged. This was the first time that an economic system encompassed much of the world with links that superseded national or other political boundaries. The new world economy differed from earlier multiple political framework of the earlier *empire systems* because it was not a single political unit. Empires depended upon a system of government which, through commercial monopolies combined with the use of force, directed the flow of economic goods from the periphery to the centre. Empires maintained specific political boundaries, within which they maintained control through an extensive bureaucracy and a standing army. Only the techniques of modern capitalism enabled the modern world economy in allowing the process of capital accumulation, and at the same time unlike earlier *empire systems*, to extend beyond the political boundaries of any one empire and at the same time to maintain a secular trend of stratified modern state system.

The new capitalist world system is based on an *international* (the core-periphery) *division of labour* that determines relationships between different regions as well as the types of labour conditions within each region. In this model, the type of political system is also directly related to each region's placement within the world economy. As a basis for comparison, Wallerstein proposes four different structural categories of the world-system: *core, semi-periphery, periphery,* and *external,* into which all regions of the world can be placed. These structural categories describe each region's relative position within the world economy as well as certain internal political and economic characteristics. The tendency of capitalism to expand has caused the multiple political framework in the state system to expand, so that now the entire globe is both capitalist and statist. But more important, the competitive dynamic of capitalism has also made it such that local entrepreneurs find the creation and maintenance of increasingly strong state structures to be their major organizational strategy to compete in the market. Finally, the operation of both capitalist accumulation and the multiple political framework of state system in these four structural categories represent the basic structure of stratification of the world-system, organizing, more than any other feature, the basic inequalities of territory, commodities, political power and wealth.

Basic Structures of World-System: The Concepts of Core and Periphery

Core : The core regions benefited the most from the capitalist world economy. For the period under discussion, much of northwestern Europe developed as the first core region. Politically, the states within this part of Europe developed strong central governments, extensive bureaucracies, and large mercenary armies. The trading nations of Western Europe (Spain, Portugal, England, France, Holland) encouraged trades and this permitted the local bourgeoisie to obtain control over international commerce and extract capital surpluses from this trade for their own benefit. As the rural population expanded, the small but increasing number of landless wage earners provided labour for farms and manufacturing activities. The switch from feudal obligations to money rents in the aftermath of the feudal crisis encouraged the rise of independent or yeoman farmers but squeezed out many other peasants off the land. These impoverished peasants often moved to the cities, providing cheap labour essential for the growth in urban manufacturing. Agricultural productivity increased with the growing predominance of the commercially-oriented independent farmer, the rise of pastoralism, and improved farm technology. And at the same time their state-centred mercantilist policies gradually gave way to capitalism as states shifted alliance to increasing reliance on the merchant class who functioned under state authority, and had developed more resources (by making profit internationally) than the monarchies so that the influence of the aristocracy diminished with the growing power of the merchants. The major functions of the states became the support of capitalists through controlling labourers' demands, protecting property rights, safeguarding foreign markets and trade routes, and opening new geographical areas to exploitation.

The *core* first formed as a small group of Western European nations intensively involved in foreign trade. After the industrial revolution, the world-economy of the core expanded by *incorporating rather bequeathing the remainder of the world in the order: Latin America, Caribbean, North America, India, East Asia, Africa, Australia, and finally the Pacific Island* so that the entire world is covered by about 1900 except few countries like China, Iran, Arab countries, Argentina. These incorporations took several forms. The simplest form of bequeathal was *"plunder"—this could be a short-term process to be supplemented by more productive activities involving new settlement of European colonies by destroying the native citizens, as in North America and in Australia. Alternatively, existing societies remained intact but they were peripheralized in the sense that their economies were reoriented to serve wider needs within the world-economy. This could be achieved through political controls as in India or indirectly through "opening up" an area to market forces as in China. The end result of these various incorporation procedures was the eventual elimination of the external arena.*

The world-economy use the terms "core" and "periphery" as complex processes and not directly to areas, regions or states—the latter become only core- or periphery-like because of a predominance of core or peripheral processes operating there. This is "not a semantic point but directly relates to the way in which the spatial structure is modelled over time to produce a single space-time framework. The core came to include all the fully industrialized nations. These nations are characterized by wealth, capital-intensive production (increasingly focussed on high-tech manufactures, technology, services, and communications) and a relatively small proportion of workers in agriculture.

The Periphery : On the other end of the scale lay the peripheral zones. These areas lacked strong central governments or were controlled by other states, exported raw materials to the core, and relied on coercive labour practices. The core expropriated much of the capital surplus generated by the periphery through unequal trade relations. Two areas, Eastern Europe (especially Poland) and Latin America, exhibited characteristics of peripheral regions. In Poland, kings lost power to the nobility as the region became a prime exporter of wheat to the rest of Europe. To gain sufficient cheap and easily controlled labour, landlords forced rural workers into a "second serfdom" on their commercial estates. In Latin America, the Spanish and Portuguese conquests destroyed indigenous authority structures and replaced them with weak bureaucracies under the control of these European states. Powerful local landlords of Hispanic origin became aristocratic capitalist farmers. Enslavement of the native populations, the importation of African slaves, and the coercive labour practices such as the *encomienda* and forced mine labour made possible the export of cheap raw materials to Europe. Labour systems in both peripheral areas differed from earlier forms in medieval Europe in that they were established to produce goods for a capitalist world economy and not merely for internal consumption. Furthermore, the aristocracy both in Eastern Europe and Latin America grew wealthy from their relationship with the world economy and could draw on the strength of a central core region to maintain control. History cannot be denied, the *periphery structure* put in place by colonialism, as described above persist, but in changed form: the role of *export dependency,* that the underdeveloped, labour-intensive countries such as Ghana, Ethiopia etc. are carrying on to supply unprocessed agricultural goods and other primary sector (like fishing, lumbering and mining) products to the core, illustrates their peripheral character of economy: these countries cannot accumulate sufficient capital for modernization. Money and resources needed for investment flows into the hands of the core capitalists. What investment exists—for example, in mining or manufacture for export—is for the benefit of the core. Thus the concept of peripheralization implies that these areas did not join the world-economy as "equal partners" with existing members but that they joined on unfavourable terms. One of the few defences that the periphery has against such exploitation is the erection of protectionist barriers against core products and investment.

Like all core-periphery models, Wallerstein defines the process of operations of core and periphery—"the core exploits and the periphery is exploited". But *core-periphery pattern* is not a static one in Wallerstein's model because they occur through the different processes operating in different geographical zones, areas, regions or states—the latter become only core- or periphery-like due to the predominance of core or peripheral processes as "the opposite types of complex production relations" dominating in those regions respectively. This is *not a trivial semantic point but directly relates to the way in which the spatial structure is modelled. Space itself can be neither core nor periphery in nature—rather they are core and periphery processes which structure space so that at any point in time one or other of the two processes predominates.* In simple terms, *core processes consist of relations that incorporate relatively high wages, advanced technology and a diversified production mix* whereas periphery processes involve *low wages, rudimentary technology and a simple production mix.* These are general characteristics, the exact nature of which changes constantly with the evolution of the world-economy. It is important to understand that these *core* and *periphery* processes are not determined by the particular product being produced.

The Concept of Semi-periphery Category: Core and periphery do not exhaust Wallerstein's concepts for structuring space. Although these processes occur in distinctive zones to produce relatively clear-cut contrasts across the world-economy not all zones are easily designated as primarily core or periphery in nature. One of the most original elements of Wallerstein's approach is his concept of the semi-periphery. Between the two extremes lie the semi-peripheries. These areas represented either core regions in decline or peripheries attempting to improve their relative position in the world economic system. They often also served as buffers between the core and the peripheries. As such, semi-peripheries exhibited tensions between the central government and a strong local landed class. Good examples of declining cores that became semi-peripheries during the period under study are Portugal and Spain. Other semi-peripheries at this time were Italy, southern Germany, and southern France. Economically, these regions retained limited but declining access to international banking and the production of high-cost high-quality manufactured goods. Unlike the core, however, they failed to predominate in international trade and thus did not benefit to the same extent as the core. With a weak capitalist rural economy, landlords in semi-peripheries resorted to sharecropping. This lessened the risk of crop failure for landowners, and made it possible at the same time to enjoy profits from the land as well as the prestige that went with landownership.

According to Wallerstein, the semi-peripheries were exploited by the core but, as in the case of the American empires of Spain and Portugal, often were exploiters of peripheries themselves. Spain, for example, imported silver and gold from its American colonies, obtained largely through coercive labour practices, but most of this specie went to paying for manufactured goods from core countries such as England and France rather than encouraging the formation of a domestic manufacturing sector.The *semi-periphery* is neither core nor periphery in nature but includes characteristics of core and periphery, for example, by poverty, primary product reliance for trade and vulnerability to outside decision-making. Thus the semi-periphery is the "halfway house" between the other two. But it is more than that —it serves as a *buffer between the core and periphery,* keeping the world system away from disintegrating. Wallerstein (1975) suggested that the coherence of this category is derived from the fact that semi-periphery is formed by the more advanced exemplars of present-day's dependent development on the core, for example, both high levels of industrialization and large marginalized populations like Canada (or, Australia) which have more independence than periphery states and may act as regional powers. These countries serve also as a location to which production is transferred when costs of production increase in the core. They also have capital on their own, but are nevertheless dependent on the core for much of their infrastructures. Again these semi-peripheral countries are having "staples", for which they have to be still heavily dependent on the core for much of their infrastructures. However, for their *staples* because they are still heavily dependent on trade, they are forced into a delicate balancing act through exporting mixed high-tech and low-tech goods both to the core as well as to the periphery, asserting their own sovereignty while placating (*i.e.*, satisfying) the core and at the same time suffering exploitation by the core—in that way the semi-periphery is the most turbulent category, in that its members rise or fall in the global hierarchy. Potentially they tend to be closely allied with the core and might depend on it for military protection. Much of Canada's economic history has been understood in terms of these global trading pressures, shaped up by industrial expansionism, setting up demand for its different "staple" products from Canada (fish, furs, timber, wheat and minerals) with different staples shaping, changing social, economic, and political institutions within Canada as the country shifted from reliance on one staple export to another.

External Areas: These areas maintained their own economic systems and, for the most part, managed to remain outside the modern world economy. Russia fits this case well. Unlike Poland, Russia's wheat served primarily to supply its internal market. It traded with Asia as well as Europe; internal commerce remained more important than trade with outside regions. Also, the considerable power of the Russian state helped regulate the economy and limited foreign commercial influence.

Cycles of the World-System

Understanding the world-system involves also more than analyzing the processes that maintain its basic structures. In this world-system, modern political economy is further characterized by regular cyclical rhythms which are asserted to be a third parameter of the system, constitutive of it rather than merely incidental to it. Wallerstein has been relatively uninterested in short-term business cycles, but he identified three basic cyclical processes: Wallerstein was struck by the (one) regularities of the long waves emphasized by Kondratiev (from forty to fifty years in length), (two) of the even longer swings of distribution of power among core states between periods of rivalry, and (three) alterations in the core-periphery control structure between colonial control and informal control. But the data for these have not yet been treated as world-system data, and so far these three cycles are at best gross summaries of diverse movements in different accounting units (mostly states), so that they seem to be empirically correlated, such that cyclical regularities systematically affect the differentiated zones of the world-economy. The periods of expansion culminate in the rise of a hegemonic core power and their informal controls while periods of economic stagnation lead to contraction because of competition among core powers for more direct colonial-type controls so that, most peripheral and older semi-peripheral zones are hardest hit, while core and new semi-peripheral zones do relatively better.

The cyclical rhythms of the modern world-system have been of great interest in much of the research on world-systems and they have been used to explain a number of different phenomena. For example, these cycles have been used to account for the cyclical pattern in international trade between periods of relatively free trade and periods of relatively high tariff. The determination, measurement and consequences of these three cycles as a basic set of development dynamics in the world-system provide the basis for Wallerstein's periodization of modern history, and hence for the temporal boundaries of his originally projected four (now five?) volumes. The period of growth and expansion called "the long sixteenth century" (1450-1620/40) is covered in the first volume; the corresponding period of contraction and stagnation (1600-1730/50) in the second. Economic cycles give part way to political in the dating of the third and fourth stages. Stage three (1750-1917) is marked by the growing predominance of industrial over agricultural capital, the globalization of the capitalist world economy, and the rise and demise of British hegemony. And stage four (since1917) witnessed the consolidation of the global world-system, paradoxically through the growth of revolutionary socialist challenges to that system which resulted in the existence of so-called socialist semi-peripheries whose ruling groups had a stake in the continued existence of the system.

Understanding these long cycles are not only basic to the understanding in functioning of the world-system but also to the operation of capitalism. Because of the structured separation of economics from politics, the anarchy of the market, periods of expansion come to an end when production outstrips the world distribution of income and hence effective demand. In fact, the periodic crises are worsened by the initial response of individual producing units to declining profits. That response, typically, is to expand overall production in order to maintain an absolute profit level, and necessarily results in an increased problem of over supply. The ensuing severe crises are marked by increased concentration and centralization of capital, by geographic expansion, by technical change, and by struggles leading to sufficient income redistribution on a world scale to permit a renewal of capitalist expansion as well as long-term developmental changes in the world-economy. For example, the crisis of the late nineteenth century saw the rise of corporations, cartels, and trusts; the "scramble for Africa" and frontier expansion in the Americas, Australia, and Asiatic Russia; the invention of the internal combustion engine and of electrical power; and the achievements of social democracy: higher wages for core industrial workers and the benefits of the welfare state.

The resolution of periodic crises, then, furthers the movement of the *fourth parameter* of the modern world-system, which Wallerstein argues, as the *secular trends in world economy*. These trends are conceived as *both* the self-reproducing *and* the self-transforming aspects of the entire structure, with asymptotic limits that hypothetically *pre-sage its demise*. That is, capitalism is understood as a system in the process of slowly transforming itself in the direction of limits beyond which it cannot pass: three secular trends are typically distinguished—(one) the expansion of the size of the world-system, (two) the interrelated increase in commodification, in capital concentration and bureaucratization, (three) increase in mechanization—as constitutive of this moving system, these secular trends represent the cyclical rhythms, the interstate competition, and the dividing and redividing of labour. These three secular trends represent the fourth basic set of development dynamics of world-economy, since they are having finite limit, they would seem to necessarily involve the transformation of the world-system itself so that institutional competition within the world-economy may get transformed into conflict —this may result in increasing political resistance bringing a transformation of the present world-system.

Stages of Growth in World-System

The development of the modern world economy lasted centuries, during which time different regions changed their relative position within this system. Wallerstein divides the history of the capitalist world-system into four stages, which for our purposes can be simplified and divided into two basic phases:

Stages 1 and 2 (1450 to seventeenth century) : This period follows the rise of the modern world-system between 1450-1670. When all the existing Western European states attempted to strengthen their respective positions within the new world-system. In order to accomplish this move, most of the states consolidated their internal political, economic and social resources by:

(*a*) Bureaucratization: This process aided the limited but growing power of the king. By increasing the state power to collect taxes, the kings eventually increased state power to borrow money and thereby further expand the state bureaucracy. At the end of this stage, the monarch had become the supreme power and instituted what has been called "absolute monarchy."

(*b*) Homogenization of the local population: To underline state involvement in the new capitalist system and encourage the rise of indigenous capitalist groups, many core states expelled minorities. These independent capitalist groups, without deep rooted local ties, were perceived as threats to the development of strong core states. The Jews in England, Spain, and France were all expelled with the rise of absolute monarchy. Similarly, Protestants, who were often the merchants in Catholic countries, found they were targets of the Catholic Church. The Catholic Church, a trans-national institution, found the development of capitalism and the strengthening of the state threatening.

(*c*) Expansion of the militia to support the centralized monarchy and to protect the new state from invasions.

(*d*) The concept of absolutism introduced at this time related to the relative independence of the monarch from previously established laws. This distinction freed the king from prior feudal laws.

(*e*) Diversification of economic activities to maximize profits and strengthen the position of the local bourgeoisie.

By 1640, northwestern European states secured their position as core states in the emerging economy. Spain and northern Italy declined to semi-peripheral status, while northeastern Europe and Iberian America became peripheral zones. England gained ground steadily toward core status. During this period, workers in Europe experienced a dramatic fall in wages. This wage fall characterized most European centres of capitalism with the exception of cities in north and central Italy and Flanders. The reason for this exception was that these cities were relatively older centres of trade, and the workers formed strong politico-economic groups. The resistance of workers broke down the ability of employers to accumulate the large surplus necessary for the advancement of capitalism. Meanwhile, employers in other parts of Europe profited from the wage lag by accumulating large surpluses for investment. Long-distance trade with the Americas and the East provided enormous profits, in excess of 200 per cent to 300 per cent, for a small merchant elite. Smaller merchants could not hope to enter this profiteering without substantial capital and some state help. Eventually, the profits of the trans-Atlantic trade filtered down and strengthened the merchants' hold over European agriculture and industries. Merchants with sufficient power accumulated profits through the purchase of goods prior to their production. By controlling the costs of finished products, merchants could extend their profit margin and control the internal markets. This powerful merchant class provided the capital necessary for the industrialization of European core states.

Stages 3 and 4 (eighteenth century and beyond) : Industrial rather than agricultural capitalism represented this era. With the shifting emphasis on industrial production, the following reactions characterized this period.

(*a*) European states participated in active exploration for the exploitation of new markets.
(*b*) Competitive world systems such as the Indian Ocean system were absorbed into the expanding European world system. With the independence of the Latin American countries, these areas as well as previously isolated zones in the interior of the American continent entered as peripheral zones in the world economy. Asia and Africa entered the system in the nineteenth century as peripheral zones.

(*c*) The inclusion of Africa and the Asian continents as peripheral zones increased the available surplus, allowing other areas such as the USA and Germany to enhance their core status.

(*d*) During this phase, the core regions shifted from a combination of agricultural and industrial interests to purely industrial concerns. Between 1700, England was Europe's leading industrial producer as well as the leader in agricultural production. By 1900, only 10 per cent of England's population was engaged in agriculture.

(*e*) By the 1900s, with the shift toward manufacturing, core areas encouraged the rise of industries in peripheral and semi-peripheral zones so that they could sell machines to these regions.

Basic Orienting System

Wallerstein moved toward a conception of the modern world economy as the necessary inclusive "totality" from which the analysis of change in any individual country should proceed—A totality (or, mode of production) is the basic unit of analysis for study of change, because it is the arena within which the basic determinants of change are located. This world-systems theory entails (*i.e.*, necessitates) both a critique of the mainstream of social science (including both Weberian and Marxian forms) and a novel understanding of the history of capitalism. In his theory, Wallerstein critiqued mainstream social science both for its ahistoricism and for its uncritical equation of the notion of society with the spaces of nation-states. This theory is a development of dependency paradigm by drawing heavily on neo-Marxists from the Latin American dependency theorists, like A. G. Frank ("satellization" which for Wallerstein "layer within layer"). Adumbrated by Paul Baran in the 1950s, dependency theory was advanced in the 1960s by a number of prominent left-radicalists like Samir Amen ("peripheral capitalism"), Fernando Hendrique Cardoso ("associated development" to explain Brazilian exceptionalism), Arghri Emmanuel ("unequal exchange") in which the primary focus is not on individual nations or even the relationship between nations, but the workings of a single global capitalist economy, which has emerged historically four hundred years ago by the Western European nation-states through theirs practising the doctrine of mercantilism—for the latter, the fundamental principle was the emphasize the centrality to life of capital accumulation by not being confined within the boundaries of theirs own but through trade with other countries. As elaborated in a number of articles since 1974, Wallerstein's research agenda on world-systems includes five major subjects: (1) functioning of capitalist world economy as a system, (2) the how and why of its origin, (3) its relations with non-capitalist structures before it became a fully global system, (4) the comparative study of historical alternative modes of production, and (5) the ongoing transition to socialism. He claims further that in order to carry out this five-fold agenda, we require to redo our historical narratives, accumulating new world-systemic quantitative for reviewing and refining our conceptual baggage. In this context, he went by reconceptualizing the historiography with his reinterpretation. The basic orienting concepts what he did grouped are: "totality", axial division of labour, international-state system, cyclical rhythms, secular trends and antimonies (or, contradictions)—All these except *totality* apply primarily to the work on modern social change which according to Wallerstein *totality* are of three types, with a fourth possibility in the future. Karl Polanyi, who argued that capitalist development leads to constant political attempts to destroy capitalism itself, made a more fundamental contribution to world-systems analysis. For it was Polanyi who worked out the notion of three basic modes of economic organization, or types of social economy, which he termed reciprocal, redistributive, and market modes. These have become rather without modification Wallerstein's three types of totality: mini-systems, world-empires, and world-economies. Along with this great debt, however, the heritage of Marxism is at least as great. (First), the fundamental reality for Wallerstein is social conflict among materially based human groups; (Second), shared with Marxism is the concern with the relevant "totality"; (Third) is the sense of the transitory nature of social forms and theories about them; and (Fourth) is the centrality of the *accumulation* process, along with the competitive and class struggles it engenders. Here the stress is on the novelty of capitalism as a social organization, structurally based in the ever-renewed search for profit, and originating in the sixteenth century "primitive accumulation" through the expansion of Europe (colonies, precious metals, slave trade) and the reorganization of agricultural production (enclosure, capitalist ground rent). Also from Lenin, Wallerstein appropriates the centrality of both inter-imperialist rivalry—he will give it the somewhat antiseptic term "core competition"—and anti-imperialist revolution to the twentieth century. From Mao Ze-Dong, Wallerstein seizes upon the notion that class struggle continues after "socialist" revolutions, and moves toward an interpretation of Soviet "revisionism" as a consequence of the U.S.S.R.'s approaching core status in the still capitalist world-economy. At the same time, he insists upon a nuanced view of twentieth century socialist revolutions as both consolidating world capitalism and providing momentum for its undoing.

In addition, Wallerstein of course draws heavily on the neo-Marxist tendency known as dependency theory. Adumbrated in the 1950 by Paul Baran,[19] dependency theory was advanced in the following decade by a number of prominent left social scientists, among them Samir Amin, Fernando Henrique Cardoso, Theotonio dos Santos, Arghiri Emmanuel, Andre Gunder Frank, and Dudley Seers. The critical entering wedge against "developmentalism" (the idea that each national society could develop in basically the same way), dependency theory focussed on the centrality of core-periphery relations for understanding the periphery. From Amin came "peripheral capitalism," from Cardoso "associated dependent development" to explain Brazilian exceptionalism, from Emmanuel "unequal exchange," from Frank "satellitization" (Wallerstein's "layers within layers"), from Seers the "open" economy. It remained for Wallerstein to make the core-periphery relation critical for understanding the core. In any case, it is from these sources that Wallerstein derives his sense of global class struggle—with its frequently nationalist guise.

Following Polanyi, Wallerstein took three historical modes of totality: "minisystem", "world empire" and "world economic system".

Minisystem: Viewed that way, at one time all societies were having an entity that had within it a complete division of labour, a single cultural frame and a single polity in which based upon reciprocal-lineage modes of production, small kinship based groups dominate in local territories; this system he called "*minisystem*"— such minisystem ranged from primitive hunting and gathering economy to agricultural societies having sedentary practices. The minisystem as the classical subject matter of anthropology, have now been swallowed up by capitalist expansion—with increasing population, the "law of diminishing returns" started which provided an economic imperative for territorial expansion so that *agricultural hearths* got tied to the "world empires", that is, to the unite with a single division of labour but multiple cultural system. At this stage, according to Wallerstein, the minisystems got integrated into two example of world-systems: the unit (one) with a single division of labour but multiple cultural systems what he called as "world-empires" that covered extensive regions and the (other) with a common political system, there is a "world-economies"; both are world-systems. They are distinguished from world-systems in that they both involve multiple cultures, the former differs from the latter in the essential fact that they lack political integration, the former has a single political centre while the latter has multiple centres of different strength.

World Empires: World empires according to Wallerstein, are based upon "redistributive-tributary modes of production"[20] where large numbers of agricultural producers generate a surplus sufficient to allow for artisans and a military-religious-bureaucratic ruling class; these two groups are linked together by the forcible appropriation of surplus (*tribute*, hence the alternative designation, "tributary mode") and its redistribution to a stratum of rulers, if not more, for political purposes. A wide range of historical social formation is thus involved under this rubric, from the simple linkage of two tribal groups by the exchange of tribute for "protection" all the way to the far-flung, long lasting high civilisations of China or Rome. The basic pattern of change in world-economies is one of cyclical expansion and contraction, both spatially and temporally: expansion until the bureaucratic costs of appropriating tribute outweighed the amount of surplus so appropriated. On the other hand "world econom-y (-ies)" is (are) an entity based upon the capitalist mode of production based on profitability and are thus integrated through the market rather than by a single political centre. In this type of social system, two or more distinct economical and cultural regions are interdependent with respect to necessities such as food, fuel and protection, and two or more polities compete for domination without a single centre's emerging quasi-permanently. Before the modern period, world-economies tended to become world-empires: the classical trajectory from the multiple polities of Greece to the single imperium of Rome is the best known case. According to Wallerstein, the distinctive key to the dynamism of the modern world's on-going-inter-state competition within the framework of a single division of labour. This structure alone puts a premium on technical and organizational innovations that give groups the opportunity to advance their interests, prevents the total freezing of the factors of production by a single system-wide political elite, and denies to the exploited majority a focus for its political opposition. Thus structure stands in sharp contrast to that of world-empire, with their well-known technical sluggishness, "target-tribute" surplus appropriation, and vulnerability to rebellion or conquest. The European capitalist world-economy is the only one that has survived the vulnerable first stage of development and has been able to expand to eliminate other mini-systems and world empires.[21]

According to Wallerstein until the advent of capitalism in Western Europe, world-economies were unstable and tended towards disintegration or conquest by one regional group and hence, transformation into *world-empires*. What Wallerstein distinguishes, that in the government-centred mercantilist policies of the latter which gradually gave way to capitalism as the nation-states shifted to increasing reliance on the merchant class, which had developed more resources than the monarchies, the influence of the aristocracy diminished with the growing power of the merchants. The major functions of the state became the support of the capitalists through controlling labourers' demands, protecting the property rights, safeguarding foreign markets and trade routes, and opening new geographical areas to exploitation.

The system works through the market and not through states. Although there is only one world economy, there are numerous political units. It is the competition among sovereign nation-states that maintains the economic system which is based on "capitalism" and capitalism is simply a system in which the owners of the means of production—states, private companies, individuals—compete with each other in order to maximize profit and this capitalist economy did not remain confined within the boundaries of nation-states. Thus though the world economy systems existed also in ancient empires of

Hindu, Roman and China which commanded extensive regions but the major changes which have taken place over time in the ebb and flow of power, that what distinguished them from the European ones is that in contrast to the latter, the wealth was collected within the boundaries of their empires and appropriated by a military-bureaucratic ruling class of the overarching imperial states which undermined the growth of their territories. While the exchange between producers and artisans remained reciprocal, but because of the appropriation of the wealth, there had been a large-scale material inequality in all such "civilizations" so that there was the same material base to all the societies in spite of political constructs between them.

World-System: A modern capitalist world economy (or, world-system, the *third type* of totality) in Wallerstein's view is more able than a world empire to experience sustained economic development precisely since it is based upon six basic features of the world economy because (firstly) the capitalist mode of production is having a single world market; (secondly) in this mode there is no overarching political structure; (third) economic actors have more freedom to maneuver for whom the "criterion for production is profitability"; (fourth) producers exchange what they produce for the best price they can get, (fifth) the basic drive of the system is to appropriate, not for their consumption but to reinvest surpluses as *capital* for further economic gains. And, finally (sixth) competition between different units of production is ultimately controlled by the cold hand of the market so that the basic rule is either to accumulate or to perish. As the price of a product is not fixed but set in a geographically—extensive (world) market, there is competition between producers at least for periods of time. More efficient producers can undercut other producers to increase their share of total production and achieve monopoly. Such a world economy according to Wallerstein is the *capitalist world economy* which has been sustaining since the beginning of capitalist economy in the sixteenth century in Western Europe, and, even further back, to at least 1450, when international trade began to grow to shape the modern capitalist world economy.

Finally there remains a *fourth type* of totality, at this time only a possibility, but one which Wallerstein believes and hopes is in the cards. This he has called a *socialist world-government*. Presumably this form of organization would differ from a world-empire in that the production, appropriation and redistribution of world surplus would be collectively and democratically decided upon not by a bureaucratic stratum but by the world producers in accordance with an ethic of use value and social equality. Although such an alternative mode is barely on the horizon, Wallerstein finds the world currently in a transition to socialism. He has recently adumbrated its likely features in print, in *Utopistics* (2000),[22] but he found no more than prefigurings in certain unspecified practices of previously existing *soi-disant* socialist states. He roundly denounced the Soviet claim to have established a socialist world system separate from and alternative to the still existent, singular capitalist one, and the USSR's demise did not come as a great surprise to world-system scholars.

Although some of the remaining orienting concepts have meaning when applied to more than one kind of totality, they have been worked out to apply primarily to world-economies, particularly to the capitalist world economy or modern world-system. By "division of labour" Wallerstein and his collaborators mean the forces and relations of production of the world economy as a whole. They distinguish five aspects of the division of labour: core and periphery; commodity chains; semi-periphery; unequal exchange; and capital accumulation.

The central relation of the world-system perspective is that of *core* and *periphery*, geographically and culturally distinct regions specializing in capital-intensive (core) and labour-intensive (periphery) production. In Wallerstein's view neither can exist without the other: it is a relational concept describing a relational reality. The notion of "commodity chains" has been put forth to describe the production of goods as they move from raw cotton from colonies becoming Manchester textiles, peasant-grown Colombian coffee becoming Detroit labour-power, and so on. Thus twentieth century "backwardness" is seen not as the result of a late start in the race to develop but as the continued deepening of a long-standing structural relation. As the capitalist world-economy expanded over the past four centuries, new areas, formerly external, have been incorporated into the system, almost all as peripheries. At the same time, new productive activities have developed and the location of different core and peripheral production processes has shifted. The crucial element is not the product itself (although for many purposes the famous "raw materials exchanged for manufactures" is a convenient shorthand), but rather the capital intensity and skill level of the production processes themselves (U.S. wheat and Indian wheat are not the same). Hence textile manufacture was largely a "core" process in 1600 involving a relatively high capital/labour ratio and some of the most skilled workers of that era. By 1900 it was becoming a semi-peripheral process and by 2000 is well on its way to being a peripheral one, whereas computer software production has become emblematic of core economic life.

If the idea of core-periphery relations derives in large part from the dependency theorists, the notion of "semi-periphery" is original with Wallerstein. At once economic and political, it is a conception around which there is much fuzziness and disagreement, like that of "middle class". On the political side, semi-peripheral states are said to stabilize the world-system through sub-imperial practices, deflecting and absorbing some of the peripheral opposition to the core. On the economic side, there are two overlapping and not incompatible aspects of semi-peripheral zones. As implied above they are on the one hand recognized as *intermediate* between core and periphery in terms of the capital intensity, and skill and wage levels of their production processes, as with sharecropping in sixteenth century Italy or textile mills in early twentieth century Japan. On the other hand, they are characterized by "combined" development, the coexistence of some core-like and some peripheral production, with their trade flowing simultaneously in two directions as they export little-processed materials to the core and simple manufactures to the periphery. Movement into and out of semi-peripheral status is possible for a region or state, both from above and from below. But upward movement is difficult, and it is further argued that the upward movement of some is largely (though not wholly: the system does expand) at the expense of the downward movement of others. Thus one is led to anticipate such phenomena as the recent rise of some semi-peripheral states or regions (Taiwan, South Korea) along with the stagnation or decline of others (Iraq, Colombia).

By "*unequal exchange*" Wallerstein refers to the processes or mechanisms (*e.g.*, "transfer pricing") that reproduce the core-periphery division of labour. While there is much continuing debate over the precise nature of the mechanisms, they do result in the systematic transfer of surplus from the subsistence and semi-proletarian sectors located in the periphery to the high-technology, more fully proletarianized core. In this way one finds continuing and increasingly higher levels of living in the core, where high wages, worker political organization, and surplus capital combine to produce pressure for ever greater technical advance. This in turn tends to increase the differentiation between core and periphery.

By *capital accumulation*, finally, Wallerstein refers to the basic process expounded by Marx, with two important differences from the mainstream of received Marxist opinion. First, the accumulation process is seen as a *world* process, not a series of parallel national processes. Second, it is necessarily involved with the appropriation and transformation of peripheral surplus: geographic expansion and on-going primitive accumulation are not incidental to, but necessary and integral to the very constitution of capitalism. Here Wallerstein argument is formally and substantively isomorphic with the socialist-feminist position on housework. This is the source of Wallerstein's controversial assertion that not wage-labour alone, but the combination of waged and non-waged labour is the essence of capitalism. "When labour is everywhere free, we shall have socialism".

So much for the division of labour, the basically economic side. But what of the political? A second fundamental family of concepts revolves around the structure of the international state system. A basic parameter of the capitalist world-economy, the competitive system of states provides an escape from stagnation by providing for capital the freedom to escape political restrictions threatened or imposed by other social forces. Wallerstein and his collaborators have singled out three aspects of the state system as crucial: imperialism, hegemony, and class struggle.

Imperialism refers to the domination of weak peripheral regions (whether they are states, colonies, or neither) by strong core states. Semi-peripheral states are typically in between. Since state strength is a function of productive base, military organization, diplomatic alliances, and geopolitical situation, an occasional newly incorporated territory may have a sufficiently strong state to enter the world-system in a semi-peripheral position; such was the case with both Russia in the eighteenth century and Japan in the nineteenth. States are used by class forces to distort the world market in their favour and may also initiate such distortion by the use of force or diplomacy. They enforce and reinforce mechanisms of unequal exchange. As for the internal structure of states, core-states are far stronger than peripheral ones, as the ruling classes of the latter typically want to preserve open economies and to enjoy local power. But to say strong states is not to say necessarily strong or highly centralized executives (monarchs, presidents).

The very strongest states are those which need the least administrative and military machinery to do the jobs of internal coordination and external depredation. Also, there appears to be some regularity in the alternation of informal empire and formal colonization as modes of imperialism. *Hegemony* refers to the recurrent situation of one core state temporarily outstripping the rest. A hegemonic power is characterized by simultaneous supremacy in production, commerce, and finance which in turn support a most powerful military apparatus. The Dutch enjoyed such supremacy in the middle of the seventeenth century, the British achieved the same in the nineteenth century and the USA in the twentieth. Interestingly, the hegemonic power is characterized by a rickety and rather decentralized state apparatus in contrast to rival core states, having to overcome the least internal resistance to an aggressively expansionist foreign economic policy. Hegemonic powers take responsibility for maintaining a stable balance of power in world politics and for enforcing free trade, which is to its advantage so long as its economic advantage lasts. Each hegemonic power thus has contained the world financial centre of its time, a pattern which goes back to late medieval Venice, and, passing through Genoa and Antwerp, takes its characteristic modern form in Amsterdam, then London, and at present New York—which ironically started its modern career as New Amsterdam. The rise of a hegemonic power to its accepted place is typically preceded and accompanied by world wars (1618-1648, 1756-1763 and 1792-1815, 1914-1918 and 1939-1945). And precisely because the hegemonic state is never strong enough to absorb the entire world-system but merely to police it, the phenomenon of hegemony is a necessarily temporary one, due to three processes. (First), class struggles raise the wage level, lessening the competitive advantage, so that the hegemonic power can no longer undersell its rivals in the world market. (Second), technical advantages are diffused through imitation, theft, or capital export. (Third), world-wide technical advance makes possible the effectiveness of larger political units.

The final crucial aspect of the state system is the prominence of *class struggle* as the stuff of politics within and across state boundaries. Here the unconventional implications of the world-systems perspective are three. (First), much attention is directed to the existence of class alliances across state boundaries, as bourgeois in several locations collude to protect surplus appropriation even as they compete over relative shares. (Second), states themselves are conceived as mediating actors within the grand drama of a world-economy-wide class struggle. (Third), Wallerstein is impatient with descriptive occupational/class terminology (*e.g.*, slave owner, professional, peasant) that reifies social types at the expense of understanding the consequences of their activities for the operation of world capitalism. Wallerstein insisted on a high degree of interconnectedness between the division of labour and the international state system. Hence the phrase "political economy of the world-system".

Theoretical Reprise

The capitalist world economy, as envisioned by Wallerstein, is a dynamic system which changes over time. However, certain basic features remain in place. Perhaps most important is that when one examines the dynamics of this system, the core regions of northwestern Europe clearly benefited the most from this arrangement. Through extremely high profits gained from international trade and from an exchange of manufactured goods for raw materials from the periphery (and, to a lesser extent, from the semi-peripheries), the core enriched itself at the expense of the peripheral economies. This, of course, did not mean either that everybody in the periphery became poorer or that all citizens of the core regions became wealthier as a result. In the periphery, landlords for example often gained great wealth at the expense of their underpaid coerced labourers, since landowners were able to expropriate most of the surplus of their workers for themselves. In turn in the core regions, many of the rural inhabitants, increasingly landless and forced to work as wage labourers, at least initially saw a relative decline in their standard of living and in the security of their income. Overall, certainly, Wallerstein sees the development of the capitalist world economy as detrimental to a large proportion of the world's population.

Through this theory, Wallerstein attempts to explain why modernization had such wide-ranging and different effects on the world. He shows how political and economic conditions after the breakdown of feudalism transformed northwestern Europe into the predominant commercial and political power. The geographic expansion of the capitalist world economy altered political systems and labour conditions wherever it was able to penetrate. Although the functioning of the world economy appears to create increasingly larger disparities between the various types of economies, the relationship between the core and its periphery and semi-periphery remains relative, not constant. Technological advantages, for example, could result in an expansion of the world economy overall, and precipitate changes in some peripheral or semi-peripheral areas. However, Wallerstein asserts that an analysis of the history of the capitalist world system shows that it has brought about a skewed development in which economic and social disparities between sections of the world economy have increased rather than provided prosperity for all.

Geographic expansion is the first and most obvious of modern capitalism's secular trends. Essential to the transition from feudalism by making available to capitalists the labour and commodities of an arena stretching from parts of the Americas to parts of Africa and Eastern Europe, expansion has been a regular if discontinuous feature of modern history. With a few semi-peripheral exceptions, territories controlled both by mini-systems and by world-empires have been incorporated as peripheries with specialized productive roles. By the early part of the twentieth century, the capitalist world-economy covered the entire globe extensively. But there has also been a process of "inner" expansion into regions previously included within colonial or national political boundaries but theretofore existing as "subsistence redoubts," supplying primarily able-bodied migratory workers to other zones. The expansion process is particularly important in times of contraction, with the press of need for cheaper raw materials, for new markets, and for expanded resources in the intensified interstate competition of such periods. The globe itself sets a finite limit to the expansion process; it is not accidental that in the present moment we hear much about seabed mining, aquaculture, even space colonies.

Commodification : This the second secular trend that is, the conversion of everything from use value to exchange value. Markets for land and labour are the critical ones: the sub-processes are termed "commercialization" and "proletarianization." The *commercialization* of land has already reached a near-limit, as little usable space remains in the hands of tribal or communal groups. As for *proletarianization*, it has clearly not yet reached its possible limits. For most households, the bulk of their income is derived from the core; in the periphery, due to low wage levels the costs of reproducing the work force are borne by subsistence producers. In a stunning inversion of conventional wisdom, Wallerstein notes that "the slowly developing, slowly eroding, marginal, largely subsistence sector of the world-economy, within which live [a large part] of the world's rural populations .... do not pose a problem to the capitalist world-economy. But if the limits of proletarianization have yet to be reached, surely the world is closer to it now than at any past time. "The world is probably halfway, more or less, in the process of freeing the factors of production".

Mechanization of production is the third secular trend which is, increasing industrialization, that is, the growing ratio of constant to variable capital.

Bureaucratization is the final secular trend, in Weber's, it is "the strengthening of all organizational structures.......*vis-a-vis* both individuals and groups. Wallerstein asserts that this trend has both stabilized the system by increasing the material capacity of ruling groups to repress opposition and weakened it by decreasing the ability of rulers to control the bureaucracies themselves, which can translate into weakness in enforcing "politico-economic will." Again, just what limits this trend is up against remain to be specified.

A final set of orienting concepts includes the *three primary contradictions* in the modern world-system, what Wallerstein has sometimes called "antinomies." The *first* is that between economy and polity, creating dilemmas of action for politically organized class groups. Since the world-economy is characterized by multiple polities, economic decisions are primarily oriented to an arena distinct from the bounded territorial control points where state power resides. This contradiction results in renewed pressure to bureaucratize individual nation-states, and, we might add, in the growing pressure to unify the world politically. Socialism in one country is in this view impossible.

The *second* contradiction is between supply and demand. Since multiple polities make the regulation of supply impossible, anarchy prevails. Demand on the other hand results from the temporary resolution of class struggles within the various states. Capitalists seek both to extract and to realize surplus-value, an impossible balancing act. Never perfectly harmonized, the cyclical rhythms of modern capitalism speed up over time and in addition become more synchronized throughout the entire world market. Regularly, crises push forward the secular trends which constitute capitalism's approach to its limits.

Third contradiction is between capital and labour. Here Wallerstein has argued that there are three kinds of workers in the world system, those who work for subsistence, those who work for wages, and those who are "part lifetime" proletarians, the most common type in the world as a whole. He has further asserted that the process of proletarianization reduces the well-being of most of the world's workers—Marx's accumulation of misery, and that the so-called "new working class" of professionals, technicians, and bureaucrats has been the main exception to and beneficiary of this process. But free labour, especially when urbanized, brings with it heightened labour organization and political struggle. And so the world bourgeoisie is hamstrung both by having to share increasing amounts of surplus with the intermediate strata.

Wallerstein's WST (world-system theory) identified historical cycles accounting for economic "booms and busts" of various durations: the main economic periods for WST are the long-wave *theories of Kondratiev cycles* (named after the Russian economist who discovered them in 1920) : each cycle or *long wave* lasting about 50-60 years, representing a phase of economic downturn, stagnation and profit squeeze, the latter pushes capitalist economy to seek a new and innovative way to a qualitatively different phase of global capitalism, not just an adjustment from the previous order. The most recent Kondratiev cycle began in 1945, expanded until 1967-73 and then contracted through the 1980s—for this recent cycle, key economic roles and structures include the US dollar as the global currency, and supranational bodies such as the World Bank, IMF (International Monetary Fund), and G7 group of countries. The turn of the twentieth century finds the world in an inter-cycle transition phase—institutions and rules such as WTO (World Trade Organization) and Free Trade, that are designed to stabilize and ensure continued profitability and global power for core countries in a new Kondratiev cycle, are being contested.

The Basic Elements of the World-System Economy: Three Geographical Tiers of Nation-States

The Temporal cycles of systematic integration, order, turbulence, transition and reconstitution of the global economy play out variably across geographic space. The world-system is very unequal. Because of several centuries of worldwide economic integration of countries through trade into spatial duality of core-periphery, the differences into spatial duality of core-periphery, the differences in per capita income separating the richest and poorest countries which was 3:1 in 1820 and 35:1 in 1950 became 72:1 in 1992, as par by UNDP's Human Development Report of 1999. Within this highly unequal world order, however, there are place-specific dynamics. At times, regions rise and fall in terms of power, development and economic potential. WST has mainly described this globally-differentiated space as the basic elements of world capitalist economy (or, world economy), which Wallerstein (1984)[23] identifies into *three groups*. These three basic elements of world capitalist economy (or, world-economy) are : (*i*) a single world market, (*ii*) a multiple state system, and (*iii*) a three-tier structure.

(*i*) A single world market—the world-economy system consists of a single world market which is capitalist in the sense that production is for exchange rather than use: producers do not consume what they produce but exchange it on the market for the best price that they can get. These products known as commodities whose value is determined by the market—it is for this reason that the capitalist market is a price-setting institution (unlike pre-capitalist markets which are based upon traditional fixed prices). Since the price of any commodity is not fixed there is economic competition between producers—in this competition the more efficient producers can undercut the prices of other producers to increase their share of the market and to eliminate rivals. In this way the world market determines in the long run the quantity, type and location of production. The concrete result of this process has been uneven economic development across the world.

(*ii*) A multiple state system—in contrast to one economic market there has always been a territorial division between a number of political states in the world-economy. This division both predates and grew along with the geographical expansion of the world economy. In the present context, this is part of the definition of the system because its most important feature is that if one state came to control the whole system the world market would transform into a world-empire. At the same time the multiple state system provides through tariff, trade quotas and financial incentives for the protection and stimulus of developing (or protecting) "infant industries" by restricting access to domestic markets of foreign-made goods and encouraging development of domestic production. *It is the competition among this sovereign inter-state system is a necessary element of the world-economy.* If there were a single political centre, competition would disappear and capitalist system would collapse.

Again through this, single states are able to find a way for distorting the markets in the interest of their national capitalist group to protect their interests collectively within their own boundaries and powerful states can distort the market well beyond their boundaries for a short time. This is the very stuff of "international politics". The concrete result of this process is a competitive state system in which each state attempts to have the best of its abilities to insulate itself from the vagaries of the world market while trying to turn the world market to its advantage so that the states become involved very quickly in various allocation and stabilization activities that have progressively grown in scope and a variety of "balance of power" situations may prevail.

(*iii*) A three-tier geographical structure —Wallerstein argues that the exploitative processes that work through the world-economy always is based upon a functional economic division of labour in three geographically differentiated zones of a "three-tiered geographical format"—*core, semi-periphery* and *periphery*—this geography is defined by global division of labour at any particular time and tied together by world market trade in bulk commodities that are necessities for everyday consumption. This third essential element of world-economy is also "political" in nature, but is more subtle than the previous one. Geographically, the early world economy was consisted of Europe and those parts of South and Central America under Spanish-Portuguese control—the rest of the world was an *external arena* essentially outside the workings of the world economy that time. Later by means of plunder, European settlements and the re-orientation of local economies to the world economy, the rest of the world got transformed into a periphery like Latin American colonies. The world economy thus came to consist of a "core" (Western Europe first, joined later by USA and Japan) and a "periphery". Geographically, the most interesting example in Wallerstein's concept is that of the "semi-periphery", which separates the extremes of material well-being in the modern world-economy which Wallerstein terms the *core* and *periphery*.

Dimensions of World-Economy: Its Spatial Structure in the Historical Context

World-economy of Wallerstein is having a systems approach providing a theoretical language which combines sociological conflict theory with economic and historical data considered on a world scale. In conceptualizing social change—such changes are described in terms of societies that are equated with countries —thus we talk of occidental (or, western) society, oriental society, and the like. This position of social change occurring region-by-region, country-by-country is accepted by *orthodox social science* and we may term it the "multiple society assumption". But Wallerstein's world-system analysis (1984) rejects this orthodox assumption as a valid starting point for understanding today's world—he postulates world-economy system is a global one and it accepts a "single society assumption" rather than multiple society—in this concept, many "societies as countries" become merely parts of a larger whole that is, a particular social change can be understood only within the world context that is the world-system. For instance, the decline of England since the late nineteenth century is not merely a "British phenomenon", it is part of a wider world-system process which should be termed as "hegemonic decline". Trying to explain this specific social change by concentrating on England alone will produce only a very partial view of the processes that were unfolding at the end of the nineteenth century in which production became more based on "machinofacture" that is based on increased mechanization, automation, and intensively used skilled power which took place in USA enabling to take over from England as the leading core nation. Thus within world-economy systems analysis, we can identify basic forms that social changes can take. It is worth reiterating the key point that it is these entities that are the objects of change and these entities are the *"societies of this historical social theory"*.

According to Wallerstein, the world economy is the basic economic entity comprising a single economic division of labour, in which it is natural that there are different countries with multiple sovereignties who perform with separate histories, having parallel societies and carrying on different economic tasks, but no one of which can subsume and control the entire economic system being "parts of a whole reflecting that world system." To understand the ways in which the various socio-economic profiles and distinctive political economy of a particular country, according to Wallerstein, we require to first situate that country in the world economy. In the dialectical perspective of world-system analysis, what is required is the demand of the knowledge of the complex texture of social reality by the historical concreteness within a long-range perspective that observes trends and forces of the world-system, which can explain what underlies and informs the diverse historically concrete phenomena. Thus, for example, in nineteenth century India was organized to provide the Lancashire textile industry with cotton and Australia to provide the Yorkshire industry with wool and both were producing raw materials for the textile industry in the core so that their economic function within the world economy was broadly similar. *Nevertheless, the social relations embodied in these two productions were very different, with India being "an imposed peripheral process"* and Australia a *"transplanted core process"*—thus the outcomes for these two countries have clearly depended on these "social relations" and "not the particular type of product". Another example can be of similar products leading to contrasting outcomes due to "production relations" in the contrast between the tropical hardwood production of central Africa and the softwood production of Canada—the former combines expensive wood and cheap labour, the latter cheap wood and expensive labour.

The semi-periphery is interesting because it is the dynamic category within the world-economy. Finally, much restructuring of space *in fits and starts* during long-term cyclical epochs (or, long-wave rhythms) of economic and political fluctuations (that is, of alternate expansion and contraction) in world-economy. These trends within the context of the long-wave rhythms of the world-economy correlate highly with the emergence of distinctive *technological systems* (known as "Kondratiev long wave cycle" in 1920)[24] which involves "distinctive economic epochs" of core economy rising and sinking through the *semi-periphery*. Political processes are very important here in the selection of "success and failure" (that is, expansion in production and supply increasing to meet increased consumer demand followed inevitably by overshoot when overly optimistic producers and suppliers overestimate the rate of increase in demand and then collapse as overcapacity and excess inventory leads to falling profits, reduced asset values, pessimism, and reduced levels of investment) in case of the world-economy. The mechanics of the epochs are associated with the development and maturation of a distinctive *technology system* consisted of new clusters of energy sources, transportation and communication and key industries. A fundamental consequence of the succession of Kondratiev cycles is that a new phase economic development, based on new technology system and requiring new resources and new markets, initiates a round of "creative destruction" that leaves an indelible imprint on economic landscapes by their pattern and intensity of inequality. Superimposed on Kondratiev cycle if the free-dynamics that is, the rate of economic growth per capita, one gets a 11 to 15 years of duration cycle of regular changes of acceleration followed by deceleration and *vice versa*, this cycle is known as "Kuznets cycle". [25] The rhythm of this Kuznets cycle is important particularly, for semi-periphery, since it is the "crucial middle zone" in the spatial manifestation of Wallerstein's three-tiered structure, its role is more political than economic which never disappears.

The World-System (WS) as a whole is relatively stable and a nation-state's position in WST is historically path-dependent, but not deterministically as nation-states can move between categories over time, depending on their developmental success, international alliances and foreign aid, and the nature of current accumulation regime. For example, the three nation-states which world-system theorists consider to have been dominant in the past four centuries—the Dutch, the British, and the Americans—all began as semi-peripheral to the world-system.

Once major powers such as Spain, Portugal, and The Netherlands have become minor players in the core. The British Empire maintained hegemony for a hundred years but lost it in the twentieth century. Japan moved from the core to the periphery after its defeat in World War II, then rapidly returned to the core. The states such as Taiwan, South Korea, Singapore and Hongkong are rapidly claiming their place from the poor countries of the South to the semi-periphery status. The position of the Soviet bloc was always ambiguous, but with the collapse of socialism, some countries seem to be moving into the core, others into the periphery and some others into semi-periphery status. Indeed, WST is quite useful for analyzing the upward and downward movement of countries over time.

Wallerstein speaks of three mechanisms for *moving from periphery to semi-periphery* (and even *vice versa!).* The (first) is *"seizing the chance"*, when an aggressive state "takes advantaged positions of core countries and weakened economic position of domestic opponents of such policies". The (second) is *"by invitation"* when a transnational corporation simply moves into a less developed part of the world, and in doing so brings the area into the semi-periphery. The (third), riskiest approach is *"self-reliance"*—distancing one's economy from the world system, perhaps by nationalizing a resource, thereby claiming both the loss of foreign investment and core pressure for re-incorporation into the periphery. Wallerstein actually considers role of the semi-periphery more political than economic but which serve the economics.

The Political Economy in World-Economy Context

Political economy in its basal structures of capitalism, deals with spheres of production as well as with spheres of exchange, and with crisis and conflict rather than equilibrium. The distinguishing feature of capitalist mode of production involves the extraction of economic surplus for accumulation within the world-economy. This surplus, the distinguishing feature of the capitalist economy ethically is expropriated in *two* related ways—(one) *expropriation via the price-fixing market,* and (two) the traditional appropriation through the military-bureaucratic ruling class of world-empires of bygone ages. Though, of course, this second method of expropriation could not dominate the world-economic system (or else the world-economy system itself would be transformed into a new mode of production), but it has not been entirely eliminated—known as *mercantilism* or the mercantile economy it should not be undervalued as a process in the world economy from the initial historic "plunder" which took place on the colonized countries in the eighteenth century to today's support of MNEs' interests by the "home" country, usually the United States. The important point is that these two methods of expropriation should not be interpreted as two separate processes—or two "logics" —one "political" and the other "economic" but they are the aspects of the same overall political-economy logic since "the logic of the accumulation process in terms of appropriation *includes* the logic of state building and geopolitics". Thus political economic processes lie at the heart of world-economic system analysis but they are located there separately in isolation since politics are equated with activities surrounding states. While state-centred politics are crucial for understanding the world-economy they do not constitute the sum of political activities. If we equate politics with use of power then we soon appreciate that political processes do not begin and end with states: all social institutions have their politics.

Classical Political Economy Theory

Classical political economy started with Adam Smith's *The Wealth of Nation* in 1776 who stressed the role of free market and individual liberty in economic success. Individuals, unfettered by state inference, would use their ingenuity to the greatest extent. Division and specialization of labour would allow resources to be used in the most efficient and productive manner possible. If all individuals pursued their narrow self-interests, all of society would benefit inadvertently. State interventions to relieve poverty would inhibit initiative, and would stifle investment because they would rely on increased taxes. Therefore, the prescribed role for the state in the economy was a minimal one. Adam Smith identified only *three functions* for the state to perform: (1) defence of national sovereignty and enforcement of contracts, (2) the protection of citizen's rights in private property against violation by one another, and (3) the provision of public or collective goods. Public or collective goods are those that society needs but the market will not normally provide because the gains are so widely dispersed.[26]

The other feature of classical political economy was its conception of citizens' rights, in which it was the state's task to defend. Classical political economy, along with classical and neo-classical liberalism, conceived of individual rights in negative terms. Citizens enjoyed certain liberties from coercion. Such as freedom to practice religion, trade and economic enterprise, and these could not be violated by either the state or other individuals but this freedom has always meant simply freedom from physical restrictions imposed by another person or the state. The citizens did not, however, possess *positive rights*, that is, the rights *to* something, whether it be employment, education, housing, or the like.[27] The *price of this negative freedom is inequality*. Because people have different aptitudes, endowments, and inheritances: some will prosper and some will not. *The classical theorists insisted that it is not the state's task to redistribute resources to equalize society* and this view of classical forebearers is still uphold today—it is contended that the least well off in society benefit more from this inequality—because it speeds up economic progress, which in turn benefits them—than they do from an *egalitarian society* that inhibits economic progress. Herein lies a fundamental problem.[28]

At any rate, classical political economy saw the capitalist system as a complex and delicate mechanism that could easily break down once the state started meddling with it. Left to itself, the free market was seen to be self-regulating : even when it appeared to have broken down, it was still functioning and would repair itself naturally. Hence the term *laissez-faire capitalism* that is, precisely a capitalism that is left alone to (market forces). For example, in an economic depression there is a showdown of economic activity and widespread unemployment. The economy appears to have stopped functioning. But classical political economy "sees a silver lining to the gray cloud" : with so many people unemployed, there are more people competing for fewer jobs so they must offer to work for less than their competitors. Thus labour wages drop, and employees respond by hiring more workers. More workers with more money to spend translates into increased demand for goods and services, which in turn causes producers to expand their activity, which compels them to hire more workers, and so forth.

Modern Liberal Political Economy: In the twentieth century the monetary instability and lack of credit had inhibited trade among nations: the conditions worsened during the 1930 Depression; modern liberals wanted the governments to adopt protectionist policies when they could not pay for their imports. And in that way they favoured using the "the state as an agent of social transformation". The state, it was held, could both develop nation's economy and alter societies in such a way as to make them suit human needs. Underlying this was a belief that the state could embody "collective will" more effectively than the market, which favoured "privileged interests". In the early post-World War II period, in the development thought, like conventional economic wisdom, there existed a broad consensus that national economies needed more state intervention than they had been given in the past—in fact during this time in Latin America the right-wing authoritarian regimes that begun employing *statist development strategies*. Meanwhile the horrors of Depression of 1930s and post-war political developments had given "Keynesian economics" (the economic thinker, John Maynard Keynes who put out his *General Theory of Employment, Interest and Money* in 1936)[29] pride of place in academic and policy circles in the North.

Keynes had no problem with the market economy—he liked the machine, but judged it to be in need of improvement if it was to operate well. In particular, Keynes took issue with the conventional economic assumption that during a downturn, labour prices drop, causing employers to hire more workers and thereby mop up unemployment. The Depression led Keynes to believe that high unemployment could persist indefinitely. He advocated the use of *fiscal policy that is government spending—to deal with recessionary condition.* This was an instrument that virtually all governments were then loath to use.

By building roads and dams, for example, a government could create jobs, which in turn would create more demand for goods and services, which would cause factories to increase their output and then to take on more workers, and so on in an upward spiral. Once good times returned, the government could prevent the economy from overheating by taking money back out of it. In short, Keynes' prescription for improving the capitalist economy was for governments to save in good times, spend in bad.

Keynes was not the first to advise governments to spend in their way out of recessions. However, his innovation was to call on governments to borrow, if necessary, to pump money into the country's economy. The loans would be repaid later from the earnings generated by a newly robust economy. Such public spending would worsen inflation, as more money would chase fewer goods. But Keynes argued that this expansionary fiscal shock would not cause inflation because increased investment would occur along with increased demand. It all heralded the advent of *managed* (or, *command*) *capitalism;* this revolution in economic policymaking overthrow the "doctrine of *laissez-faire capitalism*" that the Depression had discredited.

In the late 1940s, governments in Western Europe and North America started taking Keynes' advice. By then, the USSR had begun to consolidate its hold on Eastern Europe by establishing *puppet regimes* in the six countries it had liberated from Nazi rule (East Germany, Poland, Romania, Bulgaria, Hungary, and Czechoslovakia). This solidified the *iron curtain* that Winston Churchill said had fallen across Europe, dividing it in two. It was becoming obvious that the new Soviet Bloc was not going to join the economic order prescribed at Bretton Woods Conference in 1944. The dust was slowly settling on Western Europe, though, even if the future looked uncertain immediately after the war, especially with communist parties threatening to take power in Italy, France and Greece. Capitalism only firmly re-established its hold on Western Europe when the United States instituted the Marshall Plan,[30] whereby it injected 21 billions of dollars into the reconstruction of Western Europe's ravaged infrastructure. At the same time, liberal democratic parties committed to a more equitable social order came to power in Western Europe.

What emerged in the politics of Western Europe, and indeed in virtually all the developed capitalist countries, has come to be known as the post-war *Keynesian consensus.*[31] Not only did this innovation safeguarded capitalism, it also won the support of the Western World's working classes. Western governments made full employment a top priority, alongwith improved social benefits such as public education, housing and health-care. Post-war capitalism was both *to be redistributive and managed.* Western governments, through nationalization of declining or important private companies, regulation of the economy, public spending, and other means, involved themselves far more deeply to the arrangement of their economies than ever before. In this new version, *capitalism was to be not only more efficient, but indeed more humane. It was a "recipe" for social peace* like none seen before: investors would grow richer and so too the workers, and *poverty would become a thing of the past. Scholars proclaimed that correct economic management would prevent there ever being another Depression, and that the high growth rates that followed in the 1950s were a permanent feature.* All of this was possible because the ingredient missing from earlier capitali*sm—an appropriate interventionist role by the state—was now in place.*

This was the political and intellectual climate into which the Southern Countries were born at the end of the World War II. The industrial world had polarized between capitalism and Soviet communism, while a new form of "statist liberalism" had taken hold in the capitalist west.

The Neo-Classical Political Economy

In 1944, when the World War II entered its final phase and allied victory was just a matter of time, the allied leaders met at Bretton Woods, in New Hampshire to give a shape to the world economy so that a favourable international trading environment get created. To this end the Bretton Woods Conference gave rise to the IMF (International Monetary Fund) and the International Bank for Reconstruction and Development (which became known as World Bank). In 1947 the Bretton Woods system was rounded out by the General Agreement on Tariffs and Trade (GATT). All were designed to create as stable and freely flowing an international trading environment as possible. This influenced the academics in the Southern Countries too, whose confidence in the state was further reinforced by the emergence of "structuralist economics". Aware of the imperfections in the market and global capitalism, and confident that the state could overcome them, development theorists proposed models that assigned the state a leading role in the economy. Pretty soon, the left came to espouse these models. Many Countries of the South, many of which had just won their independence eagerly accepted the models, for they seemed to promise a rapid journey into the industrial age.

At first, the statist (or, state-led) models seemed to deliver just that. With the post-war world economy booming, demand for the products of Southern Countries rose. This provided the governments of the Southern Countries with the capital they needed to develop their industry and infrastructure. However, as time went by, problems in these strategies came to light. The economies of Southern Countries were growing more slowly than required to continue improving the standards of living of the world's poorest citizens. The industrial development that took place consumed more resources than it generated, "a waste exacerbated by inefficient states". And when the post-war boom came to an end in the 1970s, the shortcomings of the state-led development became plain: capitalism got convinced that "welfare state" had become so generous that it robbed individuals of discipline and initiative, and started believing that the growing intrusion of the state in the economy hobbled private enterprise. In 1979 Margaret Thatcher led the Conservative Party to victory in Britain's general election and within the year, Ronald Reagan won the US presidency— these events heralded a shift to the right all over the West; further conservative victories were to follow in other countries and where leftist parties won or retained power, they nevertheless moved right or formed coalitions with right-wing parties. They started the *Reaganomics Era* with which came the "neo-classical political economy" and there started the application of *supply-side economies* to the management of the US economy in the 1980s. The central tenet of this supply-side economics is individuals who behave as rational utility maximizers that is, people are self-interested: they know best what they want, and they also know how to get it. In the pursuit of their goals, people act rationally and efficiently. Thus if individuals are left to pursue their self-interests, society as a whole benefits, whereas if individuals are compelled to pursue collective interests, society as a whole suffers. For example, creating a business in order to generate wealth for oneself nevertheless creates jobs for others, whereas taxing that business in order to redistribute its profits will discourage the owner from expanding it further and creating any more jobs. Accepting this "doctrine of unintended consequences", neo-classical economists conclude that *free-market economies* enable individuals to pursue their self-interest to the benefit of society, whereas *command economies* stifle self-interest and initiative and thus slow society's progress. As argued by one of the doyens of neo-classical economists, Friedrich Hayek of the Chicago School of Economists, "individualism ensured that more things would be tried, and greater the number of things being tried, the more innovation and progress there would be". But he maintained, individuals would only incur the costs of trying something new if they knew they would reap the benefits of any success they had; people are not "altruistic" : taxing the rich to feed the poor hindered the most well-off, reduced initiative and thus innovation, and so hurt all of society.

Interestingly, the forerunners of the contemporary neo-classical theory emerged at the same time as Keynes put out his General Theory of Employment and Money in 1936. But so dominant was Keynes's thinking that time that neo-classical ideas remained confined to academic circles for a few more decades. It was only in the post-1960s that criticism of Keynes moved out of the margins of academic community and since this post-Keynesian period, it was argued that monetary policy is a more useful instrument to manage "*capitalism's boom-and-bust cycles*" . It is contended that by tightening the money supply during bouts of high inflation, and loosening it during times of recession, governments could regulate aggregate demand and maintain economic growth. Money supply can be loosened by lowering interest rates, and tightened by raising them. When interest rates are high, people prefer to invest rather than spend their money, and the high cost of loans discourages people from buying on credit. Economic activity thus slows, less money chases after the same supply of goods, and prices rise more slowly or even fall. In times of recession, lower interest rates have the opposite effect : people withdraw money from savings and spend it, they even buy on credit because it is no longer expensive, and activity resumes. This was a more effective means to deal with the boom-and-bust cycle than Keynes's proposed control of government purse strings. Compared to traditional political economy of Adam Smith, this neo-classical theory though assigned government a greater role in economy, but this theory implied a reduction in the size of the state by removing many of the government's levers on fiscal policy like "government investment and nationalization" and such things went against the post-war Keynesian consensus. In neo-classical theory, the task of the government was merely to create the right environment for business and individuals to maximize their potential. Government should concern itself merely with stabilizing monetary growth, which would "provide a monetary climate favourable to the effective operation of those basic forces of enterprise, ingenuity, invention, hard work, and thrift that are the true springs of economic growth".

At first, the neo-classical economic impact was modest. That changed in the 1970s, when *stagflation* hit the developed economies of the North. As the decade progressed, the people in the North became more concerned with inflation than with unemployment. The latter, a devil the post-war generation had feared, had led people to see in Keynesian economies a powerful "exorcist". By the 1970s, however, due to post-war economic affluence, the working classes had diminished as a proportion of the population in the Northern Countries and the middle classes had emerged to become the prominent electoral constituency. They feared unemployment less than the high inflation that was eating into their standards of living and raising their mortgage payments. *The monetarist recipe of tightening the money supply in order to reduce inflation appealed to them.* By this time there had emerged an even more radical economic theory, known as "rational expectations", whose essential claim was that people had learned to anticipate government policies and thus could effectively derail government attempts to make adjustments in the economy. The proposed solution was even more extreme than monetarism's "hands-off approach" : *a complete retreat of the state from economic life.*

Along with neo-classical economics there arose a separate but related school of thought in political economic theory: *neo-classical liberalism.* This revived the idea of classical liberalism idea of Adam Smith who stressed individuals above all else, seeing individuals as the building block of the society: this neo-classical liberalism believed the minimalist state produced not only a better economy, but a better society as well. Left with maximum freedom, people would not only realize their potential and pursue those things in life at which they were best, but would also become more responsible and self-reliant. They would form *institutions*, such as families and neighbourhoods that would then look after the young, elderly and weak. Expanding the state not only deprived people of freedom, but by usurping many of the tasks performed by society—as, for instance, social agencies replaced families and community associations—it robbed them of initiative and responsibility.

In the nineteenthth century, classical liberalism of *laissez-faire economy* gradually gave way to "modern liberalism". This judged that society was riven by so many historical inequalities that only state intervention could level the playing field to give all the same degree of freedom and opportunity to realize their potential. However, especially after the 1960s, classical liberalism went through a renaissance in the North, resulting in "neo-classical liberalism". This school of thought did help to push the political agenda of the North away from statism, profoundly influencing the "Reaganomics", the politicians such as Margaret Thatcher and Ronald Reagan. This in turn pushed the agendas of the donor agencies to the right, and prodded many Southern governments to re-examine their statist development practices. It was held that paring back the state would improve the operation of not only the economy but the state itself. Reducing the state's resources would at the same time reduce opportunities for corruption; eliminating civil-service jobs would encourage educated people to create their own opportunities for enrichment in the private sector rather than look to the state for advancement. Making the state "leaner and meaner" would improve its operation while at the same time releasing resources into the private sector.

Each of these three major zones of the world economy has their each characteristic form of "labour control", for example, skilled wage labour and tenancy in the core, semi-skilled in the semi-periphery and unskilled in periphery. These three different zones are differentially rewarded by the world capitalist economy, with surplus flowing disproportionately to the core areas. Moreover, the economic structure of each zone supports a given sort of dominant class oriented toward the world market, as well as states of a certain strength (strongest in the core and weakest in the periphery) that operates in that class.

Finally, according to Wallerstein, the differential strength of the multiple states within the world capitalist economy is crucial for maintaining the system as a whole, for the strong states reinforce and increase the differential flow of surplus to the core zone. This happens because strong states can provide "extra-economic" assistance to allow their favour on the world market. Hence the existence of differences in wealth among the states is not an anomaly but rather a natural outcome of the fundamental processes driving the world economy, otherwise if the gap between the rich and poor nations disappear ultimately, then the capitalist world-system that has been in place since the sixteenth century will disappear.

Thus this capitalist world-system of Wallerstein provides a theoretical paradigm which combines sociological conflict theory with economic and historical data considered on a global scale. It analyzes particular problems of development and underdevelopment in terms of their history, sociology and economics, and in the light of the broadest happenings in a world society, in particular the development of capitalism. The world-system is composed of parts which make it work in a particular way, enabling the spread and development of capitalism. For example, in social system of the Southern countries, in small-scale peasant farming or pastoralism, there is no employment of labour and the system is not capitalist but in their relations of dependency between societies they contributed to the continuation and growth of capitalism by providing very cheap inputs to capitalist production in contrast to the large-scale capitalist farming which because of its bigness is not inherently more efficient (than the former) but they provide the muscle to coerce a labour force into being, as well as the discipline and authority necessary to extract surplus value from that labour. Thus the "otherness" which the labour experience in this case can now be seen as the outcome of their location within long-term historical changes taking place throughout the world. Hence the material basis of the world-economy system is defined by the myriad commodity chains that criss-cross global space. Each chain consists of a series of production nodes from the point of initial expropriation of nature as raw material to the final points of consumption. At each node the means of production are brought together and, through a specific set of relations of production, value is added to the commodity and passed on down the chain. The social relations at each node determine how much of the additional value added to the commodity is retained at that location and how much seeps through the next node. Basically the operation of this mechanism can be reduced to two processes, one associated with the core of the world-economy, the other with the periphery. As their name suggest, core- and periphery-making processes are spatially polarized across the world and have generated two distinctive zones. All peoples are part of a single capitalist world-economy whether resident in the "core" or in the "periphery" —both are integrated into the "metropolis-satellite chains" which facilitate the extraction of surplus produces at each stage of production and its transfer and consumption by the social classes in the core. Wallerstein (1984) theory focuses primarily on historical forces giving rise to a trade-induced world division of labour, arguing that they produced an international structure of UNEQUAL NATION STATES which by perpetuating an uneven pattern of accumulation at the core brought about capitalist development at the core and underdevelopment in the periphery. Thus, WST has undoubtedly deepened our understanding of the global political economy. It is also a satisfying antidote to the reductionism, ahistoricism and superficiality in most popular interpretations of economic change.

Changing Perspecitve of World-Economy System

Wallerstein's this (world) capitalist system of world-economy is an *open system* that evolves over time along an obvious *path dependence to its motion* so that through its expansions and their fluctuations through recessions, it returns to its initial advantage that is produced through "increasing returns to scale" making (the world-economy) as a whole relatively stable, but there is also some movement within it, some locational reversals and shifts. Over time, the dynamics of the capitalist mode of production which has forged the contemporary world economy, the loci of economic activities can keep changing due to many factors like *ecological exhaustion*, socio-economic consequences and the impact of new technology. Hence some areas "progress' and others "regress". Once major mercantile powers of the nineteenth century such as Spain, Portugal, and the Netherlands have become *minor players* in the core. The British Empire which maintained hegemony over the world for a hundred years but lost it in the twentieth century after the World War II by the ascent of United States—this relative decline of Britain is interpreted as *climacteric eclipse* of core economies (Knox and Agnew, 1994).[32] Semi-periphery states such as Taiwan, South Korea and Singapore are rapidly claiming their place in the core. The position of the Soviet Bloc was always ambiguous, but with the collapse of socialism, some countries of the Second World seem to be moving into the core, others into the periphery, and others into semi-periphery status. All these reflect the shifts in the operation of the world-economy, especially the recent slow disintegration of Fordist/organized capitalism that was important in the core for much of the twentieth century.

The creation of a world-economy has momentous implications for our understanding of the world in which we live. Development and underdevelopment are global processes, and it is to them that we must look if we are to begin to understand the human condition and to act to improve it. Drawn by these forces into convergent activities, people of diverse origins and social makeup are driven to take part in the construction of common world which include the European sea merchants and solidarity of various nationalities as well as native Africans, Americans and Asians. In the process, the societies and cultures of all these people undertake major changes which singles out not only the peoples as the carriers of *real* history but also what the social anthropologists have called *primitives* and have often been studied as *pristine survivals from a timeless past*. The global processes set in motion by European expansion constitute their history as well. There are thus, according to Wolf, 1982[33] : "no contemporary ancestors, no people without history and no people—whose histories have remained cold".

Thus, in one sense, the capitalist economy poses a unified system on the world. Yet, each individual culture had a different starting point, and each was transformed in its own way and at its own rate. Thus, we should not expect a single proletariat culture but a multitude of cultures, all subsumed under capitalist domination. Here, "in a sense, we are thus back full circle with a diverse world which is more than merely fascinating. It is a world constructed by a global set of social relations, developed and underdeveloped by the same processes—the precise mechanism of which cause controversy in theory as well as in practice—and divided by ideologies sustained by the terrifying control of the means of violence in the system of nation states. It is a world which we fail to strive to understand at our own, very real, peril" (Lee, 1989).[34]

But the fact that the above-named particular countries change their position over time in the world-economy, from core to semi-periphery or to periphery, or *vice versa*, does not in itself change the nature of the world-economy system. These shifts will be registered for individual countries as "development" or "regression". The key factor to note is that within a capitalist world economy, all states cannot "develop" simultaneously *by definition*, since the system itself functions by virtue of having unequal core and peripheral regions. In the mid-twentieth century, this core shifted from Europe to the United States, which replaced Great Britain's primacy (as the "first" industrial power) and since then it holds the *hegemony* of the core. Today, the United States increasingly shares its economic power with Japan and European Union for quite the same way.

Wallerstein insists that any theory of social change must refer to a "social system" —that is, a 'largely self-contained" entity whose development dynamics are "largely internal". For self-containment to obtain, he reasons, the entity in question must be based upon a complete economic division of labour for which he takes the large scale social system of early modern Europe. Hence in the structural consolidation of this capitalist world system, he has dealt with historical evidence primarily in terms of a pre-conceived model of a world capitalist economy which opted over the past three centuries by three basic developments.

The (first) has been the capitalization of world agriculture, meaning the ever more use efficient use of the world's land and sea resources in large productive units with larger and larger components of fixed capital. Over time, this has encompassed more and more of the earth's surface, and at present we are probably about to witness the last major physical expansion, the elimination of all remaining plots restricted to small-scale, so-called "subsistence" production. The counterpart of this process has been the steady concentration of the world's population as salaried workers in small-to-large, dense-to-denser pockets—*i.e.*, *proletarianization* and *urbanization*. When transportation costs were high and communication difficult, places were protected from competition by friction of distance and locations could depend upon a relatively high degree of monopoly power. But diminished transport costs have made production, merchanting, marketing, and particularly finance capital much more geographically mobile than heretofore: The monopoly power inherent in location is much reduced now and this allows much freer choice of location which in turn permits capitalists to take more rather than less advantage of small difference in resource qualities, quantities, costs and amenities between places. This also means that the models of economic activities that identify, for example, as the key factor earlier in determining location are potentially misleading now in contemporary circumstances so that "static models" need to be adjusted or supplemented by more dynamic models of the transition from "organized capitalism" based upon close coordination between business, government and labour which dominated throughout much of the twentieth century to "disorganized capitalism" with new "emerging sectors" representing the new modes of organization and production that now appears underway. For example, multinational capitals has become much more sensitive to the qualities of places in its search for more profitable accumulation. Thus, the world-economy as an open system is a complex and changing intermeshing of institutions and markets producing different effects in different places, rather than a closed global *isotropic plane* governed by the same invariant determinants everywhere and always. *And this is the central message which we can have now.*

The (second) major structural change has been the development of technology that maximizes the ability to transform the resources of the earth into usable commodities at 'reasonable" cost levels. This is *industrialization* which is far from over—the spread of industrial activity which in the past remained largely concentrated in the temperate core areas to the tropical and semi-tropical peripheral areas. Industrialization too has hitherto tended to consolidate the world's capitalist system in providing a large part of the profit that makes the system worth the while of those who are on top of it, with a large enough surplus to sustain and appease the world's middle strata. More extension of industrial activity will not change a peripheral area into a core area, for the core areas will concentrate on ever, newer, specialized activities.

The (third) major development, at once technological and social, has been the strengthening of all organizational structure—the states, the economic corporate structures, and even the cultural institutions—*vis-a-vis* both individual and groups. This is the "process of bureaucratization", by which within a world economy, the state structures function as ways for particular groups (semi-periphery/ periphery) to affect and distort the functioning of the market in favour of their advantage. While the bureaucracy has been uneven, the stronger the state machinery, the more its ability to distort the world-market. The core states have stronger state machineries than the peripheral statės, the head of the state of a core country has the power today that monarchs of yesteryears sought in vain to achieve.

The capitalist world economy developed through long cycles of "expansion" and "contraction", is based on an *international division of labour* between the *core*, the *periphery*, and the *semi-periphery* of the world economy. "Expansion" which is the upward cycle, occurs as the major changes which have taken place over time in the ebb and flow of power, when the totality of world production is less than world effective demand, as permitted by the existing social distribution of world purchasing power, and that "contraction" occurs when total world production exceeds world effective demand—the downward (cycle of contraction) is only resolved by a political realocation of world income that effectively expands world demand. Since, however, political boundaries of the countries (states) are smaller than the economic whole, they will reflect different groupings of economic tasks and strengths in the world-market. Over time, some of these differences may be accentuated rather than diminished.

Three-Tier Structure and Geographical Scales : A Trilogy

Geography employs traditionally three scales of observation for analysis—international (or, global), national (or, state) and an intra-national (usually urban or metropolitan scale). Though unlike the world-economy paradigm of Wallerstein's three-tier structure, to justify this trilogy of geographical scales are merely given—there has been no articulation of theory, but the recognition of trilogy of geographical scales appears to have been implicit in many social science studies (beyond geography) as representing a particular way of viewing the world as a "subtle form of developmentalism" (Taylor,

1992).[35] If this trilogy of geographical scale is pivoted around the *nation-state,* then the three scales represent: a "national scale" *of ideology*, a *"local scale"* of *experience* (at which we live our daily lives, whether urban or rural) and a "global scale" of *reality* (filtered through particular aggregations of nation-states, in which the communities are located). This model of geographical scale does not posit three processes operating at three different scales but just *one process* which is manifest at three scales. In general, this process takes the following form : the needs of accumulation are experienced locally and justified nationally for ultimate benefits organized globally (for example, by MNEs paying less tax). This geographical scale is thus a "single process" in which ideology separates experience from reality but there is "one system" which is world-economy. Thus at local scale, "individual cities" possess economic bases to play functional roles which stem from their handling of the factors of production which at a *regional scale* the urban system of southern Ontario and Quebec region characterize a *heartland* in serving the hinterlands as well as in shaping the regional geography of Canada but at *world-system scale* this industrial heartland becomes paradoxically, both an "*economic hinterland*" of USA and a "*quasi-metropolitan state*".

World-system economy occupies a given geographic space: its core (a central pivot) about which there is the semi-periphery and finally comes the very wide peripheral areas, which, are not having political maturity or industrial diversification within the international division of labour characteristic of the world economy, hence they are the subordinates rather than true participants, so that the spatial landscapes there resembles "purgatory or even hell", like the countries of Ethiopia.

The "world system" analysis theory of Immanuel Wallerstein (1984) is a development of dependency theory, but differs from it inasmuch as it pays considerable attention to the specific histories of different regions of the world, and does not generalize in quite the same way. It provides a theoretical paradigm which combines sociological conflict theory with economic and historical data considered on a global scale. It analyzes particular problems of development and underdevelopment in terms of their history, sociology and economics, and in the light of the broadest happenings in a world society, in particular the development of capitalism. The world-system is composed of parts which make it work in a particular way, enabling the spread and development of capitalism. For example, in social system of Countries of the South, in small-scale peasant farming or pastoralism, there is no employment of labour and the system is not capitalist but in their relations of dependency between societies they contribute to the continuation and growth of capitalism by providing very cheap inputs to capitalist production in contrast to the large-scale capitalist farming which because of its bigness is not inherently more efficient (than the former) but they provide the muscle to coerce a labour force into being, as well as the discipline and authority necessary to extract surplus value from that labour. Thus the "otherness" which the labour experience in this case can now be seen as the outcome of their location within long-term historical changes taking place throughout the world. Hence the material basis of the world-economy system is defined by the myriad commodity chains that criss-cross global space. Each chain consists of a series of production nodes from the point of initial expropriation of nature as raw material to the final points of consumption. At each node the means of production are brought together and, through a specific set of relations of production, value is added to the commodity and passed on down the chain. The social relations at each node determine how much of the additional value added to the commodity is retained at that location and how much seeps through to the next node. Basically the operation of this mechanism can be reduced to two processes, one associated with the core of the world-economy, the other with the periphery. As their name suggest, core- and periphery-making processes are spatially polarized across the world and have generated two distinctive zones. All peoples are part of a single capitalist world economy whether resident in the "core" or in the "periphery"—both are integrated into the "metropolis-satellite chains" which facilitate the extraction of surplus produces at each stage of production and its transfer and consumption by the social classes in the core. Wallerstein (1984) theory focuses primarily on historical forces giving rise to a trade-induced world division of labour, arguing that they produced an international structure of UNEQUAL NATION STATES which by perpetuating an uneven pattern of accumulation at the core brought about capitalist development at the core and underdevelopment in the periphery.

Critiques of Dependency Theorists

The main assumptions of dependency theory of A. G. Frank are the spread of underdevelopment in the peripheral world and the need for greater self-reliant development whereas the world system theory of Immanuel Wallerstein considers the spread of development and the need for greater world interdependence. Thus though these dependency theorists are being criticized as their theories are derivations from the theories of Marxism and the processes which they deal with are mechanistic and are on the sweeping world scale analysis, but their attempts in stressing the unity of the economic relations of production and distribution, and the social-political dimensions of the modern state, have been successful in separating the "modern" from the "traditional", the capitalist from the non-capitalist.

The main assumptions of dependency/underdevelopment theory of A. G. Frank are underdevelopment. They are neither ascriptive nor an original stage for the Countries of the South but rather the contact with and oppression by Europe which "stagnated their economy and underdeveloped them". In exemplifying his points, Frank points out to the British deindustrialization of India as well as to the negative effects of industrialization which plagued the Latin American countries by the 1950s and he proposes the need for greater self-reliant development by dissociating from the developed nations whereas the world system theory of Immanuel Wallerstein is a quasi-Marxist one since Wallerstein gives credit to capitalism for creating the world system and the need for greater world interdependence. However, on the negative side, capitalism (as accumulation of wealth), according to him, "has been exploitative through extraction of surplus labour value and destruction of economy of vast majority of persons located within the boundaries of world-economy, than any previous mode of production in world history. Thus both these theories are located in the historical dynamics of the expanding capitalist mode of production of the modernization theory (originally propounded by Max Weber) which is not just a theory but an "action programme", aimed at making the rest of the world "just like us". Thus both the views: dependency/ underdevelopment and the world-system enable to provide a *new theoretical paradigm* to guide our investigations of the emergence and development of capitalism, industrialism and nation-states. This dependency paradigm thus has a number of strengths : For (one thing), it recognized that the dynamics of the countries of Global South are very different from those in the North—Simplistic prescriptions of modernization paradigm based on evolutionary development from "tradition" to "modernity" which took place in Europe and the United States are not applicable to poor Southern Countries. Early modernization theorists tended to see poverty and wealth in terms of production: those countries that produced more would develop faster (thus, development was measured by increases in GNP). The dependency theorists added an important dimension by viewing the issue not as a matter of production but of structural relations of capitalist exchange—they put the modernization approaches to task for their ahistorical ideal types of "tradition" *versus* "modernity" elaboration and then their application to national cases. Also, the dependency idea of a world economic system based on a division of labour between core and periphery is a valuable one. Both Frank and Wallerstein placed a strong, and badly needed, emphasis on the history of the Countries of Global North and the effects of that history over hundreds of years. Finally, despite the significant heuristic values of the dependency theories, discussed above, a cornerstone of these theories, is since peripheral-capitalist relations of production persisted and countries were unable to withdraw from the world economy, underdevelopment remained a terminal condition of the peripheral world, "neo-colonialism" argument comes.

Modernization and Dependency—A Reprise: Towards A Synthetic Theory

Thus on the primary source of underdevelopment, there have been two schools of thoughts: those of the *modernization school* tend to emphasize internal constraints, while *dependency theorists*, including those of the world-systems approach, stress historical processes and international structures. During the 1970s and 1980s, however, these dependency theorists are being criticized as their theories are (one) derivations from the theories of Marxism, and (two) the processes which they deal with are mechanistic and particularly, (three) the world system approach are on the sweeping world scale analysis so that it is difficult to apply in specific cases or to meaningfully test empirically. But their attempts in stressing the unity of the economic relations of production and distribution, and the social-political dimensions of the modern state, have been successful in separating the "modern" from the "traditional", the capitalist from the non-capitalist. And by the 1970s, these theories of *dependency paradigm including those of the world-systems approach* as the opposite of *modernization paradigm* were well received by most social scientists that they addressed themselves to certain crucial variables in providing insight into the nature and dimensions of development process. Though both the theories of development are usually argued as *antithetical,* both can be defended, while economists have tended to reject or ignore dependency arguments as an explanation only at the "structural" level—for them, David Ricardo's "theory of comparative advantage" among nations remains a virtual article of faith, although it is largely repudiated by dependency theorists. There is probably enough truth—and enough that is questionable—in each perspective to keep the opponents arguing for some time.[36]

The dependency paradigm has a number of strengths. For one thing, it recognized that the dynamics of the South are very different from those in the North; simplistic prescriptions based on development in the United States and Europe are not applicable to poor countries. Early modernization theorists tended to see poverty and wealth in terms of production; those countries that produced more would develop faster (thus, development was measured by increases in GNP). The dependency theorists added an important dimension by viewing the issue not as a matter of production but of the structural relations of capitalist exchange. Also, the dependency idea of a world economic system based on a division of labour between centre and periphery is a valuable one. Finally, such theorists as Wallerstein placed a strong, and badly needed, emphasis on the history of the South and the effects of that history over hundreds of years.

Critics have pointed out much theorizing—especially that of the world-system approach—tends to be at such a high level of abstraction that it is difficult to apply in specific cases or to meaningfully test empirically. Dependency theory tends to focus almost exclusively on economic factors (an inheritance from Marx) to the exclusion of political, military, social, and cultural factors. Critics also charge that dependency theorists ignore factors internal to the various countries, overemphasizing the international domain.

Though often argued in categorical terms, modernization and dependency perspectives may not be as mutually hostile as they first appear. Though the dependency perspective argues that the characteristics which define the Countries of the South are not intrinsic but were developed through interaction with the West, however, there is also the added assumption in this paradigm that underdevelopment persists today because of *ongoing* structures of inequality. One might argue that though such structures certainly exist and are formidable, there are *also* powerful forces internal to various countries. The fact that such forces were developed historically from Western expansion does not make them any less real or less internal now. An example would be the continuity of elites in Latin America. These elites who date back hundreds of years to the conquest and colonial periods, are more interested in the maintenance of their own power and position than in supporting or benefiting the core states (though they may use core investment and military support to their own advantage). Another example is the emergence in many Southern Countries of *huge informal economies*—shanty, towns, gypsy transportation, and illegal markets—that are relatively divorced from dictates of international capitalism.

Some recent theorists have sought to reclaim the state, rather than the economy, as the focus of analysis. These scholars see the Southern Countries not as a mere servant of international capitalism but as an autonomous entity in its own right, with the power to direct (or mis-direct) development through policy decisions and through acting as a primary investor in indigenous production.

One thing is certain: there will be no agreement among theorists about the causes of national poverty or the best road toward development. However, each skirmish in international battle leads to new and valuable insights into this single most critical issue of our times.

Endogenous Economic Development Approaches in the Post-1950s: Monopoly Capitalism and Neo-Colonialism

Early generation of dependency theory was inclined to claim that

Table 5.1 : The Modernization and Dependency Paradigms

Focus of Analysis	*Modernization Paradigms*		*Dependency Paradigms*	
	Early Theorists	*Later Theorists*	*Dependency Theory*	*World-System Theory*
	The internal systems of individual Nations		*International system historically developed*	
Causes of Under-development	Nations start out under-developed.	Vicious cycle of poverty. Internal Colonialism	Capitalist Exploitation. Unequal terms of trade.	Historical emergence of capitalist world system since sixteenth century.
	Lack of modern technology,	Lack of social differentiation		
Entrepreneurial	Spin-off effects of spirit, modernizing elites.	International division and social mobilization.	*Comprador elites*; Lack of primary export production.	Division of labour, within which each country has its function.
Structure of the System	Internationally : Developed and Underdeveloped Countries. Internally : Traditional and Modern Sectors.	Internationally; Developed and Underdeveloped Countries. Internally : Dual Traditional and Modern Economies. "Colonial" dominance by a major city or region.	Developed centre and Underdeveloped periphery.	Developed core, underdeveloped periphery, partially developed semi-periphery.
Geographical Path to Development	Underdeveloped countries must follow the paths	Foreign aid and foreign investment to overcome	Import Substitution to develop the domestic economy.	May be necessary to "delink" from the capitalist world-

Table 5.1 : (Contd.)

Focus of Analysis	*Modernization Paradigms*		*Dependency Paradigms*	
	Early Theorists	*Later Theorists*	*Dependency Theory*	*World-System Theory*
	The internal systems of individual Nations		*International system historically developed*	
	traversed by developed countries. The latter can help by transferring technology, expertise, capital, etc.	vicious cycle of poverty. Open trade. Export production—at first of primary resources, then of manufactured goods.	Acceptance of foreign investment but with strong controls to break dependency.	system for a time. Countries can move from periphery to semi-periphery core.
Key Theorists	W. W. Rostow	S. N. Eisenstadt	Andre Gunder Frank Johan Galtung	Immanuel Wallerstein

Southern Countries would remain locked into "classical dependence" on the Countries of the North but later generation theorists maintain that since so long as dependence of Southern Countries remain locked to the First World, so they could never break free of their dependence and poverty. So what needed were the policy of autonomous national development strategies and in doing so for insulating a rational economy from the world-economy the dependency theorists placed their faith in the state as the "engine for development". Thus following this ideology, in the meantime the *adoption of endogenous economic development* strategies pursued by certain Southern Countries (like China, India, since their de-colonization) on policies of near self-sufficiency (*autarky*), or through indigenous development models known as, Import Substituting Industrialization", *ISI* (designed to limit reliance upon certain classes of imports) got weakened because of various factors —the policy led the volume of world trade considerably lower by the 1970s, the Southern Countries' inability to maintain fixed currencies and the disruption caused by official devaluation resulted in the early 1970s led to the abandonment of currency controls. With the introduction of floating currencies, new opportunities for speculation emerged. The apparent successes of the export-led development policies of the East Asian economies, which were recording high rates of growth and experiencing rapid industrial transformation led many countries practising statist interventionism to believe themselves that "integration could be a preferred strategies to that of isolation". Many poor countries were plunged into massive indebtedness and when they approached the International Monetary Fund and World Bank for loans and other assistance in rescheduling debt payments, support by the latter was made conditional upon a standard medicine of undertakings about changes in economic policy for removal of export restrictions and to expand the role of private enterprise in production and trade. As a result these organizations came to play a critically important role in the emerging global economy by "imposing a singular economic model throughout the countries of Global South". This New World Order is a capitalist order, and thus, the world system model described by Immanuel Wallerstein (1984) has grown immeasurably stronger and more integrated. The model of open trade, based on (Ricardo's) idea of comparative advantage, is that which the First World countries believe is the most beneficial, has to be also for countries of Global South. To protect the First World's financial interests, the severe conditionality restrictions on debtor nations, repatriating capital through squeezing exports and retrenching public expenditures, what these organizations imposed, could have been the envy of the big-stick diplomats of the heyday of gunboat diplomacy.

Justification for this policy (of the role of private enterprise in production and trade) lies in the gross distortions and inefficiencies which frequently arise when political and bureaucratic controls impede the normal working of market mechanisms, examples of which are readily found in industries operating at very low capacity utilization, in ever-increasing price differentials of domestic goods over their imported equivalent, and in uncompetitive industries. Despite widespread evidence that efforts by central bureaucracy to control and integrate the many complex process of an economy are often seriously deficient, there are voiciferous critics of the monotheistic views of international bodies which favour free enterprise. Thus finally despite the significant heuristic values of the dependency theories, (discussed above), got stifled as an alternative process of development, and a "neo-classical capitalist form of development" was imposed on the Southern countries for their revival. Using terms such as *neo-imperialism and neo-colonialism*, Baran and Sweezy (1966)[37] noted that economic neo-imperialism and neo-colonialism have "effortlessly replaced colonial (political) control and brought about sustained under-development. Surplus continues to be drained off, principally through the repatriation of profits from foreign investments" in the Countries of South. Even the class structures of such peripheral societies, which are primarily proletarian, are products of neo-imperialism. Capitalists in imperialist societies can readily expand their influence to less developed countries by overwhelming the *petit bourgeoise* (the local producers or owner-workers) who are driven out by competition, fail financially, and end up in the working class, or *proletariat*, this further increases the ratio of workers to capitalists, thereby producing more surplus value for each capitalist— all these happen not only by investment, but by their control of technological expertise and of money through the World Bank, International Monetary Fund. And such neo-colonialism, accompanied by uneven economic development, is the essence of *Monopoly Capitalism* and the result is capitalism has now become Wallerstein's worldwide system of economy with some nations at the core and others on the periphery.

The last important issue is "monopoly capitalism", that is *imperialism* and an important feature in present day imperialism is the *multinational* or the *trans-national corporations*, MNCs or TNCs —with the improvement in transport, and the technological form of communication through the computer, making the flow of capital, easier than ever for the centralization and internationalization of capital as well as core dominance by economy and by military over the periphery. Manufacturing where there is cheap labour, and distribution close to markets, make it rational to locate enterprises in more than one setting, sometimes in several advanced societies, and sometimes in less developed countries. Nation-state in economy has become less relevant since the capital that operates internationally needs the support of a home state to protect its interests.

Hence as a cornerstone of the dependency theories, the peripheral-capitalist relations of production persisted and countries were unable to withdraw from the world-economy and underdevelopment remained a terminal condition of the peripheral world, and the "neo-colonialism" argument comes through the operation of the multinational corporations who do not address underdevelopment in the periphery but they undermine domestic entrepreneurship as well as the autonomous domestic development by their capital-intensive technology-oriented production. Thus the relationship between domestic and foreign capital allows for a disproportional concentration of wealth and productive capacities so that there is an internal as well as an external conditioning that leads to dependency of the periphery and a general theory of Frank and others as underdevelopment being caused by dependence on centre (industrialized) economies failed.

Notes and References

1. Friedman, John, 1963, *Regional Development Policy : A Case Study of Venezuela,* MIT Press, Cambridge, Ma.
2. Friedman, John, 1972, *Urbanization, Planning and National Development,* Beverley Hills, California.
3. The term, "Region", refers to a relatively large area within a country, *e.g.*, North Eastern India. Sometimes large world regions are often implied. More often than not the precise foundaries are left unspecified. Arguments usually can be applied to all scales of regions, from world-regional to a metropolis'

suburban region, implying that the same processes producing regional differences are to work at all scales. The major concern in defining a region has been to provide spatial frameworks for thinking about the mobilization of resources and the spread on concetration of resources and the spread on concentration of economic activity.

4. Massey, Doreen, 1984, *Spatial Divisions of Labour*, Methuen, London.
5. Krugman, Paul, 1995, *Development, Geography and Economic Theory,* MIT Press, Cambridge, Ma.
6. Krugman, Paul, 1991, *Theorization of Uneven Regional Development,* MIT Press, Cambridge, Ma.
7. Martin, Ronald Peter Sunley, 1998, "Paul Krugman's Geographical Economics and Its Implication for Regional Development Theory : A Critical Assessment", *Economic Geography* 74.
8. Frank, A. J., S. Amin, G. Arrighi and I. Wallerstein, 1982, *Dynamics in Global Crisis*, Monthly Review Press, New York
9. Myrdal, Gunar, 1957, *Asian Drama : Rich Lands and Poor,* Harper and Row, New York.
10. *Op. cit.,* 32, in Chapter 3.
11. Slater David, 1992, *Internal Atomization,* Monthly Review Press, New York.
12. *Op. cit.,* 6.
13. Slater, David, 1975, Special Organization in Africa's Development.
14. Habib Irfan, 1969, "Problems of Markist Historical Analysis in India", Reprinted in S.A. Shah, 1973 (ed.) *Towards National Liberation; Essays on the Political Economy of India,* Montreal.
15. For Wallerstein, it is world-system and not world systems—with a hyphen in between, rather the two words constitute a single concept. Since world is not attribute of the system, he meant one word in plural sense, because for him there were many world system, Frank's system is a world systems in an attributive sense, in that it has been tending over time to cover the whole world.
16. Wallerstein, Immanuel, 1975, "The Present State of the Debate on World Inequality" , Cited in, Seligson, Mitchell A. and Passe-Smith, John, 1992, *Development and Underdevelopment* : *The Political Economy of Inequality, Lynne Rienner Publishers,* Boulder, U.S.A.
17. Wallerstein, Immanuel, 1984, "*The Politics of World Economy : The States, The Movements and The Civilizations*", Cambridge University Press, New York.
18. Wallerstein, Immanuel, 1974, *The Modern World System I : Capitalist Agriculture and the Origins of the European World-Economy in the Sixteenth Century,* Academic Press, New York, subsequent two volumes are : *The Modern World-System II,* subtitled "Mercantilism and the Consolidation of the European World-Economy, 1600-1750," was published in 1980, and Volume III, bringing the story into the 1840s, appeared in 1989.
19. Paul Baran is looked upon as the Father of Dependency theory for his publication in 1957, "The Political Economy of Growth", *Op. cit.,* Chapter 3.
20. Refer to the surplus of production taken from the producers/artison forcibly by the ruling class as "tributes". See also *Glossary.*

21. Walter, L. Goldfrank, 2000, "Paradigm Regained? The Rules of Wallerstein's World-System Method", *Journal of World System Research*, Vol. VI, No.2, New York.
22. Wallerstein, Immanuel, 2000, *The Essential Wallerstein*, The New Press, New York.
23. Note the variability in world economy across the geographic space as identified by Wallerstein.
24. The boom and bust phases in world economy as noted by N.D. Kondratiev in 1920. See also *Glossary*.
25. The rhythm in world economy as indicated by economic growth in G.N.P. of a country, first identified by Simon Kurnets. See also *Glossary*.
26. *Op. cit.,* 33, Chapter 1.
27. Supply and demand are basic to the definition of capitalism : production increases automatically to meet demand, and once the market is satiated, production will be reduced. The result is a degree of efficiency that is impossible if demand is controlled through price subsidies or supply is determined by government edict. The argument is pervasive. A difficulty lies in the nature of demand and demand is determined by purchasing power that is, *money* which has very little to do with *need.* For example, in the United States, there is substantial need for low-cost housing, but very few entrepreneurs are investing in such construction. The people who need housing the most do not have the money to pay for it. Therefore, almost all new housing is devoted to middle- and upper-middle class buyers. In Countries of the South, where 40 per cent or 60 per cent of the population may be extremely poor, this dilemma is greatly amplified. In El Salvador, the poor people desperately need food. However, poor people do not have money to create effective demand, so virtually all the most fertile agricultural land is turned over to crops such as coffee for export. The NEED is in El Salvador, but the DEMAND is in the United States. There is no assurance in both the cases within an economy based upon the logic of supply and demand that production will be oriented to supplying the needs of the large majority of the people. *Scarce resources* may be turned over to producing or importing automobiles for the wealthiest 5 per cent of the population. *The burdens of growth tend to fall most heavily on the people at the bottom.*
28. The response of the neo-liberal theorists is that as the countries develop through free-market policies, new jobs will be created so that the income level of the poor will increase. As the masses get money, their power in the market will grow so that more and more will be produced for them. It is important that money be spent on new production that will create new jobs, rather than on social benefits that produce nothing. *SHORT-TERM AUSTERITY IS NECESSARY FOR THE LONG-TERM PROSPERITY.*
29. *Op. cit.*, 35, Chapter 1.
30. Aid plan for war-ravaged Western Europe.
31. Known also as Keynesianism, See *Glossary.*
32. *Op. cit.,* Knox and Agew, 1944, 4, Chapter 1.
33. Wolf, Eric R., 1982, "Europe and People without History", California-University Press, Berkeley.

34. Lee, 1989, "Social Relations and Geography of Material Life", in D. Gregory and R. Watford (eds.), *Horizons in Human Geography,* Macmillan, London.
35. Taylor, Peter J., 1992, Second edition, *Political Geography : World Economy, Nation-State and Locality,* Longman Scientific and Technical, London.
36. David Ricardo, *Op. cit.,* 34, Chapter 1.
37. *Op. cit.,* 19.

6

THE ECONOMIC APPROACH FOR DEVELOPMENT THROUGH CAPITAL FORMATION

Harrod-Domar Model

The beginning of modern economic growth theory started with the independent work of *Roy, E. Harrod* and *Evsey D. Domar* published in 1946.[1] They assigned a crucial role in the process of growth to capital accumulation which generates income through the multiplier process (*i.e.*, through productive capacity). The basic idea of the Harrod-Domar model is that economic prosperity and growth occurs through a reinforcing process of higher rate of savings and a lower capital-output ratio whereby capital is accumulated. Given the assumptions of the model, the rate of an economy depends on two things: the propensity to save for investment and the capital-output-ratio, a constant that turns capital into production, provide handy tools for triggering growth in time-structure of production. This model, initially developed in 1939, has a strong Keynesian economic flavour, both indicating that the economy does not automatically achieve its potential.

It is based on the notion that actual income determines the amount of saving, which again determines investment, and that is what affects the rate of economic growth. If saving is not enough, the potential growth rate will not be achieved. This model is the simplest and best known production function used in the analysis of economic development. In advanced capitalist societies, this model explains the relationship between the growth and unemployment. However, the Harrod-Domar model is used in developing nations as an easy way of looking at the relationships between growth and capital requirements.

It does explain the differences in growth performances between countries and it allows one to predict an estimate of growth for a nation which can be compared to predictions of growth for a different country.

The model suggests that the economy's rate of growth depends on:

- the level of saving,
- the productivity of investment *i.e.* the capital-output ratio.

For example, if $ 10 worth of capital equipment produces each $ 1 of annual output, a capital-output ratio of 10 to 1 exists. A 3 to 1 capital-output ratio indicates that only $ 3 of capital is required to produce each $ 1 of output annually.

And it concludes that :

- Economic growth depends on the amount of labour and capital.
- As LDCs often have an abundant supply of labour it is a lack of physical capital that holds back economic growth and development.
- More physical capital generates economic growth.
- Net investment leads to more capital accumulation, which generates higher output and income.
- Higher income allows higher levels of saving.

Implications of the Model: Harrod-Domar

The key to economic growth is to expand the level of investment both in terms of *fixed capital* and *human capital.* To do this policies are needed that encourage saving and/or generate technological advances which enable firms to produce more-output with less capital *i.e.* lower their capital-output ratio. Investment I is a very important variable for the economy and investment has a dual role: it represents an important component of the demand for the output of an economy as well as the increase in capital stock. For equilibrium there must be a balance between supply and demand for a nation's output. In simple case this equilibrium condition reduces to $I_t = S_t$, where I_t and S_t are investments and savings at time, t.

The Harrod-Domar growth model gives some insights into the dynamics of growth. We want a method of determining an equilibrium growth rate g for the economy. Let Y_t be GDP and S_t be savings at time, t. The level of savings is a function of the level of GDP, say $S_t = s\,Y_t$.

This model is constructed on the basis of assumption that the country is having low investment or savings, which combined with limited factor substitutability, prevent the economy of the country from expanding rapidly enough to absorb its labour force. For any country, with its land and labour existing in space, the stable economic growth what the country needs is its capacity for the capital stock what it can invest for increasing its output. If we divide the total stock of capital, K by the total output, Y for a period, t, we get the average capital-output ratio, σ. Thus the level of capital K needed to produce an output Y is given by the equation $K = \sigma Y$ where σ is the capital-output ratio. It is also possible to estimate the marginal capital-output ratio, which is the increment in stock capital from one period, t to the next time period, $t + 1$ and the corresponding increment in output and then divide the first by the second ($= \Delta K / \Delta Y$ as in the Eq. 2 below). In this context, the basic assumptions, which are needed for, are that (1) aggregate supply and demand would be balanced when net investment, I_t in any period, t *i.e.*, rate of change in the *capital stock,* per unit of time equals the change in national income capital $Y_t - Y_{t-1}$ times the marginal or incremental capital-output ratio, σ, that is, investment, $I_t = \sigma(\Delta Y_t)$ where σ indicates the value of capital required to produce one unit of output by value in a single period, and (2) that at equilibrium in a closed economy, intended investment would equal intended savings, S_t, that is, $I_t = S_t$. Note that since the first assumption being an accelerator process, it involves a (3) assumption which is that new capital comes into full production instantly, so that the model assumes no depreciation in capital stock on one hand and no technical innovation on other hand. These assumptions are involved in the Harrod-Domar model which uses the capital-output ratio, σ to relate a country's economic growth rate to its propensity to save and to its effectiveness in using its capital stock to produce output; accordingly this model thus produce the following equilibrium condition that is :

$$I_t = S_t = \sigma.(Y_t - Y_{t-1}) = \sigma.(\Delta Y) \qquad \text{.......Eq. 1}$$

in which the changes from one period to another are represented by Δ, so that the growth in capital and national income are ΔK and ΔY. Now the marginal value of capital required to produce one unit of output by value in a single period is:

$$\sigma = \Delta K / \Delta Y \text{ which again is equal to } I_t / Y_t \qquad \text{.......Eq. 2}$$

Now by dividing Eq. 1 through Y_t to produce the saving rate, that is the average and marginal propensity to save, s $(=S_t / Y_t)$ which means the ratio between savings S_t and output Y_t. So as an unit of capital can add I / K unit to the output, which in turn gives rise to s/σ unit of savings. Thus the stock of capital grows at the rate of s/σ . Again, since the rate of growth of output (or, national income rate), g, equal to $(Y_t - Y_{t-1}) / Y_t$, we obtain the equilibrium rate growth equation, g as follows :

$$s = \sigma \,.\, g \text{ or } g = s/\sigma \qquad \text{............Eq. 3}$$

Also, since $\sigma\Delta Y = sY$, therefore the equilibrium rate of growth g is given by

$$g = \Delta Y/Y = s/\sigma \qquad \text{.............Eq. 3a}$$

In words, thus this Eq. 3 or Eq. 3a defines that the equilibrium growth rate of output is equal to the ratio of the marginal propensity to save and the capital-output ratio, that is, growth as a function of the savings ratio divided by the marginal capital-output ratio which itself becomes the equilibrium conditions for steady economic growth, this equation Harrod calls the "fundamental equation". Thus the stock of capital grows at the rate of Eq. 3 (or Eq. 3a). Since we have a fixed capital-output ratio, output also grows at the rate of s/σ Or, to be more precise, this is the warranted rate of growth, in the sense that if this rate of growth is expected and investment is planned on the basis of this expectation, the rate of growth will in fact be achieved. Steady growth will result from this unity between expectation and achievement. If a different rate of growth or output is expected, the actual rate will be different from both the expected and warranted rate and instability will ensue.[2]

This model, thus suggests savings provide the funds which are borrowed for investment purposes and it tells us how the economy can grow such that the growth in the capacity of the economy to produce is matched by the demand for the economy's output.

A numerical illustration can be considered. For example, the economy is currently operating at a capacity production level of 1000 per year and has a capital-output ratio of 3. This means the capital stock is 3000. Assume the marginal propensity to consume out of GDP is 0.7 so the marginal propensity to save is 0.3. This includes business and public saving as well as household saving. The Harrod-Domar growth model tells that the equilibrium growth rate is $g = 0.3/3 = 0.1$, *i.e.*, the economy can grow at 10 per cent per year. We can now check this result. At the current GDP of 1000 the level of saving is 0.3*1000 = 300. The growth in GDP is 0.1*1000 = 100 and with a capital-output ratio of 3 the additional capital required to produce the additional output is 3*100 = 300. This is the investment required in order to increase capacity by the right amount and, sure enough, this happens to be equal to the amount of saving available in the economy.

But we must made sure there is adequate aggregate demand next year to absorb the production of 1100. At that level of income the consumer demand is 0.7*1100 = 770. The level of investment the next year under the assumed equilibrium growth conditions is derived as above. The 10 per cent growth on production of 1100 is 110 which with a capital-output ratio of 3 requires an increase in capital stock of 330. Thus next year's investment will be 330. This, added to the consumer demand of 770 gives an aggregate demand of 1100. Thus everything balances.

For contrast, let us consider what would happen if the level of current level of investment were to be higher, say 350. This, combined with consumption demand of 700 generates more demand than the capacity of the economy to produce. That excess of investment of 50 induces a demand for an additional 3*50 = 150 units of capital which the economy cannot achieve. There is an irresolvable excess of demand in the economy.

On the other hand, suppose the investment demand fell short of 300, say 250. Now the aggregate demand is only 950, less than the capacity of the economy. If production falls to 950 there is an excess of capital and no need for any investment. Thus aggregate demand fall as investment dropped to zero and consumer demand would drop along with. There would be an irresolvable deficiency of demand. The equilibrium of the Harrod-Domar model is a razor-edge equilibrium. If the economy deviates from it in either direction there will be an economy calamity.

- *Problems of the model : The law of diminishing returns would suggest that as investment increases the productivity of the capital will diminish and the capital to output ratio rise.*

The immediate attraction of this equilibrium model is that it relates one of the primary objectives of development, *the rate of economic growth, g*, to either by expanding the proportion of national income saved or what is often considered its major limiting factor, *investment* through lowering the capital-output ratio, σ. The latter is equivalent to increasing the effectiveness with which the capital stock is used to produce output. The growth rate Eq. 3 also serves as a basis for estimates of planned investment. Attempts have been made to employ this fundamental Harrod-Domar model (with appropriate allowances for *capital depreciation*), to provide a useful, although crude answer to such questions as : (1) given the assumed capital-output ratio, σ of a specified country, what rate of savings (hence investment) is necessary to achieve a target growth rate? Or (2) what savings and investment patterns will maximize consumption pattern of a country over time?

Again, if we want a formula for the rate of growth of per capita income, the above growth rate Eq. 3 has to be slightly modified in which all we have to do is to subtract the rate of increase in population from the rate of growth per capita income. Thus with a savings propensity of 12 per cent, a capital-output ratio of 3, and an increase in population of 2 per cent per year, the economy can grow at a rate of $0.12 / 3 - 0.02 = 0.02$, or 2 per cent per year per capita. Similarly, on this context of σ as 3, increase in population by 0.02 to have a 3 per cent increase in income per capita, the annual investment in economy of what is required is 15 per cent of the national income. This is what the approach outlined in Harrod-Domar's model of capital-output ratio tells us.

Alternatively, we can begin with an estimate of the actual rate of savings and, if we also know the capital-output ratio, σ, *we can deduce the rate of growth of national income that can be achieved on this basis.* We can compare this with what we regard as a desirable rate and so arrive at some idea of the "gap" that has to be covered in order to attain certain goals. The mechanism involved in this theory is that rate of growth of national income is a simple function of the "stock" of capital, hence the main factor governing the optimum rate of investment is taken as the growth potential of the economy of a country: savings are transmuted into investment which, by creating additional productive capacity, leads to more income in the following period (as shown by incremental capital -output ratio). The recognition that higher income in the period leads to still higher amounts of savings, investment, capital formation in the next period's source of output, i.e., that investment creates the capacity for increased production in the future, which in turn affects the size of equilibrium output and income, becomes the important contribution of the Harrod-Domar model.[3]

Simple model always sacrifice realism: this model relies largely on a capital theory of value: capital stock and output in contrast to Ricardo-Marx labour theory of value.[4] This model overlooks two factors: the (first factor) is savings mean postponement of present consumption, in case it becomes normal for an individual, it is irrational from the point of view of society as a whole which, so to speak live eternally; it is not merely that future consumption is underestimated because of temporal uncertainties, but the future is not felt with the keenness of the immediate present. The argument is, in fact, that more we provide for the future the less will be the marginal utility of increased provision for the future and the greater, therefore, will be the claim of present consumption. The (second) factor is that a higher rate of savings and investment may, beyond a point be associated with a greater amounts of capital which will be required to produce an additional unit of output. The limits to growth are not set simply by the possibilities of an investible surplus.

A simple extension of this economic growth model is by considering it in an open economy context (as against the closed economy context earlier). This is typically done by introducing the foreign trade context by rephrasing the model such that,

$$g = s/\sigma + b/\sigma \text{ where } b = (\text{imports} - \text{exports}) / Y \quad \text{......Eq. 4}$$

The significance of this "opening up" in the economic growth model is for exports exceeding import which creates additional investment for expediting the growth rate. The income-generating effect of this additional investment would be assumed to be the same as would occur from investment financed by domestic savings.[5] On other hand, if export-import trade is in balance on current account, *b* will equal to zero and higher volumes of trade will have no beneficial effect upon economic growth. Again, the Eq. 4 can be significant only if imports exceed exports (and *b* be positive) if a country is in receipt of foreign investment or credit or aid the latter will enable a capital transfer for the additional investment to be put to the economy.

Statist Liberalism : State-Led/Statist Development in Practice

Prior to 1980, many countries quite deliberately adopted policies that were designed to insulate their economies from the world market in order to give their domestic industries an opportunity to advance to the point so that they foster their domestic industrial activity where they could be competitive for their catching up with the world countries. The policy of development via import-substituted industrialization, (ISI), for example, was often associated with protective tariffs and subsidies for key industries. Performance requirements on foreign investment were also common. These measures often required foreign investors to employ native workers in skilled positions, and to purchase inputs from domestic producers, as ways of ensuring technology transfers. It was also common for developing countries to sharply restrict capital flows. This was done for a number of purposes: to increase the stability of currencies, to encourage both foreign corporations and citizens holding large amounts of domestic currency to invest within the country, and to use the allocation and price of foreign exchange as part of an industrial or development policy.

The origins of this approach are not new. Since at least the late eighteenth century when most western countries including USA sought to "catch-up" with England, the centrepiece of industrial development used much the same approach, with the basic principles laid out by

Adam Smith (1776). Every rich nation today considered it essential to possess national capacity in so-called "essential" industries—Inward FDI and trade were largely controlled and limited in their scope, unless they met conditions that promoted the self-sufficiency view by enhancing the host country's domestic sector. This view considered government intervention as a necessary tool by promoting *infant industries*. There was also a lot of protectionism and intervention in technology transfer. There was an attempt to provide some sort of equality, education, health, and other services to help enhance the nation.The industrialized nations have understood that some forms of protection allows capital to remain within the economy, and hence via a *multiplier effect*, help enhance the economy. Thus these countries have become developed because in the past their governments took major responsibility to promote economic growth. Thus, as seen in the structural adjustment initiatives and other western-imposed policies, the developing nations are effectively being forced to cut back these very same provisions that have helped the developed countries to prosper in the past. Europe is a good example of this. While the phrase "Welfare State" probably conjures up negative images, with regards to globalization, it is realized by the European communities that protecting their people when developing helps the societies, economies and cultures to thrive. It may be that for real free trade to be effective (which itself is a subject of intense debate noted in Chapter 5 earlier!) countries with similar strength economies can reduce such protective measures. However, to try and get developing countries to compete in the global market place at the same level as the industrialized nations and before their foundations are stable, is almost economic suicide.

An example of this can be seen with the global economic crisis of 1997/98/99 that affected Asia in particular. An UN report looking into this suggested that such nations should rely on domestic roots for growth, diversifying exports and deepening social safety nets.[6]

As far back in the 1950s (or, even earlier after the World War II), once the political and economic chaos that plagued Europe, got settled, international conditions favoured growth. And the Countries of Global North were all poised for an economic boom and on top of this the countries like USA and Canada experienced a *"baby boom"*. Demand and investment rose throughout most of the world so that world economy were growing at average annual rates of between 5 to 6 per cent and there did not appear to be any end in sight to this growth. Such high expectations led to greatly expanded social welfare states. GIVEN SUCH RISING DEMAND, it was not surprising that initially, things went well for many newly independent countries of South. It would have taken some doing for a government to implement a strategy that did not produce healthy growth rates. The successes of the time actually obscured what were there, in most cases, insufficient performances, however, this would not become obvious that time. *Keynesianism* (the statist liberalism approach of the leading economic thinker of the age, John Maynard Keynes) in 1936 began finding its way into the work of development theorists, on the other hand "Soviet central planning" of the 1930s which surged ahead (at a time when Western capitalism was on decay) lent further support to the principle of an expanded state role in the economy. In this environment, a *"statist liberalism"* approaches of Keynesianism and structuralism ascendant, the quarter-century that followed World War II, witnessed a degree of state intervention all over the world on a scale in the economy hitherto unseen : the state was held, could both develop economies and alter societies in such a way as to make them suit human needs. Underlying this, there was a belief that state could act as an agent of social transformation so that the statist development strategies could embody collective will more effectively than the market, which favoured privileged interests.

The post-war statist liberalism approach by adopting the innovation of *Keynesianism* brought a favourable international environment for economic growth in the Countries of Global North—Western governments made full employment a top priority, alongwith improved social benefits such as public education, housing and health care. Thus post-war capitalism was to be both *redistributive and well-managed.*

In the Countries of Global South, this statist development strategy started with modern Turkey which got created in 1923 out of the crumbling of Ottoman Empire—the bold venture of Kemal Ataturk went for statist liberalism for industrializing Turkey. Since the World War II, particularly 1950s, the quarter-century that followed, witnessed a degree of state intervention in economies all over the world on a role hitherto unseen: "development thought", like conventional wisdom was really neither left nor right; there existed a broad consensus among the development theorists that assigned state intervention a leading role in the economy. Thus at the end of the World War II, state intervention took the form of generous welfare legislation, nationalism of private industries, and immense public programmes.

Pretty soon, the left came to expose the statist models. This influenced the Countries of the South: in Latin American countries, there existed a broad consensus that economies needed more state intervention than they had been given in the past and the right-wing authoritarian regimes began employing statist development strategies.

Many of the countries in South Asia and Africa who had just won their independence, eagerly adopted the statist models, for they seemed to promise a rapid journey into the industrial age. For most Countries of the South, they were under the traditional development concept, the latter were formulated in the forties or fifties when in the developing countries the industrialization was very rare and when the typical rates of capital accumulation were quite low. The typical way of life in these countries is rural and agriculture-based in which families and households work together to produce crops and livestock to support for themselves or to sell for cash, or more likely a mixture of activities. As far back in 1958, Hirschman (1958)[7] identified "rural underemployment (including so-called "disguised unemployment") which resulted from the immobility of its subsistence agriculture in India during the British period and its highly inelastic character during that period thwarted the industrial growth in the country. Hence for acceleration of capital accumulation, Hirschman (1981)[8] suggested the *traditional development concept* of an "activist state and for planning to overcome the disadvantages of lateness" through what calls for "a deliberate, intensive and guided effort".

Statist Model Strategy : "Import-Substituting-Industrialization" (ISI)

In addition to the aura of intellectual sanction which the statist approach got world over, there were practical factors too that made "statism" appealing as development strategies to the governments of the South : colonial powers had hindered the development of industries, industrial bases as well as the entrepreneurial class. So when the Southern Countries sought to industrialize rapidly, but lacked the *bourgeoisies* (the local capitalists or entrepreneurs) upon whom to devolve the task, the obvious agent for this transformation was the "state". Furthermore, in the early days of industrialization, the industrial firms grew from by amassing the capital from small *ateliers* (*i.e.*, workshops) and cottage industries to immense factories of the modern days. The costs of establishing a new venture in the present days being immense, the development planners had three options, that is :

(One) The first option was to cut the national economy off from the world economy, and try to take it through its own process of indigenous development, a model known as "autarky" which has been more popular historically in theory but in practice, seldom proved possible.

(Two) The second option was for the state to accumulate the necessary resources by restricting imports of goods in question by way of *tariffs i.e.*, taxes on imported goods, or of non-tariff barriers such as *quotas*, in terms of quantitative restrictions on the goods being imported as their qualitative restrictions through convenient restrictions or through *content regulations*, such as like 50 per cent of the given product to be locally produced and/or a *quality controls* (which can create a list of requirements that local producers are able to meet but that importers have a difficult time for satisfying). Such restrictions raise the prices of imported goods to local consumers, either by adding a surcharge to the world price, as tariffs do, or by reducing supply and thereby causing buyers to bid up the price, as non-tariff barriers do. Either way, the local investors who could not normally compete with foreign suppliers find the market suddenly benign. Provided they can get hold of the start-up capital, they can import the production machinery and begin to produce the goods locally.

Because the domestic market is relatively small, producers will operate at lower volumes than does the foreign competition. This means that they will not be able to take advantage of "economies of scale" which means that as the volume of output increases, efficiency of labour is maximized so that *unit production cost* decreases. Because the countries operating in an ISI regime cannot "exploit economies of scale", the prices of those goods will be higher than those on the world market. Nevertheless, provided these prices remain below the administratively inflated prices of imports, the indigenous venture can turn a profit.

(Three) The third option that the governments can go further to guarantee profits—they can establish licensing schemes (*License Raj* in India) that limit the number of firms allowed to produce a given product or import a needed input. Some governments have even allowed only one firm to produce a given product, this in effect giving it a legal monopoly that in combination with import restriction, provides an almost "watertight guarantee of profits".

With prices kept high and costs low, the attractions to invest are enough to persuade even the most conservative of investors. If a local entrepreneur cannot find capital to start up a venture, a foreign firm probably will. Import barriers may have closed off an export market to a foreign firm, but by setting up a "branch plant"—a capital-intensive-assembly plant (but none of the research and development capacity), it can sneak in under the wire and realize even greater profits than it had been earning when it was selling goods from its plant under this arrangement, or when it license state industry, protectionism and Soviet-style Five-Year Planning became the hales for a local firm to use its technology to produce its product, it typically allows the branch plant/license from ever competing with the parent company in export markets and thereby eroding any of its scales.

Strategies of Industrial Development in India: Statist Model of Mahalanobis

The pursuit of industrial development for a country essentially involves a choice between producing goods *either* for the domestic market *or* for the overseas market (export-oriented growth). Obviously the two strategies are not mutually exclusive, and in reality the question of strategy relates to the degree of emphasis attached to serving one set of markets rather than another. If a country goes for import-substituting–industrialization (ISI), for its investment of domestic savings as well as foreign exchange flows (either through export-import balances and/or multinational transfer/government loans), it can have a choice out of a number of options open to it. These are: importing investment goods for (*a*) for manufacturing low-capital, low-risk "wage goods" or simple consumer products for mass consumption, or, importing capital goods for (*b*) making intermediate goods and develop domestic raw material supplies, or, for (*c*) making capital (or, investment) goods out of the import of machine tools. Thus the ISI persuasion model for a country involve a choice between different options (variations) in these three vertically integrated sectors like (*a*), (*b*), or (*c*) or as a sequence, that is, beginning with the first stage (*a*) and moving on to stage (*b*), and eventually to stage (*c*). In the ISI model, if variation (*a*) is pursued, the strategy is, the importation of investment goods in order to produce consumer goods (previously imported), for the domestic consumer market. The countries like India having industrial base already in the country (Bagchi,1982)[9] adopted the type (*c*) that is, making capital goods out of the import of machine tools under the dominant role of the central government.

For overcoming colonial deprivation and neglect, after its independençe in 1947, India was the first country of the Global South to articulate a consistent and intellectually exciting post-colonial development strategy which sought to break the historical (and colonial) patterns of trade and production through democratic means of "national plans", the leaders considered the "trade as a whirlpool of international capitalism". In the 1930s and 1940s, the issue of "planning *versus* the market" which incorporated a central role to the national government in respect of the control and direction of economic activity, had been widely celebrated around the world. In India, the Industrial Policy Resolution was adopted in 1948 followed by its revised version in 1956, set down the fundamentals for planned economic development with a strong ideological orientation, for promoting a balanced regional development. P.C. Mahalanobis in 1955 provided the draft plan frame for the Indian Planning Commission's Second Five Year Plan of India for 1956 to 1960, in which a multi-sectoral version was employed in establishing two-sectoral targets for the national plan. In it following the "Fiedman's 1928 Soviet model", Mahalanobis put into practice the socialist ideas of investment in a large public sector at the expense of private sector, with heavy industry emphasis (at the expense of consumer goods) and a focus on import substitution (at the expense of export promotion). The plan frame proposed a two-pronged attack on problem of breaking through India's static economy: the creation of a large machine-building industry as a basis for expansion of secondary consumer goods industries; and decentralized cottage industries to assist the development of small scale and ancillary industries and to create jobs for India's army of unemployed. The latter was expected to produce consumer goods at a low cost and create employment with minimum social dislocation—like once a machine-building industry was established, the handloom would be replaced by powerlooms, and low-cost cottage industry would be able to compete with factory manufactured goods. The plan became a joint venture between Keynesian macroeconomics, Stalinist public investment policy, and Gandhian rural development and thus, it was in the spirit of high adventure and profound confidence that India embarked upon the programme of planned development to release the economy out of centuries of backwardness". Industry under public sector, centralized planning and protectionism became the hallmarks of the Indian development model and though this state-led ISI model did originated under Kemal Ataturk in Turkey in 1923, India is often considered to have been the paradigmatic case of this development strategy (Rapley, 1996)[10].

Mahalanobis model shares the assumptions of the economic growth model of Harrod-Domar that output is a function of capital only, that at equilibrium in the economy the output of any one sector (the *i*-th) will equal its stock of capital times its output-capital ratio ($b_i = 1/\sigma$), and that there is no international sector. Mahalanobis assumed for India, its $b_i = 1/\sigma$ *would lie between* 1:4 and 1:5; a population growth rate of India was assumed to be 2.5 per cent per annum, then the investment equal to 10 per cent of national income would be necessary to exert a "critical minimum effort" which would enable the economy in maintaining a "balanced-growth" for escaping the "gravitational pull of population increase".[11] The model can be applied in two sectors only, which are defined as producing consumption and investment goods respectively. In the investment goods sector two types of machines are assumed to be made, those required to produce consumption goods and those required to make more machines. Since all savings are invested in machines it follows that :

1. It pushes the questions of intersectoral development choice to the forefront. The rate of growth of consumption goods output depends in the short run upon that sector's share of the new machines produced in any period (i.e., its share of new investment).

2. It draws attention to the capital goods sector and argues that its capacity needs to be expanded if higher growth rates are to be attempted—the larger this sector becomes, the greater the economy's capacity to produce machines for consumption of goods manufacture. The rate of growth of consumption of goods output ultimately depends upon the share of new investment in the capital goods sector. In fact, the investment goods share of investment, y_k is the key variable in the model, and at equilibrium $g = f(y_k)$.

If the marginal rate of savings (s') is raised above the average rate (s), the share of new investment in the heavy (or, capital) goods sector can be increased at the expense of the consumption goods sector since the marginal growth rate in consumer demand will be correspondingly reduced. Thus, if a country can raise its savings rate (which implies $s' > s$), it is in a position to increase its rate of investment in productive capacity to permit future consumption levels to exceed what they otherwise would be.

When the average and marginal rates of savings are equal ($s = s'$), an equilibrium, in which machine capacity is fully employed, is achieved when the rates of growth of national income, investment and consumption are equal to the investment goods sector's share of investment times its output-capital ratio. In the Mahalanobis planning exercise, the output-capital ratios are adjusted for capital depreciation. This adjustment involves adding replacement requirements to the capital side of the ratios.

This chosen strategy of *planned economic development through an activist state* was perhaps the right response to the conditions prevailing in 1945 or 1950, when the private investment in basic industries and infrastructure was not forthcoming for various reasons such as lumpy investments, long gestation period and low returns. Also the private sector that time was found weak and ill-equipped to face the challenges of the time and this sector was not willing to establish industries in backward regions of the country. The public sector (whether under the centre or the state) in the country includes all its activities funded out of government's budget. In this period, due to the hereditary state of dependence on erstwhile imperialist power continued, so large number of foreign collaboration agreements took place with UK, USSR and other European countries for technologies which were becoming obsolete (but none of their research and development capacities)—so technological inefficiency remained with most of the public sector enterprises since their early days. And spatially there was an effort in emphasizing on fostering development of "enclaves" of modern enterprise at the cost of failing to develop the productive forces in the countryside. In the establishment and functioning of public sector enterprises, their existence was justified on grounds of meeting social and equity objectives but the national interest of profitability through productivity as a relationship between the output generated by a production unit and the inputs provided for the purpose, was not given much importance. Thus as compared to just 5 public sector enterprises in 1950-51, according to Public Sector Enterprises Survey 1998-99, on 31st March, 1999, there were 248 central public sector enterprises in the country. Of these total of 248, about two-third, 160 enterprises were in manufacturing/producing goods, about one-third, 75 enterprises were in services sector, 5 in construction and 8 in finance: By 1998-99, capital invested in public sector enterprises became worth of Rs. 2.737 trillion (or, about US$ 60 billion) in which manufacturing firms were having a two-third and the service enterprises an one-third share. Thus public sector occupies a key sector in Indian economy: in 1998-99, with its capital input of Rs.1.831 trillion in 160 manufacturing enterprises, generating 2.0 million jobs, it produced 90.71 per cent of petroleum, 89.95 per cent of coal and lignite, 79.00 to 87.00 of non-ferrous metals, 43.61 per cent of aluminium, 32.07 per cent of finished steel and 28.50 per cent of fertilizers (Agarwal, 2001).[12]

With the large domestic market to support just about any factory's or industry's existence without having to look for export markets, the phase of ISI could continue in India till the mid-1980s, during this three decades, there has been a short boom period till the late 1960s. Determined government action in ISI in India also led to the establishment of intermediate (iron and steel) and capital goods (non-ferrous metals, locomotives, ships and the simpler types of machine-making) industries in India. But except for the oil sector, this huge public sector remained inefficient, having excess manpower and having a large number of sick units. Due to poor financial performance of a large number of public sector enterprises in the country, it is perceived that the public sector as a whole in India has neither been able to improve productivity levels and efficiency, nor this sector as a whole become competitive in the domestic market.

Public Sector Enterprises: The Issue about their Privatization

The public sector firms created by many governments in the Southern Countries had become poor performers and unlike the private sector firms, they do not meet the standard measures of quality of operating productively and more efficiently financially because of their bloated staffs, high budgets, unused production capacity, heavy debts, and consistent loses on their operations have all too often stacked up poorly.

Unlike the structural adjustments in fiscal austerity which are useful when imposed in a discriminating manner like cuts in some budgets, increases in others, but the same structural adjustments cannot be done by privatization of the public sector enterprises, since these measures often fail to capture some of the particular tasks taken on by public firms since the public sector firms may also confer "beneficial externalities" on the economy so that the costs of a product or service are concentrated in the firm while its benefits are spread throughout the economy. A common example of such an externality is *human capital formation*, for example, the best way to develop a pool of engineering talent is to create an engineering firm; technological capability can be improved by creating a firm that specializes in research and development (R&D). In the less developed economies of the South, the costs of such firms often exceed their revenues. However, if in the meantime a pool of engineering or scientific talent is built up, which can then be exploited by the private sector, the net gain to the economy may well outweigh the investment. This occurred in power equipment industry in Brazil, where poorly performing public firms helped create technological capability, and in Taiwan, where they helped foster industrial development and diversification by building up new industrial sectors.

Thus there can be legitimate economic reasons for a government to maintain inefficient, loss-making firms and what seems to govern the quality of a firm's performance is less who owns it than who runs it, the condition under which it is run, and the structure of the industry in which the firm is located. In most cases in which public sector firms perform poorly their performance can be improved without privatization—for example, a public sector firm run poorly, can be improved by cleaning the administration; similarly, if a public sector firm's mandate is so extensive, or its hands are so tied by such things as price controls, that it cannot hope to recover its cost, then deregulating those firms and allowing them to operate as private agents can improve their performance.

In public sector firms, laxity occurs on the firm's administration, if there is a practice of "soft budgeting" so that the state covers the losses of a firm out of public revenue, which eliminates the fear of bankruptcy of the concerned firm, but if this inefficient firm in question is in a monopoly market, it can enjoy the laziness afforded by this monopoly—in such a case, creating a rival company shifts the monopoly from one agent to another so that competition is injected into the industry or in the sector of it.

Hence in all the above cases, public sector reform seems to be the measures at least as likely as privatization to improve the performance of the public sector firm in question. Where reform has been used instead of privatization, the results have been positive. Not only does privatization do less than hoped to improve firm performance and accelerate economic development, it seldom raises much money for the governments selling them to reduce their fiscal deficit. Public sector firms are sometimes sold cut-rate for political reasons, perhaps to favour friends of the government; this might encourage rent seekers and at the same time worsen income distribution within the economy. But, money-losing firms must be sold at a loss; profit-earning firms may earn the government a good price, but less than they might have earned over the long-term in dividends. Thus privatization of public sector firms may hinder resources that could be used productively for other purposes—for example, attracting investors into the purchase of public sector firms may crowd out investment in private sector firms when capital is in short supply. Thus instead of releasing public sector firms into a market economy, it is better that *public sector reform coupled with policies to encourage new private investment for creation of new firms, seems the best policy.* Meanwhile, the immense bureaucracy of the state-industrial sector could not always be trusted to cooperate in any effort to reform the public sector and thus undermine its own power base. Hence even in the situations the state-industrial sector needs fresh influxes of capital in order to complete their public sector reforms, but are unable to obtain this capital without selling some or all of their shares, however, it is best to reorganize public sector firms, turning them from state- to market-oriented enterprises before selling them off. *Relying on the private sector to do this may be a mistake.*

On the other hand, in export front, the country has to go on depending on agricultural products—replacement of products like jute, cotton and sisal were being partially or completely replaced by their synthetic materials in the countries of Global North and the demands for such beverages tea and coffee reached saturation points in the world market. Indian exports as a proportion of world exports fell from 2.2 per cent in 1948 to 0.5 per cent in 1983 and 0.7 per cent in 1987.[13] There is no widely accepted single quantitative factor that captures the depth and degree of integration of a given country with the world's economy, but the ratio of exports plus imports to GDP is a commonly used measure. India's ratio was a mere 16.6 per cent in 1980, compared with China's 22.3 per cent, Malaysia's 112.5 per cent, South Korea's 75.5 per cent and Thailand's 54.4 per cent. Thus India was withdrawing from international trade while international trade was increasing substantially, "doubling every eight or nine years" (Jalan, 1999)[14]. The country had to import foodgrains under PL480 of the USA in the sixties. Thus, ISI of this type results in segmental development "pauperizing vast numbers and leaving a broad swath of population out in the cold" (Bagchi, 1982).[15]

Apart from the big (investment) push, import substitution industrialization, ISI and the "balanced growth" approaches, Mahalanobis' thinking on the Second Plan was logically extended to the damaging Industrial Policy Resolution in 1956, which reserved seventeen industries exclusively for the public sector. The reserved industries included iron and steel, mining, machine tool manufacture, petro-chemicals and heavy electrical plants. The overall logic behind "reservations of these capital-using high cost industries through interventionist policy of the centre" was that these were industries of "basic strategic importance" and "required investment on a scale which only the state could provide." Gradually deregulation in form of partial reform got started since 1975s which left the door open for private entry into the reserved list, but in practice the door was closed for 35 years when there were hardly few activities except for those in agricultural sector and the unorganized sector (that could be undertaken without government permission), so that whatever surge the domestic economy had couldn't increase India's international competitiveness : the country's growth performances became unsustainable, the gross fiscal deficit of the government which averaged 3.2 in 1970-75, 4.3 per cent in 1975-80 and 7.8 per cent in 1980-91; in 1990-91, current deficit rose to a record of 3.2 per cent and debt-service payments amounted to as much as 35.3 per cent of the country's foreign exchange receipts and her foreign exchange reserves dwindled enough to sustain for 2 and half-a-month, over and above the country's short-term deposits amounted to a dangerously high level of 146.5 per cent of foreign exchange reserves the reservation remained sacrosanct until the reforms of 1991, when the country came to a "turning point".

Licensing Policy and the New Political Economy: Even more damaging than placing the public sector at "the commanding heights of the economy" were the creeping controls on the private sector. The most bizarre and damaging was the restrictive "licensing system" for control of production of consumer goods. It began with the Industries (Development and Regulation) Act of 1951, which required an entrepreneur to get a license to set up a new unit, to expand it, or change the product mix. The purpose of licensing was (*a*) to create the planned pattern of investment, (*b*) to counteract monopoly and the concentration of wealth, (*c*) to maintain regional balance in locating industries, (*d*) to protect the interests of small-scale producers and encourage the entry of new entrepreneurs, and (*e*) to encourage optimum scale of plants and advanced technology. All these were good intentions but the way that the bureaucracy went about administering the licensing system created a nightmare for the entrepreneur without achieving any of its social objectives. This led to a curb on labour-using industries, such as engineering and consumer goods—their low output could not support the capacity created for the production of capital and other intermediate goods. While during this period of early fifties, in several countries of Taiwan, South Korea, Singapore and China, government intervention was positive and gave competitive inducements to efficiency. In India such intervention through the "License Raj system" became negative in the 1960s—it fostered monopolies, encouraged uneconomic-scale plants, brought delays, arbitrariness, and corruption, and achieved nothing.

The neo-classical theory is rooted in the assumption that humans are rational, self-interested and utility maximizers. When this rational action of human beings is applied to "import licensing system", Krueger (1974).[16] got a new dimension of political theory of development, a *new political economy*. Krueger studied the effect of quotas on the behaviour of firms. In any situation in which a government restricts the supply of a given good to a level that is below demand, the local price of that good will be bit up above the world price. The difference between the price paid by the importer (*i.e.*, the world price) and the price the importer charges local buyers (the local price) is called *economic rent*. Because quotas create the windfall for importers, import licenses become hot commodities that are sought after for their own sake, not just because they offer access to needed inputs. Krueger found that with licenses being assigned to reflect firm's capacity, plant managers would invest to expand their plant even when they had idle capacity. The problem with idle plant that it generates no income, its owners must continue paying mortgage and other bills on it which enabled them to obtain bigger import licenses, which they could then sell to other managers at a profit. However, in the process their productivity dropped even further, as an even larger share of plant capacity went unused. Plant managers also tried to obtain licenses through bribery, hiring the relatives of officials in return for licenses, and so forth; such rent-seeking behaviour consumed resources that could have better spent elsewhere in the economy.

Jagdish Bhagwati[17] in 1982 expanded this to look at tariff evasion, tariff seeking and revenue seeking. These were, all he said, directly unproductive, profit-seeking activities made possible by government controls. They were profitable, but produced no goods or services, and thus wasted valuable resources. Capital-gains tax treatment, for example, led to the overbuilding of apartments or uneconomic oil exploration. The policy implications of this new political economy were clear: less government control. If some kind of protection were required, tariffs were better than quotas, because tariffs created no opportunities for rent. In this regard, neo-classical theorists distinguished between discretion and rules. Quotas and licenses were applied in a discretionary manner by bureaucrats or politicians, who could abuse their powers to favour themselves or their friends. Tariffs, on the other hand, were rules : they applied equally to everybody, and so could not create opportunities for rent seeking. Neo-classical theorists tended to favour rules over discretion whenever some form of state intervention was deemed necessary.

The new political economy was further elaborated in the work of Robert Bates (1984)[18] on Sub-Saharan Africa. The static comparative advantages for African countries lay in agriculture, but Bates found that governments in Africa seemed biased for the urban and that also at the cost of the farm sector. Currency overvaluation and pricing policies kept export prices low on farm products on one hand, and subsidizing the urban population's food bill internally on the other hand. At the same time, currency overvaluation also kept the prices on imported industrial inputs low, while protectionism kept profits high, which made life good for the industrialists. Marketing Boards, in turn, skimmed off revenue from the primary sector to fuel urban development. All in all, Bates found that cities were squeezing the rural sector in order to fuel their own growth, dampening the dynamism of agriculture which should have been the economy's "engine of growth for Sub-Saharan Africa"—import substitution industries were gobbling up foreign exchange and earning none in return for agriculture, so that agriculture, the sector of the economy that did garner foreign exchange, was contracting. The unattractive prices prompted many farmers to resort to subsistence production or to pack up altogether to be a wage earner in the city.

Puzzled by the apparent irrationality of this self-defeating policy, Bates turned to "interest-group analysis" in trying to find an answer—*Interest-group analysis has a long history in the study of industrial politics* : a "small group" with a common interest is more effective than a "large group" in terms of getting a benefit from the government, to which they are deprived of by voicing their protests, because large groups are saddled with the problem of dispersed benefits and the free-rider effect, so though a large group by its sheer weight of numbers may appear daunting in securing a policy that offers everyone an equal share of the gains through public demonstration, the relative incentive for action is lost. Thus the interest-group politics in liberal democracies, often lead to undemocratic outcomes, because small groups work to secure desired policies while large groups remain largely ineffectual.

Since individuals are self-interested, so though the development policies were counter-productive, they nonetheless served the interests of the urban elite of skilled labourers and the industrialists. This class alliance underpinned the power of modern Africa's regimes, and no government could afford to antagonize it—industrialists through their wealth and personal connections get a desirable support base, the working class, living in densely packed neighbourhoods, can easily take to the streets and threaten stability if it feels that it has been pushed too far. Bates considered this "urban bias" a key factor in Africa's underdevelopment. It had to be overcome and African governments had to be prodded to realize that their static comparative advantages, which for the most part lay in ågriculture. Thus raising the prices for peasant farmer's produce would lead them to increase their output and would bring more foreign exchange into the country and this producer prices could be easily raised through currency devaluation. Soon after Bates published work in 1984, "getting the prices right" became a guiding concern of the World Bank in Africa.

Another factor which affected a country's economic behaviour badlý was its *tariff policies*—all countries in their macro-economic policies intervene in international trade through tariff measures. However, such intervention should not be such as to make exports less profitable than production for the domestic market, or provide indefinite protection to inefficient and high cost firms from internal and external competition. During the British period of India, trade was used an instrument of economic exploitation and there was substantial drain of resources from India to England. "Free trade" was used as a mechanism to transfer raw materials cheaply to England, which in turn processed them into final goods for export back to India. The inequity inherent in this type of trade led India to be highly suspicious of trade as a positive instrument of development policy—thus after Independence the country put heavy import duties which became among the highest in the world: the tariffs in India on imports became four or five times as high as in other countries, thereby keeping the profitability of monopoly domestic sales high. The fact that this "export pessimism" of fifties and sixties continued too long, even after the conditions of international trade had changed fundamentally, cost the country dearly. Thus *the more the country tried to industrialize through ISI phase*, the more it had to import intermediate goods and more sophisticated capital goods which led the country to frequent balance of payment crisis and to become dependent on Countries of Global North for technology, product development and managerial expertise without developing the productive forces of any but few *modern enclaves* here and there in the country, all these acted as severe constraints on the long-run growth of its economy. Had India been able to adopt a policy akin to that of Japan (which rigidly excluded the entry of foreign firms and foreign private capital while permitting the purchase and imitation of foreign techniques in every field), after a fairly fast rate of industrialization along the path of ISI , India might need not have more or less stumbled into the situation of serious balance of payments deficits which it faced in the 1990s.

The *new political economy* reached the following conclusions: Given that people behave in a self-interested manner, they would seek the available opportunities to maximize their gains. If these opportunities lie in the market, their self-interested behaviour would create spin-off-benefits for others—new jobs, more productions and so forth. However, if those opportunities lie in a large and interventionist state, people will neglect the private sector and engage in activities that are detrimental to the welfare of society as a whole, such as corruption, rent seeking, and nepotism. The solution for all these as well as the recommendations for development strategy was obviously pointed in one direction : Reduce the size of the state and its role in the economy, so as to bring less government intervention, abandon ISI in favour of *outward orientation.* Free up the market and make it attractive to entrepreneurs, and at the same time Remove opportunities for corruption, rent seeking, and other economically harmful activities.

Post-Keynesian Neo-Classical Agenda

By the 1980s, a formidable corpus of literature had come together that hobbled Keynesian economics and reasserted the primacy of new political economy over the statism of the post-war generation of development political economists. By "outward orientation" neo-political economy specifically mean not only export-led growth (rather than domestic-led growth) but a minimum of state control in this process. At any rate, underlying neo-political economy was a sort of "trade optimism" that trade could be relied for growth. Economic planning was not needed to alter the structure of production: agriculture should be left free to flourish, and trade with the North was a boon, not a hindrance. Thus in the politics of Countries of the North, the post-war Keynesian consensus was about to be shattered by an all-out assault on statist development theory through the rise of conservative governments and the rightward shift of virtually the entire political spectrum.

Neo-Classical Recipe for Development in the South

Many developing nations are in debt and poverty partly due to the policies of institutions such as the International Monetary Fund (IMF) and the World Bank. Their programmes have been heavily criticized for many years for resulting in an increased dependency and poverty by the developing countries upon the richer nations. This is despite the IMF and World Bank's claim that they will reduce poverty. They impose Structural Adjustment Policies (SAPs), to ensure debt repayment in such a way that social spending and development must be cut back and debt repayment must be made the priority. In effect, the IMF and World Bank demand that these poor nations lower the standard of living of their people.

SAPs—The Root of the Problem

In Countries of the South, neo-classical theory has been embodied in *Structural Adjustment Programmes* (SAPs). Since 1976, SAPs are the package of IMF- and World Bank-designed macro-economic conditions which developing countries must comply with before they can receive resources from those institutions. Essentially, SAPs seek to make both the state and the market more efficient in such a way as to accelerate growth and eliminate waste. "Structural adjustment" embodies the goals of neo-classical economy: It places the market at centre stage, assigns the state a secondary role in development, and puts its faith in the potential of unfettered individual initiative, creativity and ingenuity. Sensitive to the obstacles placed in the way of such individualism by an interventionist state, *Structural Adjustment Packages,* SAPs aim to remove perceived structural blockages to the efficient operation of markets. To this end SAPs have usually included such elements as *fiscal austerity* and *disinflationary policies*, the *privatization of state-owned enterprises, trade liberalization, currency devaluation*, and the general *deregulation of the economy*, including *financial* and *labour-market deregulation.* SAPs also try to attract new private foreign investment in industry. SAPs seek to increase the powers and freedoms of entrepreneurs and investors, increase pecuniary incentives and competition, lower costs, restore macroeconomic stability, and make the state *leaner* and *reduce its presence in the economy. All these represent a decisive shift away from the state and back toward the market what has come to be seen as a "market-state dichotomy"*. All in all, SAPs are based on economic prescriptions contained in the 'Washington Consensus', which recommends a set of privatisation, liberalisation and economic stabilisation measures. The Consensus is a recipe for globalisation, or 'policy coherence' facilitated by the WTO, IMF and World Bank. Many people see the Consensus as epitomizing the kind of ideological prescriptions divined in Washington [19] and exported to the rest of the world.

Many of the objectives of the Consensus are valid, only if they arise out of a homegrown "national domestic consensus" (NDC) through consensus-building processes based on deep and broad participation from civil society and other domestic interest groups which balances competing objectives with attention to the needs of vulnerable people, local producers, and the natural environment. As it is, adjustment programmes, which implement the Consensus, often conflict with homegrown development agendas—which feature the priorities of social development (*e.g.* poverty reduction, equity, popular participation) and environmental protection. The virtues of such a national approach (that is, NDC) are multiple :

- it enhances the accountability of developing country governments to their citizens and democratic institutions and builds broad-based ownership for the development objectives of the nation.
- the NDC serves as a kind of "business plan" for the country, which external actors can opt into, or not, as the case may be.
- poverty is treated as integral to long-term strategies for accumulation and distribution.
- the process of developing a NDC is an opportunity for broad-based national dialogue and consensus building, essential for

democracy and national unity, especially in weak post-colonial states.

The results of SAPs have been especially dismal in low-income countries where the NDC initiative is being implemented. Low-income countries in Africa, Asia and Latin America have experienced economic and social stagnation or decline and, essentially, lost two decades of progress. To some extent, Africa is a "forgotten continent." Because SAPs often conflict with homegrown development agendas, they are, according to the World Bank's own account, implemented in secretive and non-transparent ways. The Bank allies itself with powerful partners to pursue the adjustment agenda. It finds other relatively powerless allies (*e.g.* NGOs) to mitigate the negative effects of adjustment and, on the periphery of its operations, achieve some aspects of the popular agenda. This 'command and control' style of doing business has undermined the evolution and nurturing of democratic processes. In country after country, the SAP agenda has undercut government-civil society relationships and, hence, progress toward democratization. Furthermore, in their fervour to promote SAPs, the IMF and World Bank have backed governments that abuse human rights, including the right to equitable, sustainable and participatory development. Some nations have been forgotten, or left behind, because they lack access to international capital markets. Many other nations that are integrated into the global economy have suffered dramatic economic reversals. Some of these nations were the IMF and World Bank's star performers. Bad advice from the institutions heightened the vulnerability of some countries to the rip tide of volatile international capital flows. The power of governments and regulatory bodies has been totally outstripped by the awesome and explosive growth of these flows. As a consequence, we have witnessed financial debacles accompanied by increasing levels of instability and cascading adverse economic and social effects in country after country. SAPs often promote production and distribution patterns that contribute to poverty creation. Instead of ameliorating adjustment with new safety nets or more welfare programmes, the institutions should re-conceive the programmes in ways that benefit people and the environment. Mitigation strategies, such as social safety nets, offer poor and vulnerable populations too little assistance, too late. The IMF and

World Bank place higher priority on the protection of social sector budgets than they did years ago. Still, these budgets are often squeezed by efforts to reduce fiscal deficits and dampen inflation. Mitigation efforts alone are inappropriate responses to bad economic policies. At the root of the criticisms of the International Financial Institutions lies concern with their use of SAPs to export a liberalising agenda, both in terms of the economic content and impact of the programmes, and their imposed nature. Noted economist, Ricardo Hausman, has compared structural adjustment programmes to military tanks that roll through a nation creating collateral damage. He compares welfare programmes (*e.g.* social safety net programmes) to ambulances that attempt to rescue or mitigate the harm to casualties.

Fiscal Austerity: Fiscal austerity (or, "belt tightening" as it is sometimes known) refers to government reductions in spending—it has been an important component not only of SAP, but of the government-retrenchment programmes all over the Countries in the North in the 1980s and 1990s.

The logic of fiscal austerity is straightforward—the more money the government spends, the more money it takes out of the economy. This money is removed directly, through taxes, or indirectly, by borrowing. When governments increase their borrowing they compete with private borrowers, such as banks and corporate bond issuers, for scarce capital. The quickest way attract lenders is to raise the interest rates paid to them. When interest goes up, not only do businesses and consumers cut spending—because the cost of credit, by which so much spending is done, gets too high—but people with money to spend are attracted to put it in the bank, where returns are high, rather than spend it or invest in lower-yielding securities (stocks).

Furthermore, whereas government spending can be productive over the long-term, for political and other reasons it often prompts *inflation.* Much government spending takes the form of short-term transfers including salaries, welfare payments, subsidies, and grants. Salaries, in turn are often increased regularly to retain the support of civil and defence services, often important underpinnings of a government in the South. Although this money is pumped back into the economy, it is spent rather than invested it contributes to inflation that is, when the amount of money in the economy is increased more rapidly than the economy's productive capacity, buyers bid up the prices of goods.

So the combined effects of excessive government spending are as follows: by withdrawing money from the economy, through taxes and borrowing, and by driving up the interest rates, the government "crowds out" private investors. Businesses find it hard to attract savings, and so must restrict their investment. *Economic activity therefore declines.*

High inflation rates beyond 4 per cent can further inhibit investment, because they reduce business confidence and make profits unsure. When potential profits seem likely to be eroded by inflation, investment in new technology becomes unappealing. Investors are then more likely to prefer investments that promise high returns in the short run but may contribute little to long-term development, such as property speculation and trade. Under such conditions big investors often find it safer to export their money to havens where the value of their investments is less likely to be eaten into by inflation.

The solution to all of these problems appears simple. By reducing spending, governments enable interest-rate cuts. By capping pay rises and slashing budgets, they reduce inflation. Private investment thus becomes cheaper, and the environment for business more attractive. Economic activity should therefore resume.

In addition to lowering inflation and borrowing costs and encouraging investment, fiscal austerity should achieve another goal: government-spending cuts and caps on salaries and transfers should lead to a fall in real wages, which in turn should reduce overall consumption in the economy. This so-called demand compression should leave a surplus of unsold goods that will then be available for export. Ideally, more foreign exchange should flow into the economy as a result, stimulating economic growth and rectifying any imbalances in the current account (that part of a nation's balance of payments that covers income and trade flows).

Privatization : The idea behind privatization is self evident. Any economic vision based on the virtues of a private market economy tends to frown upon the state performing the functions that can be taken on by private companies. The severe abuses and inefficiencies often associated with public firms in the Southern Countries provide added impetus to privatization. It is also believed that the owners of a private firm have a greater interest in maintaining its efficiency and profitability than do public-sector managers, who operate more like civil servants and so might be given to such strategies as "empire building". In theory, privatization should raise money for cash-starved governments, enhance the normal operations of the market economy, and improve the efficiency and financial performance of the firms privatized. It is worth noting, however, that the argument for privatization has often been expressed more strongly by the political wing of the neo-classical school than by its economists.

Trade Liberalization, Currency Devaluation, and the Abolition of Marketing Boards : Trade liberalization refers to the effort to reduce hindrances to trade, thus maximizing the free flow of goods and services. At a general level there are two types of trade liberalization: (First) there is the liberalization of foreign trade by eliminating or reducing qualitative and quantitative restrictions on imports, especially quotas, streamlining taxes on imports and devaluing overvalued currencies. Then there is (two) the liberalization of domestic markets through the elimination of price controls and marketing boards. In addition, because it raises the price of export goods in local terms, devaluation has often been promoted as a means to give producers of export goods an incentive to increase production.

As a rule, ISI regimes limited imports of consumer goods but favoured industrial producers when it came to the allocation of her currency, whose price was kept down by overvaluation. This hard currency is then used to import the inputs and capital goods needed in the production process. When a currency is devalued, its purchasing power on international markets declines. Therefore, trade liberalization and currency devaluation doubly hurt firms that formerly relied on imported goods to produce consumer goods for a protected market. Their import costs jump just as imported consumer goods start entering the country, stiffening competition. These firms must find ways of lowering their costs, or else go out of business: wasteful firms go under; efficient survivors then pick up the slack and thrive. In sum, trade liberalization and currency devaluation should stimulate an economy to realize its "static comparative advantage". In other words, an economy should specialize in those industries in which it has the lowest opportunity costs, abandon those that are expensive for the economy to maintain, and rely on imports to fill the gap. This will ensure that the economy's resources are used with maximum efficiency.

In the Southern Countries, especially in a less-developed one, much of the *static comparative advantage* lies in the agricultural sector. Devaluation boosts this sector by giving export-crop farmers a leap in income, because even if the world prices on their crops remain constant, the new exchange rate generates a greater amount of local currency. Their improved position should ordinarily encourage farmers to augment their output. This practice of "getting the price right" is a key concern of the new political economy. Given this school's belief that producer prices on primary goods were artificially distorted downward by an interventionist state, it follows that rolling back the state should, all other things being equal, lead to increased prices and thus output. Although domestic market liberalization is intended to improve the functioning of all domestic markets, in practice the concern of new political economy has been to improve agricultural markets.

One way to liberalize domestic markets is to abolish marketing boards. This should introduce competition into local markets, thereby increasing the bargaining power of farmers and enabling them to obtain better prices on their sales. If a marketing board is to remain, it can be pressurized by donor agencies into paying farmers better prices. However, given the sometimes terrible abuses wrought by marketing boards in Africa, where prices designed to extract maximum revenue from producers were so low they simply drove producers out of the market, dismantling market boards altogether which made sense. Africa is hindered from fuller participation in the global economy by a "standards divide"—a combination of inadequate capacity to meet world trade standards for goods and limited opportunities to help shape these standards to ensure that they are consistent and fair.

Retrenchment and Deregulation: At a general level, government retrenchment and deregulation should free up the market and reduce the inhibitions on private entrepreneurs. Deregulation should enable the market to function more effectively, reducing price distributions and allowing them to find levels that encourage efficient resource allocation. Wages may drop, encouraging investors to hire more workers and use more appropriate labour-intensive technology. Bankers will find the business environment more conducive, and will expand their operations and make more credit available.

An added concern is the battle against corruption in the Southern

Countries, in which retrenchment is said to be a useful weapon. Paring back the state reduces channels to resource accumulation in the public sector. Opportunities for rent seeking diminish, there are less patronage appointments to be used to gain political influence, and there are fewer chances to use public firms or marketing boards to skim resources from the economy. Ambitious individuals will therefore turn to the private sector to seek upward mobility. Similarly, trade liberalization should allow highly skilled managers who formerly lobbied for quota shares to turn their attention to productive endeavours.

A Spiralling Race to the Bottom

The IMF and World Bank prescribe cut backs "liberalization" of the economy and resource extraction/export-oriented open markets. To be attractive to foreign investors various regulations and standards are reduced or removed. Factors such as the following then, lead to further misery for the developing nations and keep them dependent on developed nations:

- Poor countries must export more in order to raise enough money to pay off their debts in a timely manner.
- Because there are so many nations being asked or forced into the global market place — before they are economically and socially stable and ready—and told to concentrate on similar cash crops and commodities as others, it is like a huge price war. The resources then become even cheaper from the poorer regions (which favours consumers in the West). Governments then need to increase exports (by further reducing costs, making the resources even cheaper etc.) just to keep their currencies stable (which may not be sustainable, either) and earn foreign exchange with which to help pay off debts.
- Structural adjustment [demanded by the IMF] is best summed up in four words: "earn more, spend less". While such advice might be valid if it were given to only a few countries at once, dozens of debtors are now attempting to earn more by exporting whatever they have at hand; particularly natural resources including minerals, tropical crops, timber, meat and fish. With so many jostling for a share of limited world

> markets, prices plummet, forcing governments to seek ever-higher levels of exports in a desperate attempt to keep their hard currency revenues stable. The "export-led growth" model on which the IMF and the World Bank insist is a purely extractive one involving more the "mining" than the management—much less conservation of resources.

"The IMF has repeatedly stated that it is not, and was never intended to be, a *development* institution." Neither was the World Bank, "The fundamental goal of creating markets for industrialized countries' exports, was written into [their] Charter "139" which means debt traps and mercantilist dependency were the goals all along.

When the IMF/World Bank/GATT/NAFTA/WTO forces other countries to devalue their currencies and reduce consumption to increase sales of their valuable resources to the developed countries, this lowers the value of their labour and commodities, and raises the relative value of manufactured products from the developed countries. The claim is made that this is due to the inefficiencies of developing countries industry and labour, but this is hardly so. The low level of wages is intimately linked not to low productivity of labour-time, as classical economic theory would suggest, but to the undervalued exchange rates and the workings of the international monetary system....The world's monetary system does not set values of the currencies on the basis of relative productivity of workers....The present system is based on balance of payments considerations and on capital flows....The *Mexican currency is valued much lower than the relative productivity of Mexican workers collectively*....While the average amount produced per unit of time by workers in Mexico, Brazil or Bangladesh is lower than in France, the United States or Japan, *the difference in wages at present exchange rates is much bigger than the difference in productivity*. This explains why the purchasing power of U.S. dollars, French francs or Japanese yen is much bigger in Mexico, Bangladesh or Brazil than it is in their countries of origin.

The product that export platform countries of the South are selling is not merely cheap labour, but highly productive labour. In Singapore...McGraw Hill produces in one year an encyclopedia that takes five years to produce in the U.S....Mexican metal workers are 40 per cent more productive than U.S. workers, electronics workers

10 to 15 per cent more productive, and seamstresses produce 30 per cent more sewing per hour than their U.S. counterparts.

Developed countries claim to be financing the developing countries, but actually the poor countries are financing the rich through low pay for equally productive developing countries labour, investment largely in commodity production for the wealthy world, and the many methods of inequality of trades. In this process, between 1980 and 1990—when measured against the dollar (not internal relative values)—wage levels in Mexico declined by 60 per cent [another 40 per cent in 1994-95],...in Argentina by 50 per cent and in Peru by 70 per cent and these were before the 1997-98 collapse of developing countries currencies.

In only seven years, the price of a tractor for Tanzania, measured by the export value of Tanzanian "sisal" (is like Rupee for Indian currency), doubled. The relative value for rubber exporters dropped 300 per cent between 1960 and 1975. Cotton exporters lost 60 per cent of their buying power in the same time span. The price of primary commodities that the Countries of the South exports have collapsed to the same price as 21 years earlier while prices for manufactured products have soared, forcing then to export more and more while importing less and less. The IMF/World Bank/GATT/NAFTA/WTO imposes these austerity measures but seldom restricts the purchase of arms, or consumer purchases for the elite from the North. The impoverishment of the Southern Countries for the enrichment of the developed countries, once the very signature of a highly successful neo-mercantilist trade policy, has evolved now into the impoverishment of both the as well as the Countries of the North for the enrichment of a reconstituted financial aristocracy Southern Countries, the *emerging corporate mercantalists*. The IMF/World Bank/GATT/NAFTA/WTO lays down and, with the military as last resort, enforces the rules. The Southern Countries are expected to lower their living standards and export more minerals, lumber, and food, all to pay debts that did little for their economic development. Typically those debts were incurred for investments to extract resources, produce agricultural exports, and build the infrastructure to ship these commodities to the developed countries. During the last 50 years of colonial period from the late nineteenth century to shortly before war when empires had absolute control of their colonies, the primary commodities price dropped 60 per cent relative to the manufactured commodities price. The discrepancy in relative values between manufactured and primary commodities highlights the reality that Adam Smith's *free trade*, as interpreted and designed by neo-mercantilists, is even more effective than colonialism in appropriating the natural wealth and labours of the Southern Countries.

To become prosperous or maintain prosperity, the developed countries knows they must educate their citizens, they must provide them with health-care, they must build transportation systems so people and goods can be moved, they must support the building of efficient industries to process their natural resources, they must pay their labour well to produce consumer buying power, they must maintain a healthy economic multiplier through a proper balance between manufacturing their own consumer products and their imports and exports, and they must trade equally with the rest of the world.

Military colossus insists that other nations reduce their education, reduce their health-care, eliminate supports for industry, reduce the wages of an already impoverished labour force, enforce their monopoly on industrial technology, and they impose unequal trades upon the Southern Countries. That the structural adjustments forced upon the Southern Countries are exactly opposite the policies under which every wealthy nation developed tells us the power brokers of the developed countries know exactly what they are doing. Their Grand Strategy is to impose unequal trades upon the world so as to lay claim to the natural wealth and the labourers of weak nations.

The Development of the Domestic Sector: Sustaining Absorptive Capacity

It seems rather facile to highlight this, but for an FDI-assisted development strategy, (first) there needs to be an appropriate compensatory domestic sector to support the FDI, and (secondly) their need to be domestic units with the capacity to learn from the MNEs—that is, MNE dependent export strategy needs a proactive element for dynamic competitiveness. Indeed, the presence and condition of the domestic sector is crucial. If no domestic sector were to exist (say, in an LDC) there can be no opportunity to absorb spillovers from FDI. In a perfectly liberalized world, MNEs have no incentive to encourage the development of domestic firms in a LDC to meet their needs because other MNEs may be able to do so more efficiently, either through imports or FDI; in an extreme case, there may actually be no FDI inflow, because MNEs will prefer to locate production in a regionally optimal location and simply import. Thus FDI in a completely liberalized milieu does not lead to growth in the domestic sector. The benefits of FDI only occur when there is domestic investment, and where the domestic investment has the ability to internalize the externalities from FDI.

The absorptive capacity supports further accumulation of technological knowledge, and technological advances support the further development of absorptive capacity in a cumulative, interactive and virtuous process during the catching-up stage. However, this virtuous circle of technological accumulation takes place only in an adequate minimum level of absorptive capacity if it is initially present. This threshold level of absorptive capacity is most significantly associated with the development of what is called "basic infrastructure" which represent "genric" location advantages. Acquiring and sustaining this threshold level of absorptive capacity if it occurs in the pre-catching-up stage. In this respect much of Latin America does quite well.

However, progress towards more technology-intensive manufacturing activities depends on the existence of "high tech infrastructure". For firms to be able to internalize and absorb externally generated technologies, and to create their own, this type of infrastructure is the key. Such infrastructure plays an important role in promoting the innovatory and absorptive capacity of firms. It also acts a mechanism to "direct" technology strategy and as a mechanism to overcome market failure. It is important to understand that while learning and absorption takes place at firm level, but the success or failure of individual firms occurs in orchestration with an "entire" system. Within any system, there exists a broader non-firm-specific knowledge base within what might best be described as "non-firm actors" that are crucial to a country-level understanding of the process of technological accumulation. Non-firm actors determine the knowledge infrastructure that supplements and supports firm-specific innovation. For knowledge level one can speak of *national technological* or *competitive advantages*, which is not simply the sum of the firms, but the synergistic effect of all the knowledge infrastructure a country is having which is not simply the sum of the educational institutions.

Absorptive capacity is also about the creation of the appropriate quality and quantity of human capital. Human capital represents a subset of absorptive capabilities. While both physical and human capital are necessary inputs for catching up, the lack of appropriate incentives for production and investment can compromise the success of the technological upgrading. The availability of a large stock of suitably qualified workers does not in itself result in efficient absorption of knowledge. This requires the presence of institutions and economic actors within industry which defines the stock of knowledge in a given location, and the efficient use of markets and hierarchies, be they intra-firm, intra-industry or intra-country. This knowledge is not costless, and must be accumulated over time. Important externalities arise which impinge on the ease of diffusion and efficiency of absorption and utilization of external knowledge.

Absorptive capacity is therefore also concerned with the efficient use of knowledge acquired. Firms need the ability to use prior knowledge in the solution of practical problems that are commercially viable. Thus, absorptive capacity implies problem-solving skills that emerge directly as a result of attempts to assimilate external knowledge. These efforts represent a potential for learning how to undertake different activities through investing in R&D, *i.e.,* to create new knowledge, and is referred to as the process of "learning to learn". Absorptive capacity accumulates only if an effort to internalize the external knowledge is exerted and in particular if the prior-knowledge has been applied to the solution of problems.

Conclusion

Neo-classical advocates of structural adjustment recognized that there would be losers along with gainers, but contended that this was not necessarily bad because the losers were gobbling up scarce resources in an inefficient manner. Their collapse would thus free up resources for more efficient producers. Losers would include large, protected industries producing for the home market and inefficient state firms. These would now have to compete with imports, lose state subsidies and protection, and pay more for imported inputs. Among the winners would be export industries, smaller firms, and farmers, especially export-crop farmers. They would benefit from currency devaluation, their goods becoming cheaper on export markets, they would gain more credit thanks to financial liberalization, and they would have fewer restrictions on their behaviour.

Foreign backing made structural adjustment all the more attractive. In contrast, the countries that had resisted pressure to implement the proposed reforms found it increasingly difficult to obtain development assistance at the time they needed it most. Thus the "neo-classical assault rolled on" through the 1980s. In both policy and intellectual circles, opposition to the assault was weak, just as opposition to the initial wave of Keynesian intervention had been. Socialist thought, which by now constituted the main opposition to neo-classical theory in the field of development studies, was dealt a severe blow by the collapse of Soviet and Eastern European communism after 1989. Yet for as long as it existed, the Soviet model stood as a reminder that it was possible to build an economy on principles other than the capitalist ones. Its collapse seemed to show that history's great experiment with socialism had in fact been what detractors such as Freidrich Hayek, the doyen of neo-classical theory had said: "it was all along a dangerously romantic delusion".

By the 1990s very few holdouts remained against structural adjustment. Many experiments with structural adjustment were less than whole hearted. India approached it hesitatingly at first in the 1980s with a reform process, though it had to move into the full weight of structural adjustment in 1991 when it was nearly bankrupt and in Zambia the government was forced to backpedal when riots broke out. Mexico shifted to a neo-classical strategy in the 1980s when development planners agreed on the need for structural reform. Similarly, in Ghana, in spite of their radical stance, there was a strong conviction that the country had no choice but to turn to the West by imposing an IMF-sponsored reform package. But elsewhere *shock therapy* which was embraced by the governments of several of these countries, signalling a complete rupture with past ways and this continues to be applied. Few if any other options were presented themselves to governments facing economic stagnation and persistent balance-of-payment crises. In all of these cases, what seemed to tip the local balance in favour of reform was the gravity of the economic situation.

In the late 1980s, the situation in the development debate was thus the "mirror image" of that which had prevailed in the late 1940s. Where neo-classical theory had once been a dissenting school, and Keynesianism and structural economics were the orthodoxy in both academic and policy circles, now neo-classical theory is the new orthodoxy. Socialism is reeling, structuralism weak, and ISI discredited.

A great many Countries of the South have implemented SAPs of one variety or another. As a result, most of the Countries of the South has become a "laboratory for a huge experiment in neo-classical theory". The results are instructive. They shed a great deal of light on the strengths, but also the weaknesses, of neo-classical theory. Just as neo-classical critiques had trickled in steadily throughout the late 1940s and early 1950s, posing questions that orthodoxy could not answer or pointing to the phenomena that orthodoxy had trouble explaining, so it goes today. Except that now it is the neo-structuralists and new schools of statist theorists asking the prickly questions.

Notes and References

1. The Harrod-Domar model is used in capital-scarce, labour-abundant developing nations as an easy way of looking at the relationships between growth and capital requirements. Being the simplest and best-known production function model used in the analysis of economic development, it is being considered here.
2. This simple equilibrium model of economic growth is subject to capital depreciation so that not all savings contribute to net addition to capital stock, as well as to the introduction of gestation lags so that the net new investment only makes its full contribution to output after a time lag.
3. Capital as sources of growth does play a major role in the expansion of contemporary developing nations. The Harrod-Domar model allows one to predict an estimate of growth for a nation which can be compared to predictions of growth for a different country.
4. In Ricardo's *Iron Law of Wages*, labour, like all other commodities can be purchased and sold, is having a natural price that enable the workers to perpetuate, this idea got revolutionaized in Karl Marx's labour laws: the labour-power of the wage-labourer can exchange itself for capital only by increasing capital, by strengthening that very power whose slave it is. Increase of capital, therefore, is increase of the proletariat.
5. This economic growth model assumed capital is infinitely flexible as to

what it can produce and that also all forms of investment are equally productive. But neither assumption is realistic, sectoral capital-output ratios are different. And it can be easily argued that additional investment financed through domestic savings is not the same as the capital transfers through transfer of foreign capital as they are not perfect substitutes for one another and that economic growth may be constrained by a *foreign exchange gap*.

6. *See Chapter 4.*
7. Hirschman, Albert, O., 1958, *The Strategy of Economic Development*, Yale University Press, New Haven.
8. Hirschman, Albert, O. 1981, *Eassays in Trespassing : Economics in Politics and Beyond*, Yale University Press, New Haven.
9. *Op. cit.,* Bagchi, A. K., 1982.
10. Rapley, John, 1996, *Understanding Development* : *Theory and Practice in the Third World*, Lynne Rienne Publishers, Boulder, Colorado.
11. Through an "early phase of import-substitution" which according to Bagchi, A. K. (1982) lasted in India until the middle of the 1950s.
12. *Source* : Public Sector Enterprises Survey, 1998-99, Department of Public Enterprises, Ministry of Heavy Industries and Public Enterprises, Government of India, New Delhi.
13. *Note* in 1948 China and Japan's share in international trade were 0.4 and 0.9 per cent respectively, lower than that of India but Japan from 0.4 per cent rose to the peak of 10 per cent in 1993 and China after its opening to the world market in 1978 rose to 4.0 per cent from low of 1.3 and 1.0 in 1963 and 1973.
14. Jalan, Bimal, 1999, *India's Economic Policy* : *Preparing for the Twenty First Century*, Viking, Delhi.
15. *Op. cit.,* I, Bagchi, A. K., 1982.
16. Anne, O. Krueger : *Cited* from Dickenson *et al.,* 1996, *A Geography of the Third World*, Routledge, London.
17. *Cited* from Dickenson *et al.,* 1996, *A Geography of the Third World*, Routledge, London.
18. *Cited* from Dickenson *et al.,* 1996, *A Geography of the Third World*, Routledge, London.
19. *Washington Consensus,* 1989.

7

NEO-CLASSICAL ANSWER TO ISI FAILURE

In 1970, the Organization for Economic Cooperation and Development (OECD), published a study that did much to push neo-classical theory to the "donor agencies" for prodding many governments of the Southern Countries. This study looked at the trade regimes put in place in the post-war period by many countries like Argentina, Brazil, Mexico, India, Pakistan, the Philippines, and Taiwan and concluded that in all these countries, ISI had done more harm than good. In their analysis, the authors made a number of observations that would form a large part of the arsenal used in many neo-classical critiques of statism. To begin with, they pointed out that in trying to build new industries, (1) ISI neglected the "comparative advantages" enjoyed by those economies. Given that these comparative advantages were often in agriculture, it was significant that industrialization had occurred at the expense of agricultural development. (2) In large part this happened because currency overvaluation had discouraged exports, both industrial and agricultural. All in all, ISI was seen to be a "wasteful strategy": Industry accounted for more investment than output, its capital-intensive nature created too few jobs, and it gobbled foreign exchange in its need for imported inputs. (3) ISI's bulky state administration created bottlenecks in the economy, further wasting resources through capacity underutilization, corruption, and sluggishness. The authors expressed the doubt common to all neo-classical theory that bureaucrats could gain access to the information needed to effectively administer the economy, and they disliked the fact that controls used in ISI appeared to curb private initiative.

As if this were not enough, the study concluded that ISI, which was often justified as a strategy that would benefit whole economy and not just preserve the wealth of a lucky few, was actually worsening income distribution. While profit earners benefited from protection and skilled labour from currency overvaluation, farmers suffered and a large share of the urban population remained unemployed, forced to seek work in the marginal or informal sectors as domestic maids, pedlars, etc.

The proposed solution was for governments to shift from ISI to "export industrialization", nurturing firms that could sell abroad rather than in the domestic market. For this purpose, the study suggested "promotional rather than protective policies" to encourage industrialization—for example, "subsidies over import restrictions". In line with its call for *export oriented industrialization* (EOI), the study advocated more openness to foreign trade, less use of controls, more use of the "price mechanism", and currency devaluation. These recommenda-tions got repeated rather many times over, by other neo-classical theorists, but in hindsight the OECD study appears relatively moderate compared with some of the later volumes. It did not oppose public ownership, it accepted some role for price controls, it emphasized the state's role in building infrastructure and in human-capital formation, and it called for some degree of state activism in helping firms to capture export markets. Nor did the study repudiate ISI outright. It merely rejected its being used for too long.

Within a year the World Bank and the Inter-American Development Bank published a trade study that strengthened the OECD study's findings. This study assessed the impact of ISI's protectionism and the currency overvaluation—or lack thereof in some cases—on the economies of Brazil, Chile, Pakistan, Mexico, West Malaysia, the Philippines, and Norway. Its conclusions were "damning" to ISI. Protection, it said, "entailed high costs" in "static (allocative) efficiencies", limited the scope for the introduction of large-scale production methods, provided few inducements to improve productivity, slowed the production and exports of primary products, and hindered the expansion of manufactured exports. ISI was, in sum, a policy that wasted resources and did too little to stimulate increases in exports. By contrast, in the countries with less-protective trade regimes agriculture and exports grew rapidly, new primary exports were developed, and exports of manufactured goods increased. Once again, while granting that protection was legitimate over set periods, the study called for devaluation coupled with disinflationary policies, the replacement of quotas by tariffs, and the use of subsidies rather than protection for the promotion of new manufacturing industries. So while the report recognized that the state had a role to play in economic development, its principal thrust was a call to roll back the state and streamline its procedures.

Stronger—some might say dogmatic—expressions of neo-classical thought were to follow, as critics gained confidence and grew convinced that their findings had throughly discredited the old statist development schools. Deepak Lal (1993) in his *Poverty of Development Economics*[1] composed a scathing indictment on what he called "development economics", saying there was no need to articulate on economics for development, as development economists had tried to do, because all the answers could be found in conventional economic theory. Lal then gave to neo-classical theory the memorable *aphorism* that "market failure was always preferable to state failure".

In 1983, the National Bureau of Economic Research, NBER[2] issued a trade study that would become so influential that some would call it the "core" of the neo-classical critique of statism. Like the OECD study, the NBER study dealt with trade regimes, reaching similar conclusions. From it emerged a focus on export-oriented industrialization, which it set against ISI. In the view of the study of authors, the latter was statist while the former, said to be practiced in the most successful of the East Asian NICs, was market-oriented. Although few more dispute the *import substituters*, a great debate soon erupted over whether or not the *export-industrializers* were free-market economies. Neo-classical theorists came to lean on the NBER study's claims that these economies illustrated the virtues of free market and liberal trade regimes, and they brandished the success of such countries as South Korea as a lesson for those such as India.

The NBER study focussed on the various market distortions caused by government intervention in the course of ISI. It argued that labour-market regulations, restrictive trade regimes, credit rationing, and social-insurance tax systems all combined to raise the domestic cost of hiring labour relative to capital. Meanwhile, currency overvaluation, the favourable treatment of capital-goods imports, and credit rationing at subsidized interest rates drove down the prices of capital services. The end result, relatively cheap capital and relatively expensive labour, clearly favoured capital-intensive production.

Although there is nothing intrinsically wrong with capital-intensive production, it normally arises in high-wage economies. When it develops in low-wage economies it excludes the mass of the population from the development process, because it creates relatively few jobs while eliminating traditional industries. The solution to this sort of problem appeared obvious: less distortion, which meant less government intervention in the economy. The trade regimes should be liberalized and there should be more "freedom" in the labour market.

The NBER study made another claim that drove to the heart of structuralist economies. Whereas structuralists had often argued that trade between the North and the South had worked to the detriment of the latter, and that interregional trade offered more hope for development, the NBER study rejected this flatly. It argued that gains from trade, including employment gains, would be maximized for countries endowed with different characteristics. In other words, the poor countries should trade with rich countries, not with poor ones. This study held that regional trade blocs in the South would do little to benefit their member states. Given that post-war approaches to development were much influenced by trade pessimism, this argument, along with neo-classical claims that the terms of trade were not going against the Southern Countries, represented a remarkable attempt to refute that pessimism. In this context, Prebisch's study in 1950 which deduced that the value of primary exports relative to finished imports had declined, had also been challenged by the neo-classical writers.

However, the golden age of statist liberalism for world economy could not last. In 1973 the world was shaken by the embargo announced by the oil-producing countries, OPEC (Organization of Petroleum Exporting Countries) on oil supplies to the Countries of Global North—this sudden cut in supply led to a fourfold increase in the world price of oil which plunged the economies of the North into recession, heralding the end of the golden age : modest growth phases returned to the world economy followed by recessions. The era of steady, high growth was now at an end. What emerged to take its place was a new phenomenon that bedevilled policy makers in their search for a cure: *stagflation,* that is, the economic stagnation, coupled with high inflation.

The success of the OPEC countries in raising the oil prices was construed as an attempt by the commodity-producing countries to increase their share of the world's wealth at the expense of the North. Whatever, flurry of truth has been there in this interpretation, the OPEC countries got much richer but this boded ill for most of the Southern Countries, not producing oil because they have to face the same jump in energy costs as did the North—the *hydra* that begun to rear its head in this oil crisis, crept into the body at once. The recession in the world economy did reduce the demand of Northern Countries for primary products which plummeted the commodity prices. But what became seemingly the incurable illness for the South was the *debt crisis,* the loan they took for the questionable projects skyrocketed because the Countries of North in their fight for inflation, hiked the interest rates to heights unknown in the post-war period so that by 1982 the large countries like "Mexico, Brazil and Argentina all announced they could not meet their current debt obligations and the debt crisis erupted. The donor agencies like World Bank, IMF relegated development projects of the South to a secondary status, and devoted their energy to trying to recover old debts and improve the solvency of their debtors".

The revenue from the sales of primary commodities was increasingly used not to fuel industrial development, but to pay off old debts, the governments had to squeeze money from their budgets to meet debt obligations, and were forced to cut their investment and social expenditure but they had little space to maneuver get steady, high growths. The difficulties faced in industrializing by the path of ISI led to the weakening of this "endogenous development" (which finally *got abandoned in 1991 in India through its adopting a* "new economic policy"). The inability to maintain fixed currencies and the disruption caused by official devaluation resulted in early 1970s (1966 in India), in the abandonment of currency controls. The inability to maintain and integrating the country more fully with international economic order centred on neo-classical economics concept of "*neo-capitalism*": the argument runs as follows: physical controls over allocation of resources and "a regime of quota restrictions accompanied with tariff on imports without any corresponding subsidies to exports" turns the domestic industries into *permanent infants*, for which the country loses the advantages of specialization in the international market and fails to reap the benefits of developing those sectors in which it happens to have a comparative advantage—this cyclical phenomena led to create an arbitrary advantage for products primarily directed towards the internal market which distort the effective relative prices in arbitrary way, misdirect resources and thwarted growth. In this respect, the apparent successes of the export-led development policies of the East Asian economies, which were recording high rates of growth and experiencing rapid industrial transformation led to believe that "integration could be a preferred strategy to than that of isolation by the doctrine of statist intervention". Further, with the introduction of "floating currencies", new opportunities for speculation emerged. So one of the first requirements for getting the country out of the regime of distorted price structures and thwarted growth is to remove the disadvantages under which industries producing exportables suffer and to *let the bracing air of international competition blow through the economy.*

Unilateral trade liberalization is opposed by those who think in mercantilistic terms that "exports are good, imports are bad" even if economic theory would prove them wrong.

In the earlier chapter on international trade, we examined two inward-oriented strategies of development : the *basic needs* model and the *import-substitution* model. There have been some short-term successes of these approaches, as well as many failures. Today, the pressure from the North and from international lending institutions is increasingly toward the New Economic Model (NEM) of *outward-directed policies*. Although, many countries follow no particular philosophy, simply exporting what they can and importing what they have to, two competing, but somewhat overlapping, ideal models have emerged.

The *neo-classical economics* dates back to the 1870s when it broke away from parent, classical economy. Theorizing became a fundamental guiding assumption in neo-classical economics in which individuals behave as rational utility maximizers: putting in another way, people became self-interested, they know best what they want, and they also know best how to get it. In the pursuit of their goals, people act rationally and efficiently. From this assumption, it follows that the most productive economy will be one in which individuals are allowed the greatest freedom to engage in activities or enter into contracts as they choose, and to reap the full benefits of their labours. Neo-classical theorists thus argue not only against government regulation, but against taxation whose aim is to redistribute wealth. And this individualism ensures that more things would be tried, and the greater the number of things being tried, the more innovation and progress there would be. But since individuals would only incur the cost of trying something new if they know they will reap the benefits of any success they had; people are not altruistic. Taxing the rich to feed the poor will hinder the most well-off, reduce initiative and thus innovation, and so hurt all of society.

The Development Approach

The development approach suggests the truism that an economy is an inherent part of society in which it operates, *i.e.,* the economy is a social construct that which is not always fully understood. An economic system, whether it is a market, planned, or mixed economy, is not an autonomous sphere that can be the subject of a *detached and subjective science, but it is a social construct, both reflects and influences the norms and values of a society, the aspirations of people, and obviously the structures and* modus operandi *of power relationships.* This means that a particular economic arrangement—a system, its positive and negative features and its overall quality—have to be assessed with social criteria, within a framework of a political and moral philosophy.

Economic conditions vary immensely throughout the world. There are enormous differences among countries and among regions in level of income, degree of modernization, use of technology, quality and density of infrastructure, capacity for saving and investment, property rights, acquiring skills, etc. Vast numbers of people struggle day by day to make a livelihood by subsistence farming or in the urban "informal sector". A few acquire money and power by speculative activities, only remotely connected to the production and exchange of goods and services. The existence of a few products of mass consumption that are available all over the world should not be taken as evidence of any significant lessening of the differences in economy and living conditions.

Yet this diverse and unequal world, more like a great *Tower of Babel* than a global village, has a few strong economic features in common—most countries have adopted some form of "market economic system", with the two basic characteristics of a regime of "private ownership" and "emphasis on freedom of enterprise". This system exists in many specific variants—(one) the mixed economies with a strong role for the state and a wide array of regulations, (two) few government agencies in countries of the North/the South apply the "concept of planning" to their efforts to influence the directions of the economic and social processes driven by market forces.

Thus the market economy as an institution and a symbol of freedom, efficiency and prosperity, has been enjoying its heyday universally. It is again the market, in its incarnation of global capitalism, that forms the overwhelmingly dominant practice and ideology in relationships between nation-states and regions. A wave of "liberalization", "deregulation" and "privatization" swept across USA and Europe towards the end of 1970s, and then engulfed the rest of the world with remarkable speed and force. Structural adjustments, reform of the economy, flexibility of the labour market have ensued. Critically important in this neo-liberal praxis and ideology is the "free circulation of capital". FDI has grown from some trifle $ 5 billion at the early 1960s to close to $ 200 billion today and the value of foreign exchange transactions has jumped from $ 15 billion a day in the 1970s to a formidable $ 1000 billion a day in the late 1990s. Obviously this free circulation of capital coupled with a very strong increase in the volume of world trade is not synonymous with a *world without border* alone and it cannot solve all disparities. Hence this approach views a broad perspective and calls for a host of measures designed to stimulate growth and development. These measures include a multidimensional intervention on part of the government to put in place an adequate industrial infrastructure constructing new roads, improving harbours, an increase in educational level of the work force, the promotion of the application of new technology, assistance in modernizing agriculture, and so on. In short, this multi-dimensional approach has to be dealt with through new economic or social initiatives.

Globalizing World Economy and the New International Division of Labour in Countries of Global South

We live at a time when capitalism has become more extreme, and is more than ever presenting itself as a force of nature, which demands such extremes. Globalization—the spread of the self-regulating market to every niche and cranny of the globe—is portrayed by its mainly establishment proponents as a process that is unfolding from everywhere at once with no centre and no discernible power structure. Global reality is one of "a fluid, infinitely expanding and highly organized system that encompasses the world's entire population," but which lacks any privileged positions or "place of power." Even Karl Marx in *The Communist Manifesto* (by Marx and Engels) supported this view of inexorable global destiny, which seemingly determines everything, but which has no manifest agent of change. Thus the World Bank quoted from in its opening page of its *1996 World Development Report*, arguing that the transition from planned to market economies and the entire thrust of neo-liberal globalization was an inescapable, elemental process, lacking any visible hand behind it :

> "Constant revolutionizing of production, uninterrupted disturbance of social conditions, everlasting uncertainty and agitation...All fixed, fast frozen relations, with their train of ancient and vunerable prejudices, and opinions, are swept away, all new-formed ones become antiquated before they can ossify. All that is solid melts into air..."[3]

"Gone—spirited away by ellipses in the World Bank quotation from the *Manifesto*—were Marx and Engels' allusions in the same passage to *the bourgeois epoch* and their subsequent reference to how the need for a constantly expanding market for its products chases the bourgeoisie over the whole surface of the globe."

"It is no doubt largely in response to this atmosphere of inevitability, in which globalization is divorced from all agency, that the movement against the neo-liberal global project has chosen to exaggerate the role of the visible instruments of globalization at the expense of any serious consideration of historical capitalism. Radical dissenters frequently single out the WTO, the IMF, the World Bank and multinational corporations—while de-emphasizing the system, and its seemingly inexorable forces".

These two distorted viewpoints, one generally in support of globalization, the other generally opposed, are mutually reinforcing in their unreality. Those who wish to intervene in these processes are thus left with no real material basis on which to ground their actions. Both perspectives have in common an emphasis on the decline of nation-state sovereignty. Now we are told that this invisible hand has been globalized to such an extent that the sovereign power of nation-states over their territorial domains themselves has been vastly diminished. Globalization is a new technological-economic system based in the microchip and ruled by an "electronic herd" of financial investors and multinational corporations, free from any nation-state or power structure, and beholden to none.

Those seeking to dispel such views might reply that capitalism with all of its contradictions remains. But most current conceptions of capitalism are too lacking in historical specificity and concreteness, and too wrapped up in the notion of unfettered competition, to be useful in countering this dominant ideology. Indeed, the very idea of capitalism is being shorn of all determinate elements. The notion of global free market hegemony without the nation-state and without discernible centres of power (only highly visible instruments of the market) means a concept of capitalism that has become virtually synonymous with *globalization*. There is, it is proclaimed, no alternative because there is nothing outside the system, and no centre within the system.

The ideological fog that pervades all aspects of the globalization debate is bound to dissipate eventually, as it becomes clear that the contradictions of capitalism, which have never been surmounted, are present in more universal and more destructive form than ever before. For those seeking to penetrate this fog at present and to understand the constellation of forces in the world today what is needed above all is a concrete and historically specific conception of capitalism that will allow us to see through such issues as globalization. Within Marxism such an analysis was provided in the twentieth century by the theory of monopoly capitalism.

The Origins of Monopoly Capital Theory

The term "monopoly capitalism" has been widely used within Marxian economics to refer to the stage of capitalism dominated by large corporations. This stage of capitalist development originated in the last quarter of the nineteenth century and reached maturity about the time of the Second World War. In *Das Capital*, like other classical political economists, Marx had assumed that the market system was characterized by conditions of free competition, in which initially capitalist enterprises were small, mainly family-run firms. Classical political economy never included such absolute fantasies as "perfect" or "pure" competition, which were to be imported into economics in its later neo-classical stage. Nevertheless, it assumed in its bedrock theory of free competition that price competition was fierce, and that no individual capitalist or firm had the power to control a significant portion of the market.

In the case of Marx, as distinct from the other classical political economists, however, capitalism was a historical system, and thus dynamic in character, passing through various stages. Although Marx himself did not present a theory of monopoly capitalism, he did point to the concentration and centralization of capital as a fundamental tendency of accumulation under capitalism. The whole development of the credit system and the stock market was for Marx "a new and terrible weapon in the battle of competition and is finally transformed into an enormous social mechanism for the centralization of capitals" (Volume 1, chapter 2, section 2). In preparing Volumes 2 and 3 of Marx's *Capital* for publication two decades later, Engels emphasized the fact that free competition had reached "the end of its road" (Volume 3, chapter 27). Marx and Engels, however, were prone to see these developments as signs of new conditions of socialization of production that would help usher in a new mode of production—not as indications of a new stage of capitalism.

It remained for later thinkers, therefore, to analyze what the economic implications of the rise of big business, and the transformations in credit, corporate finance developments meant for capitalism's "laws of motion". Within the Marxist tradition, then centred in Germany, the first important theorist of monopoly capitalism was the Austrian economist Rudolf Hilferding in his *Finance Capital: The Latest Phase of Capitalism* (1910); soon followed by Lenin in his *Imperialism: The Highest Stage of Capitalism* (1916).

Hilferding pointed to the tendency of concentration and centralization of capital to generate a greater and greater consolidation of capital, pointing eventually to one big cartel—an overly simplistic view that failed to perceive some of the countervailing influences at work. He saw these changes as mainly quantitative in character, and though his work was full of important insights, he did not explore the question of qualitative alterations in the "laws of motion of capitalism". Hilferding's perspective did, however, inspire Lenin to connect "imperialism with the monopoly stage of capitalism, and to perceive the growth of giant capital therefore as integrally related to both the expansion of capital on the world stage, and the struggle between nation-states for shares of the world market". But Lenin, like Hilferding before him, did not pursue the question of how capitalism's basic laws of motion might be modified in the monopoly stage. The concept of monopoly capitalism was to remain axiomatic for Soviet economists in the 1920s and 1930s, during which some important new departures were begun. But by the late 1930s it had been reduced to a mere dogma within the rigid orthodoxy that prevailed under Stalinism.

Historical developments, however, were pointing to such a connection. Since the turn of the twentieth century in the United States there had been a groundswell of the giant monopolies and trusts. The great merger wave at the beginning of the twentieth century was widely viewed as representing a qualitatively new reality. It has been estimated that between a quarter and a third of all U.S. capital assets underwent consolidation in mergers between 1898 and 1902 alone. The mammoth merger of the period, the formation of U.S. Steel in 1901 under the financial guidance of the investment banking house of Morgan, fused 165 separate companies. The result was a monopolistic corporation controlling about 60 per cent of the total U.S. steel industry. In 1936, Arthur R. Burns wrote his classic study, *The Decline of Competition: A Study of the Evolution of American Industry.* And in the context of the Great Depression of the 1930s it was frequently contended that the stagnation was worsened by the growth of giant corporations with a large degree of monopoly power.

The first economist to connect the theory of crisis to the theory of monopoly was the Polish economist Michael Kalecki who worked in England, developed his concept of the "degree of monopoly," which stood for the extent to which a firm was able to impose a price mark-up on prime production costs (workers' wages and raw materials). In this way, Kalecki was able to link monopoly power to the distribution of national income, and to the sources of economic crisis and stagnation. This analysis was carried forward by Josef Steindl, a young Austrian economist who had worked closely with Kalecki in England. According to Steindl's *Maturity and Stagnation in American Capitalism* (1952), giant corporations tended to promote widening profit margins, but were constantly threatened by a shortage of effective demand, due to the uneven distribution of income and resulting weakness of wage-based consumption. New investment could conceivably pick up the slack. Yet such investment resulted in new productive capacity, that is, an enlargement of the potential supply of goods. "The tragedy of investment," Kalecki wrote, "is that it is useful." Giant firms, able to control to a considerable extent their levels of price, output, and investment, would not invest if large portions of their existing productive capacity were already standing idle. Confronted with a downward shift in final demand, monopolistic or oligopolistic firms would not lower prices (as in the perfectly competitive system assumed in most economic analysis) but would instead rely almost exclusively on cutbacks in output, capacity utilization and new investment. In this way they would maintain, to whatever extent possible, existing prices and prevailing profit margins. The giant firm under monopoly capitalism was thus prone to wider profit margins (or higher rates of exploitation) and larger amounts of excess capacity than was the case for a freely competitive system, thereby generating a strong tendency toward economic stagnation.

The monopoly capitalist economy does not consist simply of giant firms, of course. Within manufacturing, for example, there are hundreds of thousands of firms, which together employ a substantial share of the work force. These smaller firms are often attached to the giants, some supplying parts, others occupying various other niches. Such firms tend to bear the brunt of an economic downturn. Conversely, during an expansion they tend to grow more rapidly than the dominant, monopolistic firms.

Thus in the development of world economy, four to five phases may be recognized, each reflecting the productive and geographical flexibility of the circuit of capital circulation. The world-wide trading and "sourcing" of raw materials constituted the initial *first phase of capitalist economy* which was better known as *mercantile economy.* This changed to *capitalist economy* in the eithteenth century ever since Adam Smith described capitalism in the late eighteenth century as a system that eliminated all need for a sovereign power in the economic realm, replacing the visible hand of the absolutist or mercantilist state with the invisible hand of the unfettered market. "The Sovereign," he wrote, "is completely discharged from a duty" with respect to the market (Book 4, section 9). And David Ricardo extolled the virtues of "comparative advantage" so that *laissez-faire capitalism* remained virtually a religion for economy of Countries of the North in the *second phase* of the world economy. The Great Depression of 1930s briefly displaced this market-dominated economy to Keynesian statist model which transformed macroeconomics in response to the depression, by remaining rooted in the age-old assumptions of *atomistic competition*. Keynes postulated the necessity of government intervention in the capitalist economy. But "Long-run development is not inherent in the capitalist economy. Thus specific *developmental factors* are required to sustain a long-run upward movement" and in the 1930s in the West, meanwhile, mainstream academic economists finally began to deal with monopoly and this Keynesian phase of economy[4] which constituted the *third phase* in the economy both for the Countries of the North as well as of the South world over was for a brief period because classical economies re-emerged in the 1970s in countries of the North and in the name of *neo-classical* (or, *neo-liberal*) *model*—it is so-called in reference to Adam Smith days' of free-market liberalism because it follows a modernization approach and pays scant attention to theories of dependency.

Since the middle of the twentieth century the world-wide flow of *money capital* in search of profitable investment through "sourcing" of cheap labour (instead of sourcing of raw materials in the mercantile days) which constituted the *fourth phase* so that today, the operation of the world economy is dominated by global production and the so-called *new international division of labour*, is based upon the operations of multi-national capital. As capitalist development has proceeded, the competitive relations between the singular units of capital has resulted in their growth, via the "process of concentration and centralization" (Roger Lee, 1995).[5] The sophisticated internal organization of these corporations and the facility of electronic communications allows the global organization of production, whereby global capital may not seek out its preferred conditions of existence throughout the world, but also take profitable advantage of economic inequalities and political discontinuities Bellamy (2002).[6]

The contemporary world economy is characterized by a single but highly dynamic international division of labour, operating over the world's complex mosaic of cultural systems and environments. None of these may, therefore, be understood simply in their own, purely local, terms or considered to be autonomous and completely self-contained. The geography of any particular part of the world must take full account of its relationships with the operation and organization of the world economy. Perhaps the most obvious example here is that of Countries of the South. The *condition* or *state* of underdevelopment may be understood (and, ultimately, rectified) only if the *process* of their underdevelopment (for example, the extraction of surplus value) are seen as a complex set of economic, social and political relationships between the underdeveloped world and the world-economy. Equally, the de-industrialization of an advanced economy, like Britain, is closely related to increasingly dominant and *hypermobile multinational capital*, which is able to take the world as "its locale" and to seek out and create cheap labour and favourable political regimes to ensure its continued profitability. In the process it helps to create the so-called Newly Industrialized Countries (NICs) which are, in truth, little more than the latest manifestation of the geographically uneven operation of global capital.[7]

For many the focal point of the New Economy has been the highflying technology stocks that carried stock market speculation to new giddy heights during what has come to be known as the "new millennium boom". The new economy is the proliferation of information technology throughout the economy that makes the current period appear so different from preceding decades. One result of the more rapid pace of Information Technology (IT) innovation has been a visible acceleration of the process that can be termed as *creative self-destruction*—the "continuous shift in which emerging technologies push out the old". Among the advantages of the New Economy is the ability of corporations to generate a flood of information in miniseconds, allowing them to "reduce unnecessary inventory and dispense with labour and capital redundancies". The spillover effects of digital technology are revolutionizing Old Economy sectors, making the New Economy a more universal phenomenon. "Computer modelling, for example, has dramatically reduced the time and cost required to design items ranging from motor vehicles to commercial airliners to skyscrapers."

The New Economy has also been associated with the development of a more "flexible workforce": non-unionized, highly mobile, just-in-time workers, sometimes embodying new job skills. Compared with the more highly unionized labour forces of Europe and Japan to that of the United States, "the relatively inflexible and, hence, more costly labour markets of these economies appear to be a significant part of the explanation as to why the New Economy has been slower to develop there. The elevated rates of return offered by the newer technologies in the United States are largely the result of a *reduction in labour costs per unit of output*. The rates of return on investment in the same new technologies are correspondingly less in Europe and Japan because businesses there face higher costs of displacing workers than in USA.[8]

The place of the socialist states within the global economy is a subject of some controversy. Whilst it is possible to differentiate between the social relations of production in socialism and capitalism, it is not easy to separate them in practice within the contemporary world economy. It is clear that most socialist states are affected, directly or indirectly, by the operation of the world economy within or around them. Even the most powerful—the Soviet Union—is indirectly involved in the world-economy through its participation in the political and material relationships within the international network of nation-states. Furthermore, although Soviet involvement in world trade amounts to less than 10 per cent of the value of its annual turnover, it has recently began to re-emphasize production for a world market, to deal with multinational enterprises for both production and trade. Thus, despite its territorial size, rich resources and potentially large internal market, the costs of developing socialism from a weak industrial base, appear to necessitate Soviet participation in the world system. Nevertheless, the integration of the socialist states into the world-economy is unevenly developed from sector to sector, as well as from country to country.

This latest phase in the evolution of world economy has been labelled under the term *disorganized capitalism,* and it has accelerated since the late 1960s. This new economic model involved the erosion of national economies so that the national economy, is no longer the sole building block of the world economy. In the 1950s when colonial countries were freed, the Countries of Global North had evolved this new techniques of production, and huge conglomerate enterprises in the name of polycentric multinational enterprises (MNEs), has been in the ascendence in the form of growth for an "immediately global market" which straddled the nations with their huge capital and closely guarded technological know-how in search of obtaining *economies of scale in production* as well as *serving a large international market.* The global shift in production has given rise to an explosion of Foreign Direct Investment (FDI), and to the emergence of trade within the multinational firms as the most rapidly expanding component of total world trade. This explosion of "disorganized capitalism" has been observed first in the Pacific-rim countries of Asia, particularly in Hongkong and Singapore or such semi-colonies as South Korea and Taiwan, where to get part of their supplies fabricated for exports, the MNEs and buying agencies based in countries such as USA and Japan have been accorded favourable treatment by these countries to utilize the low labour costs in these countries.

The East Asian Model

During the 1960s and 1970s the South Korea and Taiwan experienced periods of dramatic economic growth. This economic growth was spurred on largely by the adoption of an *Export Oriented Industrialization* (EOI) strategy for development. These two countries, along with the city states of Hong Kong and Singapore, were the *fast developers* and made up the group of export oriented industrialization success stories known as the 'Four Asian Dragons', also referred to as "Newly Industrialized Countries" (NICs). During this span of growth two things happened that made this experience even more attractive to the Countries of South who attempted themselves fully into international economy. Firstly, they were able to greatly enhance the value of their exports and diversify these exports into manufactured goods. This meant that the rise of these East Asian countries have been recognized since 1979 as NICs were less affected by the declining terms of trade suffered by commodity exporters, which requires explanation. Secondly, during these two decades the income distribution of these countries stayed at relatively the same level as before the adoption of this EOI strategy. This experience of the NICs has been seen by many to prove that the EOI strategy for development is far superior to that of the import-restraining inward looking ISI strategy. The EOI advocates have gone further and extrapolated this success of the NICs into a "general development theory for the South". This interpretation says NICs demonstrated a "twentieth century success story" and assumes that the NIC example is feasibly transferable to other LDCs, so that it is possible for poor nations to break with Import-Substituted-Industrialization, ISI and shift to export-oriented industrialization by developing in a sustained and inclusive way whilst engaging with the international economy and in the process they follow through certain domestic changes.

This type of dynamic model (*disorganized capitalism*) has been more oriented to the increased importance of the international dimension in the division of labour. From this point of view, based upon the operations of the multinational capital, New International Division of Labour (NIDL) has come into existence in which investment and production are no longer organized primarily around national economies, instead an "*offshore sourcing practice*" (*i.e.*, the practice of obtaining supplies from multiple suppliers in different countries and assembled in several) is followed which has got spreaded inexorably to Mexico, Brazil, Philippines, Malaysia and India. If product design and production activity are relatively stable, then "offshore sourcing" becomes feasible production of appliances, clothing, tools and even financial services. Such *subcontracting* on an international scale, or such *relocation of plants in low-wage economies* for export to countries of Global North as well as other countries has been seen as "the new vehicle for a variety of export-led growth" through the increased importance of the New International Division of Labour (NIDL). The actual process of production, most obviously in the automobile and electronics industries, is now global, so that increasing number of products have no obvious nationality, for example, it is difficult to distinguish some "American" from some "Japanese" cars: now that American car companies import vehicles under "their" names from Japan and Japanese companies now manufacture cars in USA in their brand names (*e.g.*, Honda in Marysville, Ohio). For a significant number of commodities, an "immediately global" market now exists which can be supplied by firms that do not organize their production and distribution (in agriculture and services, as well as manufacturing) by reference to national boundaries. This "global shift" in production has given rise to an explosion of FDI and to the emergence of trade within firms as the most rapidly expanding component of total world trade. Perhaps 30 to 40 per cent of USA imports, for example, consist of parts and finished goods from the foreign subsidiaries of US-headquartered firms —more significant is the fact that US-based MNEs are larger than the GNP of any country except the USA. For many MNEs, national markets for capital, labour and plant and office location exist only as parts of global ones. The spatial strategies of MNEs are designed to exploit national differences in labour forces, market conditions, regulatory environments and macroeconomic (fiscal and monetary) conditions so that markets are accessed in the peripheral countries and enhancing revenues become the primary motivations behind the investment strategy. Without the variation institutionalized by the territorial states into which the world is divided, the attraction to firms of shifting investments and moving production would be reduced.

The challenge to territorial states is to compete as effectively as possible in this more integrated and volatile economic environment. The individual MNEs do not always operate or compete independently. Rather, the complex permutations of financial linkages within and between companies and countries often makes them interdependent, each dependent on some of the others for certain aspects of R&D, technology, marketing and manufacturing. This new dimension of the MNEs has come to represent a serious economic threat to small nations, and it has prompted the creation of a variety of supranational organizations. Recent evidence shows that based on technological superiority, even small firms have now acquired the propensity and capability to operate globally. So the "new" conditions cannot be solely identified with giant MNEs or global corporations.[9]

Paralleling and stimulating this trend has been the emergence of these international devices for coordinating and steering capital beyond national control, for example, the "offshore" banking and financial centres which, rather like some old city-states did in their day, now service the new international division of labour (*e.g.*, Hong Kong, Singapore).[10] The circulation of capital through international circuits and in the form of investment in commercial real estate in the major financial centres (especially Tokyo, London and New York) became important features of the new world economy in their own right in the late 1970s and 1980s. Indeed some small states such as Switzerland have successfully cashed in on the new world economy to the extent that they now have median income levels higher than those of the "old" national manufacturing economies such as Britain and United States.

Superficially, these East Asian countries would appear to vindicate the methods of neo-liberalism. Over the last 15 years or so, the World Bank and IMF sanctioned *Structural Adjustment Program* (SAP) laid down an approach towards export-led growth and increased cross-border specialization and competition for carrying on international trade. SAP draws some of its inspiration from the belief that the success of the Asian NICs—derives just from the policies which these countries adopted voluntarily. But in reality, their policies are quite different in many ways. What makes the two East Asian countries, Taiwan and South Korea truly unique among the Countries of the South is not their rapid industrialization (a few other countries have experienced similar spurts) but that their development seems to have reached a point where it is self-sustaining. More importantly, all sectors of their population have benefited from economic growth and in this respect of the income equality they have got that is of greater levels than the United States and most other Western nations.

Taiwan (formerly the island of Formosa) was part of China for ten centuries before it was invaded and occupied successively by the Portuguese, the Dutch, and finally, the Japanese, who invested heavily in building an infrastructure of ports and roads. After World War II, the island reverted briefly to China. In 1948, when the communists were successful in taking over the mainland, the remnants of Chiang Kaishek's Koumintang party escaped to the island to set up what became virtually a separate country. The newcomers were an educated elite who made their escape with an army in tow and many financial resources. They quickly displaced the landed oligarchy, bought their legitimacy among the natives with an extensive land reform, and established a one-party dictatorship with absolute power (a person could be sentenced to death for publicly opposing the regime). The USA viewed Taiwan as a bastion against Chinese communism and even seems, for a while, to have divided itself that Koumintang could retake the mainland. As a result, foreign aid—both military and developmental—poured in. Other countries have received such preferential aid with little to show for it, but the government of Taiwan set out on a deliberate and systematic process of economic development. The results have been little short of amazing. Per capita income in 1950 was a mere $ 224; by 1992, it was $ 8,486 *i.e.*, an average growth rate was sustained for twenty years. After 1970, Taiwan's exports grew at an average of 22 per cent per year.

South Korea has a similar history of Japanese occupation, land reform, massive US aid, and planned economic development, with much the same results. Both countries (which alongwith the relatively city-states of Hong Kong, at present a *special area region* (SAR) of China, and Singapore, are referred to as "Four Asian Tigers".

(First), the *laissez-faire* approach opposes government intervention in the economy. But the NICs showed that extensive state administrations in the monolithic states of Taiwan and South Korea, have authority and capacity to give direction to market development, respond flexibly to private-sector needs, develop technology and human resources. The state authority in these two countries were so deeply involved in every aspect of development that their policies are referred as "state capitalism" or even "mercantilism". Initially the government owned many of the crucial industries. Private enterprises were subsidized by the state and worked closely with the state in developing export and import policies. As in Chile under Pinochet, wages did not reflect market demand but were held artificially low. Such state interventions included widespread intimidation and repression practiced in the labour markets: trade unions were kept weak, strikes banned, and welfare legislation avoided. Economic growth was planned by the state at every level and enforced through strong government intervention.

(Second), both countries employed almost two decades of import substitution in order to build up their own businesses without any involvement of foreign enterprises and their domestic labour forces. Import protections were dropped only as domestic markets were satiated and indigenous companies could turn to export. Some local industries remained protected into the 1990s. In contrast to Brazil where ISI has occurred with a very extensive, if not dominant, participation of foreign capital, South Korea effectively minimized the role played by FDI in many areas of basic and heavy industry which recall the Japanese practice to the present time.

(Third), right from the beginning, both countries Taiwan and South Korea gave priority to domestic demand. In South Korea, between 1965 and 1975, only 15 per cent of total manufactures were exported. Even as the role of exports increased, it went hand in hand with the rapid growth of the domestic economy. By contrast, the neo-liberal model gives priority to the international marketplace, with the assumption that the domestic market will follow.

(Fourth), in opposition to the neo-liberal emphasis on the sanctity of private property, both countries enacted extensive land reforms, forcibly dividing the lands of large-scale owners among the peasants. In these the then—agrarian societies, perhaps no other action was significant in redistributing wealth and bringing all the people into the national economy. *Note*, the capitalist country like USA has an ambiguous record in relation to land reform: it overthrew the Arbenz government in Guatemala in 1954 partly because of a modest land reform which threatened some property of the USA based United Fruit Company; in the 1980s, the United States promoted land reform in El Salvador at the same time it was violently opposing similar reforms in neighbouring Nicaragua.

(Fifth), redistribution of wealth and equalization of income took place *before* export-led industrialization. The neo-liberal model assumes that free-market policies will work at any level of development. This has yet to be proved. The Chilean and Brazilian examples suggest that if income equalization measures are not enacted early in the developmental process, rapid export-led industrialization will exacerbate inequities.

(Sixth), in both South Korea and Taiwan, the indigenous elites were displaced or destroyed, permitting modernizing elites to take their place. In both countries, the power of the entrenched traditional elites were at least partially broken by Japanese occupation; in Taiwan, the landed elite that had survived Japanese occupation was simply replaced by Koumintang. In contrast, the neo-liberals "trickle down" model works through existing elites, strengthening them. In Latin America, with its tremendous concentrations of wealth and power, this is a severe problem.

(Finally), the USA model encourages immediate democratization. It is unclear whether democracy really promoted development: there are arguments on both sides of the question. What is clear is that in South Korea and Taiwan, specially the periods of most rapid development took place under strong and brutal one-party dictatorships—"the invisible hand was more of an armed fist."

The East Asian model is not one policy but a sequence of government-controlled programmes, each building upon the previous ones. The industrial phase is only the last of a series of carefully planned stages, from land reform to primary import substitution to secondary import substitution to the export of small-scale manufactures to high-tech export. There is no point in the process at which one can say "This is it. This is the correct policy." The USA's neo-liberal model, on the other hand, is a static set of principles that is supposedly applicable to any country at any level of development, Taiwanese and South Korean policy has been *dynamic and evolutionary*. In most Asian countries, *e.g.*, in China (including Hong Kong), South Korea, Singapore, Malaysia and Thailand the unemployment rate was 3 per cent or lower in 1996 which could not be reached by none of the OECD countries; unemployment was highest in the Philippines at 8.6 per cent and even this rate was lower than most European Union countries.

For an increasing proportion of agricultural and manufactured commodities and for some services, production and markets of these four countries of East Asia have become world-wide. And it has been shared by a large number of scholars that the East Asian success stories do not in fact owe their growth to liberalized markets and *laissez-faire* industrial policies, but to domestic industrial development strategies that share several similarities to the ISI approach. Thus the switch from ISI to the new economic model, NEM gives a hybrid restructuring model which is a result of superimposing NEM policies on systems weaned on ISI policies. But even with the addition of some other "almost-NICs", this would not be a very long list of countries, though other countries of East Asia such as Indonesia and Malaysia, seem to be following a similar pattern of growth, though starting much later. In addition, there is some danger of exaggerating the extent to which the globalization of production lead to the rise of the MNEs. "For India's territorially extensive economies, it remains to be seen whether state administrations have authority and capacity to give direction to market development, respond flexibility to private sector needs, develop technology and human resources, while controlling labour, the latter being regulators of market" (Harris-White, 2002).[11]

Classifying Development Stages and Policies: Some Taxonomies

Comparing any two countries is a task fraught with complications.

Comparing two regions with individual countries as diverse and heterogeneous in and amongst themselves is even more complex. It is axiomatic that kinds of economic activity, domestic industrial structure and FDI associated with them, reflects the countries' economic structure and the nature of the countries' technological capabilities. We can distinguish between four stages of knowledge accumulation : (1) the pre-catching-up stage, (2) the catching-up stage, (3) the pre-frontier-sharing stage, and (4) the frontier-sharing stage. Generally speaking, pre-catching-up countries are the least developed countries with low GNP per capita and poorly developed technological capabilities, and include countries such as Bolivia, Vietnam and Paraguay. These countries are natural resource-based with commodity exports, they have little or no basic infrastructure and have yet to develop an "adequate" minimum technological capability. Catching-up countries have achieved the necessary threshold level of technological capacity, they are having "generic" basic infrastructure. They are having some level of knowledge infrastructure and growing capacity to imitate and they are engaged in low-value adding manufacturing. Growth of domestic industry in support and related sectors, little outward FDI and growing inward FDI. These countries are China, India, Malaysia, Brazil, Chile and Argentina. As any country approaches the "technological frontier", the accumulation process proceeds at a slower pace (the "pre-frontier-sharing stage). These countries are having increasingly specialized knowledge infrastructure, decline in potential to imitate and adapt increasing integration into efficiency based global production networks. The assimilation of external knowledge becomes more difficult, both because of the increasing complexity and the quantity of knowledge, and the difficulties of acquiring this knowledge. These countries are having strong domestic industry, they move towards free trade and they are having increasing use of networking to achieve modularity, rising inward FDI and rising outward FDI become also the characteristic. Pre-frontier-sharing countries are in the process of converging on the frontier countries, and they are mainly the so-called Asian NICs, and include Taiwan and Korea. A typology of the countries related to the catching-up process in terms of their motivations to FDI is given in Table 7.1.

Table 7.1 : Typology of Countries by Developmental Stages

Developmental Stage	*Pre-catching-up Stage*	*Catching-up Stage*	*Pre-Frontier Stage*	*Sharing Stage*
Technological Absorptive Capabilities	Natural Resource base; Commodity exports; No technological capabilities; Little or no basic Infrastructure. Underdeveloped Institutions. Few Domestic firms with technological capabilities: Little or no inward FDI	"Generic" basic infrastructure; Growing capacity to imitate; Engaged in low-value adding manufacturing. Growth of domestic industry in support and related sectors; Little outward FDI Growing inward FDI	Increasingly specialized knowledge infrastructure, Decline in potential to imitate and adapt; Increasing integration into efficiency based global production networks. Strong domestic industry. Increasing use of networking to achieve modularity; Rising inward FDI; rising outward FDI.	Techological opportunities primarily rest on long-term innovation and collaboration pushing frontiers of knowledge. Considerable in-house R&D activity by both domestic and foreign MNEs. Outward FDI to augment domestic capacity. Growing use of R&D alliances and networking. Strong use of knowledge of infrastructure. Growing use of outsourcing to earlier stage countries of lower-value added activities.
Economic Structure	Primary sector	Primary Sector declining ___		
	Manufacturing Sector increasing with declining growth ___			
		Service Sector increasingly predominent ___		

Table 7.1 *(Contd...)*

	Pre-catching-up Stage	*Catching-up Stage*	*Pre-Frontier Stage*	*Sharing Stage*
Motives for FDI	Resource Seeking investment, Low advantages, limited to natural resource endowments.	Resource Seeking FDI but growing low advantages, particularly unskilled labour and infrastructure attracts labour-intensive manufacturing. Growing presence of market-seeking FDI.	Market Seeking FDI and increasing efficiency seeking FDI in manufacturing as low advantages become increasingly created asset-based.	Efficiency Seeking FDI market-seeking FDI and asset-augment-FDI.

Note : Starting with primary sector only in the initial pre-catching-up stage, the successive evolution in the economic structure of the other three stages is given row-wise in the above.
(After Narula, 2002)[12]

Following the above topology of countries motivated to FDI, the East Asian experience can be identified as follows :

East Asian Model No. 1 : Autonomous Strategy : This is the model undertaken by Korea and to a lesser extent Taiwan. The primary objective was the upgreding of domestic firms by selective restrictions on FDI (in the case Taiwan), and the use of technology imports (in the case of Korea). In other words, foreign participation was largely limited in the targetted sectors, with complex industrial policies that encouraged upgrading. The yardstick for industrial development, and indeed, the objective was to promote competitiveness in exporting, and this determined which sectors were targetted, and the extent to which subsidies and incentives were provided.

*East Asian Model No.*2 : Strategic FDI dependent strategy : This strategy is best exampled by Singapore, which due to restrictions due to its limited size, could not pursue an efficient domestic industrial base. It sought instead to attract MNE activity, and then made strong efforts to upgrade the quality of FDI towards higher value-adding activities.

East Asian Model No. 3 : Passive FDI dependent strategy : In this model, FDI was also the primary driver, but instead of intervening to encourage upgrading (as with strategic FDI dependency), it relied on market forces to encourage the upgrading process. Although policies to encourage the development of "generic" location advantages were implemented (such as infrastructure development, incentives for export, skilled cheap labour), the development of complementary domestic industrial capacity was not developed in tandem with FDI upgrading. This model was followed by Malaysia, Philippines, Thailand and Indonesia.

Latin American countries like those of Asia and much of Africa followed the ISI model.The ISI approach also had as a primary objective for the upgrading of domestic firms, by selective restrictions on FDI and the use of technology imports, but focussed on ISI rather than on promoting exports. Since the mid-1980s, these countries are regarded to have adopted the "ISI restructuring model". This is a hybrid model based on a rapid transition from ISI to New Economic Model (NEM), having undertaken trade liberalization and export incentives, often as part of a structural adjustment programme. Some countries have relied on MNEs to drive their growth strategy, while others have depended almost entirely on domestic firms, and the inflow of technology through *arms length arrangements* such as licensing.

Similarities and Differences between the ISI and Autonomous Model

Latin American economies are individually very different, with different languages, geographies, histories and resource endowments. Nonetheless, they share a few common features (First), they have all pursued an ISI, inward looking policy orientation for the several decades prior to the current level towards liberalization, roughly until to the early 1990s. Indeed, the ideas and principles behind the *dependencia school* are native to Latin America, (secondly), but not entirely unrelated to the first point—there has been a long love-hate relationship between the USA and Latin America continent as its backyard, and has been the largest trading and investment partner for over a century now. In addition, USA has intervened politically and militarily on numerous occasions to maintain its economic and political dominance of the region. (Third), most countries in Latin America have been politically independent for over a 100 years, in so far as they are self-governing. (Fourth), they have had a historical dependence on natural resource and extractive industries, a trend which the import-substitution policies implemented in the 1950s and 1960s were to have helped reduce. These issues have intentionally or not coloured the attitudes of government policies towards industrial development, as well as the kinds of FDI that have been received by Latin American countries within the current wave of economic liberalization.

At the risk of oversimplifying a complex set of developments, the doctrine of import substitution hold in the post-World War II era, whereby leading economists of the day rejected the market solution as a means for the underdeveloped South to catch up with the developed North by moving away from exporting primary commodities and importing manufactures, towards developing a domestic industrial base. This, it was argued, would capture the rents that derived to the developed economies from value adding to the primary commodities imported from the South, and the resulting structural change would spur economic development, as well as promoting economic independence. The implementation of import-substitution generally involved a high degree of central planning, combined with protection. Although one of the main objectives was to decrease manufactured imports, the net effect was also to discourage exports. In both manufacturing and agriculture, *inter alia*, because of the overvalued exchange rates.

Domestic industries was to be developed by seeking capital and technology from abroad, since it was largely accepted at the time that physical capital and know-how could be transferred relatively easily through the flow of aid, turn-key projects and the provision of technical experts from the North to the South. Indeed, this view was widely held with agencies such as the World Bank promoting these technology transfer programmes.

The role of MNEs was seen as a means to actualize the process of technology transfer. Investments in most countries were permitted to targetted sectors with the explicit understanding that control, ownership and technology would gradually transfer to the domestic sector. In addition, intermediate inputs were to be phased out as domestic suppliers acquired the competence to meet the (graduated and increasing) local content stipulations that were generally included in the investment agreements. FDI was largely undertaken with the intention of supplying the local market, since neighbouring countries had implemented their own import substitution programmes. Captive markets meant that MNEs were able to pass on the costs of producing at an inefficient scale. A considerable share of productive assets were in state ownership, either as a part of the belief in central planning, or to support large capital intensive and scale-intensive projects which the private sector could not afford to maintain.

Import-substitution policies did lead to economic growth in most developing countries during the 1950s and 1960s, and even in the 1970s, although the anticipated growth of domestic manufacturing sector did not go quite as planned. This in part reflected the application of a similar ISI programme in most countries, despite the considerable differences in the countries. ISI schemes were not adjusted to reflect differences in comparative advantages but sought to duplicate the same breadth of industrial sectors regardless of their initial specialization and resource endowment. It would seem axiomatic with hindsight that the import-substituting experience of countries at different stages of economic development would necessarily be different, but this was not acknowledged at the time. Countries as varied as Argentina and Peru, for instance, in Latin America attempted to build up domestic expertise in automobiles and chemicals, despite it being the case that less developed countries have—in addition to a lower income level—have lower technological capabilities and an economic structure that favours resource-intensive and primary-sector activities. The focus on import substituting meant that little effort was made to export manufacturing output. Even as late as 1985, manufacturing export from Latin America were just 25.1 per cent of exports in the same year, less than half the level in East Asia where manufactures were about 51.7 per cent of exports in the same year (Narula, 2002).[12] It is to be noted that that manufacturing value added as a percentage of GDP was roughly equal in the two regions of Latin American as well as Asiatic countries of the South. Thus domestic acquisition and transfer of technology and managerial know-how to the domestic sector was less successful in bringing Latin America to world levels than in East Asia, and productivity of the domestic sector persistently lagged than that of the Asian countries. Furthermore, although import reduction was one of the primary goals of the ISI programmes, imports continued to be significant, as intermediate and capital goods still had to be imported. As late as 1985, manufactruing exports was 61.7 per cent of total imports in Latin America, almost identical as the same figure for East Asia and the Pacific (61 per cent). The one exception was Brazil where manufacturing exports were just 37.9 per cent of imports. Despite increasing awareness of the problems of import substitution, however, many countries continued to pursue these policies, in many instances increasing the role of state ownership as a means to increase efficiency and to promote social welfare.

The majority of East Asian economies also implemented similar infant industry programmes in the 1950s, discouraging foreign ownership wherever possible, and encouraging the development of domestic enterprise in much the same way as had Latin American and South Asian countries. While maintaining the basic industries of building up domestic manufacturing capacity, Taiwan modified the regime of ISI in the late 1950s, and Korea followed suit in the mid-1960s, seeking to encourage exports *alongside the primary goal of building domestic industrial capacity*. These included establishing a realistic exchange rate, and creating incentives to export (including subsidies, credit allocation, trade restrictions, and reduced or duty free access for imported inputs). Singapore went much further, dropping ISI policies almost completely around the same time. The new NICs Malaysia, Thailand began to move toward a greater export orientation and friendliness towards FDI from the late 1970s onwards. They were later followed by Indonesia, Philippines, China, India and eventually most of Asia. The point here is that all these countries (and later Philippines) pursued industrial policies that maintained significant elements of import-substituting regimes until (and in some cases, beyond) the 1990s, very much as Latin America has done.

Thus it is possible to say that East Asian countries adopted a *more* outward-looking, export-oriented policy orientation at a much earlier period in time than Latin America. Almost all economies in the East Asia region have actively sought to intervene to support the growth and competitiveness of their domestic sector, alongside their export-orientation. This was done through a variety of means, both by promoting domestic sectors as well as restricting imports. At 23.5 per cent tariff rates were only marginally lower in East Asia than Latin America (28.1 per cent) during 1978-80, and by the period 1981-85, were almost identical. Non-tariff barriers were in fact twice as high in East Asia than in Latin America as late as 1989-94. East Asia can therefore be said to be both *export-oriented and import-substituting at the same time.*

Latin America, however, adopted the so-called New Economic Model (NEM) only reluctantly. Unlike the East Asian's miracle economies, Latin America did not voluntarily seek to move towards promoting an outward orientation in tandem with its import-substituting regime, but was pressurized into SAPs due to problems with macroeconomic stability and the ensuing economic crises that engulfed them in the 1980s. To be sure, the large state-owned sector in many countries was highly inefficient and in dire need of reform. However, considerable industrial development had occurred in some of these countries, particularly Chile, Argentina and Brazil, and in Asia, Indonesia and India. This made the reformation and opening up of the economy an exercise undertaken with some reluctance. Nonetheless, the NEM is part of the new, received wisdom that is focussed on tackling the deep-rooted causes that underlie market distortions, explicitly reducing state intervention such that it is applied exceptionally. The emphasis thus moved away from developing domestic capability to enhancing economic efficiency and discipline that market forces were supposed to provide. They therefore moved much further away from ISI and thereby domestic capability enhancement, and much more rapidly so than did most East Asian economies. The NEM on the other hand, entailed large-scale privatization of public sector activities, rapid dismantling of import and FDI restrictions, and the termination or attenuation of state incentives and public goods aimed at enhancing competitiveness of domestic firms. Thus, the ISI programmes shared much in common with the autonomous strategy of South Korea and Taiwan, except that in Asian economies strong state intervention was clearly targetted and coordinated to enhance domestic technological capabilities and competitiveness, while at the same time emphasizing international markets as a benchmark. On the other hand, the NEM on the other hand, and the so-called ISI restructuring strategy went to the other extreme : they emphasized international markets and export competitiveness, but *withdrew the support structure that allowed firms to internalize the spillovers that derived from international competition.*

One of the goals of the ISI was to reduce the dependence of natural resources, as well as to increase economic interdependence. This particular objective was never achieved with primary exports accounting for up to 99.6 per cent of the exports of countries such as Bolivia in 1985. The NEM on the other hand, has also not been able to reduce the primary sector dependency of African and Latin American countries, with only marginal changes in the share of the primary sector in value added between 1985 and 1998. Exports of primary products had declined somewhat in favour of manufactured exports, but this probably reflects increased intra-Latin American flows in response to the regional integration schemes. This partly explains why, despite the increase in the share of primary exports, there has been also ġeneral decline in the share of value added attributable to manufacturing in almost all Latin American economies. Most primary commodities are associated with highly volatile and cyclical markets, as has been the case with coffee and copper (among others) in the last decades, and this makes countries extremely vulnerable to external shocks due to this volatility.

Although it is difficult to speak with certainty in this matter, much of the evidence suggests that ISI-to-NEM shift has resulted in a shock of transition, and has not reoriented Latin America towards the "autonomous model" or the "strategic-FDI dependent model" of East Asia, *but away from it—thus export growth for Latin American countries has not necessarily resulted to export-led growths*. Five important problems with the *ISI-Restructuring Model* (which currently enshrined the NEM doctrine) have reduced the opportunities for duplicating the East Asian success story.* These are :

1. The Attenuation role of government.
2. Unreasonable expectations from the liberalisation of FDI for industrial development.
3. The failure to sustain absorptive capacity.
4. The failure to sequence FDI and domestic capacity in tandem.
5. The failure to recognized the inertia of transition and coordination failures.

The next five sections finally deal with these issues.

The Attenuation Role of Government

At the risk of over-generalizing, the NEM has reduced the role of the state in Latin America to an extent show that the crucial point of the autonomous strategy—the role of governments in providing the proactive process of industrial upgrading—has largely been ignored. Indeed, the role of the state has been significant in the industrial development strategies of East Asian economies.

Nonetheless, given the debt crises and liquidity problems of Latin America, it needs to be acknowledged that the opportunity costs of national governments in pursuing a strategy of promoting technological development are quite high, both in terms of creating the necessary infrastructure—including cluster of related activities—and of sustaining the requisite macro-organizational policies and financial incentives over an extended period of time. In the main, the

---

* Rajnesh Narula's article "Switching from Import Substitution to the New Economic Model in Latin America : A Case of Not Learning from Asia" is an excellent one which forms the basis of the discussion in the present context here.

smaller and/or poorer devloping countries cannot afford to invest in several niches simultaneously, and as such the question of technological targetting (*i.e,* picking the "right" sectors) becomes critical.

State involvement among the Latin American countries has declined considerably with the introduction of the NEM. Perhaps the greatest change has been the reduction in state ownership and the subsequent privatization of assets. Between 1988 and 1999, $ 107.3 billion worth of privatized firms had been acquired through cross-border merger and acquisition (M&A). The share of Latin America and the Caribbean was roughly 79.8 per cent (UNCTAD, 2000).[13] In other words, during this period, about 20 per cent of the total inflows to this region was associated with privatization. During the period 1999-2000 alone privatization totalled $ 19.5 billion, of which $ 4.7 billion was directed at primary sector firms and the rest towards services. However, there has been a considerable decline in the provision of public goods by governments and this has affected domestic industry (and particularly smaller firms), which has relied on the provision of these facilities. Furthermore, the state's involvement has been also reduced in planning. Again, in sum while the change in policy orientation and the subsequent privatization of state-owned enterprises has reduced the interventionist role of governments, their role as market indicator and provider of complementary created asset-based location-specific advantages has become more critical. The lessons from East Asian economies, if anything, is not that the role of governments need to be minimized, but that it needs to be *judiciously and selectively applied.*

Governments play several important roles. There is the need for a holistic approach to selecting and leveraging sectors for dynamic growth, for stable governments, transparent policies, and the provision of basic infrastructure and skills. The provision of certain basic location advantages is perhaps most significant to note, especially for *pre-catching-up* and *catching-up-economies,* where firms (foreign and domestic) rely on governments to make available public and quasi-public goods. (First), they have a passive role in developing the appropriate public and quasi-public goods that are background to economic activity. As countries reach a threshold level of technological capabilities and become catching-up economies in earnest, governments need to provide more active support through macro-organizational policies. This implies developing and fostering specific industries and technological trajectories, such that the location advantages they offer are less "generic" and more specific, highly immobile and such that they encourage mobile investments to be looked into these assets. In other words, their role as market facilitator and provider of complementary created asset-based location-specific advantages has become more critical.

Another fundamental lesson from the Asian miracle countries tends to be lost amidst the focus on openness. Much of East Asia has enjoyed considerable stability in economic policy, and this has much to do with political stability. Political stability implies long-term continuance of economic policy since "reason returns to schooling, investment and trade operate successfully only in a peaceful, stable environment for economic activity". The basic point which is central in this context, technological capacity is central to creating a domestic industrial capacity, and that the processes underlying this are complex, slow and need careful nurturing. Knowledge may be available through markets, either because appropriate markets are thin, or because internationalization and efficient utilisation are achieved only through "learning-by-doing". In other cases, certain knowledge domains are difficult to acquire and efficiently internalize than others, and require complementary non-firm assets if they are to be useful.

The NEM de-emphasizes these issues. Knowledge acquisition is left to market forces, as is the structure of industry. Knowledge transfer is largely assumed to flow through the mechanisms of MNE activity. These are rather nuanced but potentially serious misconceptions, which the next section below will seek to clarify.

Technology and Industrial Development: The Limitations of FDI as a Driver

Although it may seem that there has been a fundamental shift in the principles of industrial development, the view that technological development and upgrading is central to growth and has not changed in more than 200 years. Indeed the concept that knowledge is easily transferable and is available throughout efficient markets is rather new, and forms the basis of the new orthodoxy reflected in the

NEM....."there has been little consideration of the potential impact of accumulation of knowledge and technological capabilities, factors crucial into sustained competitiveness in the new global economy. There is a facit presumption that the new productive structure, because it rewards efficiency, will lead to a rapid process of "learning-by-doing" and, therefore, an expanded endowments of capital, skill and technology will change each country's *comparative advantage towards higher value added.*

Technology (used here as a synonym for knowledge) is cumulative in nature and occurs on a firm-level basis. Technological capabilities are developed by the gradual accumulation of skills, information and technological effort, and firms will develop their technological capabilities in response to the market, supply and demand conditions, as well as from adapting and initiating other firms in the same or similar markets. Firms are boundedly rational, and prefer to engage in innovatory activities that minimize uncertainty of the outcome. Therefore, innovations tend to be related to its existing technological competences. Given this tendency, technology is said to be *path-dependent*, in that current technological competences are a function of its past technological competences.

Furthermore, technology is localized in nature, not only at a firm-level because of its path dependency, but also on a country-specific basis, since cooperation between users and producers in the innovatory process is often specific to a given location, and every location has different supply and demand conditions. In addition, technology has a partly public good nature: although it is relatively less costly to acquire technology than to create it, because of its localized nature and its specificity to the innovating firm, there are costs associated for the recipient firm to efficiently utilize it in its own environment. In other words, technology is only partially appropriable by other firms, and the extent to which they can do so depends on the similarity of their environments and past technological capabilities.

It is increasingly acknowledged that technological capabilities of firms (and on an aggregated level those of countries) define the competitiveness of firms in any given industry. Technological capabilities include not only the ability to search and select the most appropriate technology to be assimilated from existing ones available —what is normally referred to as absorptive capacity—but also the creation of new knowledge through investment in R&D.

These assets can be acquired by several means: through licensing; by indigenous development, and through the modality of FDI. Import-substituting programmes in most countries have sought to combine *arms length technology* imports with indigenous development (*e.g.*, Korea, India) while others sought to combine indigenous development with FDI inflows (Taiwan, China, Thailand). In a globalizing world it seems clear that there are *potentially* multiple and parallel opportunities for knowledge generation, learning and technological accumulation. This is because learning can occur through a variety of organizational means (both intra-firm and inter-firm). However, it bears repeating that learning and technological accumulation is not costless or instantaneous. Developing and sustaining a technological or a *competitive advantage is slow, reversible and highly uncertain.* Likewise, capital can be acquired through other means than FDI. However, although inward FDI does not represent the only option available to developing countries, given their urgency and limited resources it may represent the most efficient option. This is for at least four reasons : (one), the costs of acquiring technological and organizational know-how through arms length means is an expensive undertaking, and given the shortage of capital this option is not open to many developing country governments with limited resources. (Two), liberalised markets means that firms, *ceteris paribus* are likely to be more eager to maintain control of their assets and internalize the market for themselves, either through wholly owned subsidiary or in a joint venture. Exceptions exist, but only where either some strategic reason for the MNE allow for this, the host country has a strong bargaining position, or where the technology has reached the status of a commodity. (Third), infant industry protection is *de rigeur* in creating a domestic sector from scratch, and protected markets are a limited option within the framework of the NEM. (Fourth), the resources, complementary clusters and assets necessary to support a viable and strong domestic sector are also capital and knowledge intensive.

Thus, FDI is now-a-days regarded as a primary—and explicit—means by which growth can be promoted, that the availability o foreign capital and technology is an important means for economi catch-up. It is not however, a *sine qua non* for development, an ide implicit in the Washington Consensus, which largely speaking sti holds to the view that markets for knowledge are efficient, and thu that FDI is the same thing as technology imports (with the bonus of including capital flows), and that these technological imports will generate positive externalities and spillovers to domestic firms.

MNE investments in high-value added activities (often associated with high competence levels) have the tendency to be "sticky". Such subsidiaries tend to be embedded with the local milieu in terms of linkages with suppliers, customers and domestic institutions. The linkages are both formal and informal, and will probably have taken years—if not decades—to create and sustain. Firms generally dislike radical change, and will prefer to maintain the *status quo* if it does not endanger their competitiveness. When an MNE chooses to exit, it must suffer the costs of entry in another location (in terms of effort, capital and time), and these costs are non-trivial. Thus, where the level of competence of the subsidiary is high, they are more likely to maintain the subsidiary, even where an alternative location may provide a better fit to their overall strategy. It is important also, to realize that MNEs have a wide variety of operations in a variety of countries and locations. The headquarters of an MNE has a critical role within its network of subsidiaries, adding value through contributing its own expertise as well as by coordinating the flow of knowledge within the network.

FDI inflows are not a guarantee of growth. FDI and economic development are highly correlated phenomena, but both are strongly dependent on the specific resources. The kind of FDI activity a country might attract too, at different stages of development, are different. The motive of the investment is crucial in determining the extent to which linkages and externalities develop. It is generally acknowledged that there are four main motives for investment : (1) to seek natural resources : where a country possesses an absolute advantage in a given scarce resource, it is in a strong position to extract rent from the MNEs, despite the absence of infrastructure or a domestic sector. Where the resource sought is a natural one, the marginal cost of its extraction to both parties is close to zero. As such, the location is able to generate economic rent depending on the resource's rarity and accessibility to other locations. (2) to seek markets where the local or adjacent markets provide close access to significant opportunities to achieve production economies for the FDI. This requires not only a sizeable population, but also the ability of the market to support (within a reasonable time frame) the expected demand on which the investment is based. In addition though, there is often a "follow-the-leader" strategic response by other firms, whereby a market that might have supported two or three competitors is inundated with a large number of new entrants than the market can effciently support. The case of both the Chinese and the Indian automobile market represents examples of such a scenario where despite the potential for high demand levels, few participants are actually able to make a profit. This is not the case with all sectors—investments in food and personal products for instance are much more likely to achieve economies of scale, since these products have a relatively low income elasticity of demand. Indeed the automobile industry represents a special case in many developing countries, for what is now described as aggressive "market-seeking investments in developing countries" in many cases. These were only permitted under stringent conditions, but the MNEs normally expect to have access to a captive protected market in return. Much of the original FDI into Latin America prior during ISI period was primarily of this nature. Market seeking investments are the primary motive behind much of the investment in chemicals in Brazil, as well as agro-industry in Argentina, Brazil and Mexico. The dergulation of the services sector has resulted in rapid growth in FDI in financial services, energy and telecommunications. (3) Efficiency seeking, and (4) strategic seeking FDI—these two types of investment are similar in that they both normally require a certain threshold level of created assets, and are generally being regarded as being associated with countries at middle-income and industrializing countries, and in the case of asset-seeking FDI, with the industrialized countries.

For many years MNEs have engaged in export-oriented resource-seeking investment, which is *de facto* "efficiency seeking FDI". Moreover, efficiency investment—in the sense that different aspects of manufacturing activity are located in particular locations to exploit the economies of cross-border specialization and the uneven distribution of immobile created assets—is a relatively new phenomenon. In addition, efficiency seeking investment is associated with particular sectors with particular characteristics. These sectors are those dominated by firms which have integrated global networks and are highly dependent on cost-minimization on a global basis. That is, they seek to maintain their cost competitiveness by breaking up the value chain so as to locate production to seek the most efficient location and thereby the lowest overall cost. Such industries include apparel, electronics and automobiles.

In both these types of investments, the sub-national clusters and the agglomeration possesses the needful science and technology infrastructure necessary to attract asset-augmenting FDI, they are considerably different from countries which primarily affect asset-exploiting FDI. It should be noted that even where centres of excellence or agglomeration exist in a given industry, this does not imply that further knowledge intensive investments will be attracted to the same location by virtue of a single cluster existing, unless spillovers or externalities exist. Nonetheless, countries that have the basis for agglomerative economies are the ones likely to receive such FDI. This was originally the case for export processing zones (EPZs). It now applies to higher value adding activities—even in intermediate countries such as India but only where such location advantages are perceived by MNEs to efficiently complement their own core competences.

Complete liberalization and the unrestricted flow of FDI has potentially negative effects for a developing country without the appropriately endowed domestic sector. Liberalization does indeed allow for a more efficient use of a country's resources. However, it remains a basic objective of most country's industrial policy to build up a competitive domestic sector. The indiscriminate FDI policies of most developing countries has resulted in a crowding out of domestic investment. Thus, liberalization and the free flow of inward FDI can result in an FDI-dominated economy, with limited opportunity for domestic industry growth, particularly in least developed countries. From a purely economic efficiency perspective it may well be the case that FDI is "pareto-optimal". FDI-dominated production is certainly better than no production.

Can the NICs be Duplicated? An Analysis of Systemic Constraints

The NICs are the few countries which were the passive recipients of an internationally determined stimulus—the Vietnam War of 1970s as well as the Cold war created geo-political, and internal economic and political imbalances in South East Asia for USA to which USA responded by use of its power in the spheres of finance, defence, energy, and food to counter these trends, thus illustrating the fact that what is considered to be economics can only be understood in straightforward manner as *political economy* (though politics is read off as a residual effect of economic structures). Associated with these trends there was a secular decline in rates of profit in the OECD economies which helped in redeployment of industrial production goods in favour of these East Asian countries taking the advantage of their cheap labour. In theirs able to utilize to produce very rapid growth and industrialization, the latent threat before these countries because of their location must have influenced. Furthermore, the dependency perspective's insistence on a strong emphasis on the international context would appear to be vindicated. These countries, with dramatic phenomenon strengthen the basic premises of dependency which stresses the relative importance for countries of the South of extensive state intervention in the economy for protection against external competition, provided it is adjusted gradually and selectively in accordance with changing relative levels of efficiency, of the need to be concerned about the long-term consequences of FDI, of the need to consider explicitly and actively policies which will accelerate the domestic acquisition of technological capabilities and the national generation of external economies. But the two factors : (one) the lack of access to readily available external credit, diffusion of material benefits which occurred in Taiwan and South Korea in a very clear manner, and (two) a lack of effective demand in Northern markets for further Southern manufactured exports which were present for the NICs during their shift from an ISI to an EOI strategy, and in the consequent EOI period, are absent in the present global economy.

As a starting point, one must succinctly analyze the strategy and its statistical success that these two countries followed in the 1960s and 1970s. They pursued an aggressive form of EOI. Export oriented industrialization calls for a country to shift resources to export oriented sectors by means of state subsidies and access to cheap capital, "outward orientation in these three countries, South Korea, Taiwan and Singapore does not mean discrimination against production of importable but rather, discrimination against the production of non-tradable." This policy is known as a 'market augmenting' strategy while its rival development theory, ISI, is known as a 'market restricting' strategy. EOI calls for the relaxation of protective tariffs and the creation, rather than simply diversion, of demand. ISI on the other hand, calls for protective tariffs to keep imports out and is seen to not create any additional demand, but just divert it to the domestic market and producers.

Taiwan and South Korea both refused to follow a *laissez-faire*-type of EOI, as used by Hong Kong, and instead targetted certain industries within the export sector for special benefits. This targetting of industries closely followed Vernon's product-cycle theory of production that an industry goes through cycles of production. In 1980, the Korean production strategy "was also based on an attempt to upgrade the industrial structure as a whole by moving into more technology and skill intensive niches." Taiwan, recognized the problem of industrial upgrading as early as the late 60s. As its Fifth Four-Year Plan noted in 1969 the next stage was the 'transformation of the current simple labour-intensive processing industry into one that is more demanding of skills and capital.

This pursuit of a targetted EOI strategy seems to have served both South Korea and Taiwan well in terms of economic performance statistics. From 1970-1979 South Korea experienced an annual growth rate in exports of 37.9 per cent (in current prices), while during this same period world trade only grew at 20.2 per cent and other LDC exports grew at 24.6 per cent. With respect to exports, by 1982, 89 per cent of Taiwanese merchandise exports were manufactured goods and for South Korea this percentage was 90 per cent. In the late 1950s this percentage was easily below 50 per cent for both states. For the middle income oil producers, the manufactured goods percentage for merchandise exports was only 59 per cent in 1982. This huge growth and diversification of exports led to a much higher per capita income for South Korea, since in 1950 this figure was $ 146 and by 1980 it had risen to $ 1553, while Nigeria's figures were $ 150 and $ 670 respectively. Obviously South Korea could be seen as a model worthy of imitation by other LDCs.

This export oriented policy is, however, dependent on two major aspects of the world economy. These are the availability of affordable credit to support the shift from ISI to EOI for the adoption of EOI, and also a demand from external markets for these new exports. In the 1950s-70s these two elements were favourable for the adoption of an EOI strategy in the NICs. Cheap credit came from two sources. Firstly, during the 1950s and 1960s Taiwan and South Korea were both large beneficiaries of American aid money. For South Korea, aid financed nearly 70 per cent of the total imports between 1953 and 1962. Taiwan tells a similar story, of a current-account deficit of $ 1.3 billion between 1953 and 1962 aid financed approximately $ 1.1 billion or 85 per cent.

This aid money was largely garnered during the ISI periods in South Korea and Taiwan, enabling these two to build up globally competitive industries that would survive the international competitiveness inherent in an EOI strategy This aid gave the state a free stream of income, unlike the interest payment commitments of capital loans. This aid may have been especially targetted for certain areas, like military spending, but this in itself freed up more domestic capital for the state to channel independently.

Another stream of easily available credit opened up in the 1960s and blossomed in the 1970s—the international commercial loan market—and especially the large part of this market, the Eurocurrency market. The Eurocurrency market flourished in the 1970s due to the OPEC price hikes and, "the huge imbalances of the mid-1970s and early 1980s meant, in effect, that oil producers had surplus revenues to invest abroad." South Korea and Taiwan both borrowed from this Eurodollar market, where most of the OPEC money was being invested, to finance their export oriented industrialization policy during the 1970s. Taiwan and South Korea were most dependent on foreign loans during this period, at around 10 per cent of their GDP. For South Korea this influx of capital was necessary as, "their indebtedness went from $ 301 million in 1965 to $ 2.57 billion in 1970. It has been estimated that without this massive influx of foreign capital in the 1960s, Korean output in 1971 would have been smaller by a third."[14]

The Eurodollar market produced affordable credit to borrowers due to the freedom of its market. Eurodollar banks made money through loans, thus the competitiveness of this market led to lower and lower interest spreads on loans as the banks chased more profit by taking on a larger number of risky loans at lower spreads. The second external factor crucial to the success of the NICs' export oriented industrialization strategy was a sufficient level of foreign demand for their exports. For the purpose of LDC development, this entails Northern demand. In 1980, the USA, Japan, F.R.G., France, Italy and the UK controlled 82 per cent of world trade.[15]

This point is further emphasized in the view that, "the early stages of export-oriented industrialization should thus be associated with rapidly increasing export partner concentration." This demand seemed to be present in the 1960s and early 1970s. In the early 1960s and into the early 1970s the world capitalist economy was on the upswing. Core capitalist states were eager to invest, Eurocurrency was easy to borrow, and interest rates were relatively low, there were fewer trade barriers against Third World manufacturers and little competition from other labour-rich developing countries ... Reasons of timing alone make it difficult to see how the success stories of South Korea and Taiwan can be repeated in the rest of the South. The NICs were some of the first to try and succeed in EOI and thus could capture a Northern market share much easier than late EOI pursuers.

The isolation of these two factors seems to question the legitimacy of the extrapolation of the NIC's EOI experience into a general theory for LDC economic development, particularly as a theory for the Southern countries. It appears that many important multilateral institutions do see this generalization as legitimate as, "the World Bank and IMF, continue to promote the NIC path as a way for heavily indebted countries to escape the debt crisis." This policy prescription goes against the logic of the oft quoted argument that an affordable line of external credit is needed to adopt EOI. The example of Brazil in the Uruguay Round of GATT talks seems to refute the World Bank and IMF position [16] : This lack of available credit for the Brazilian state means, "even if the government wished to pursue a more liberal import policy, it is simply impossible at the moment". Brazil was making certain moves towards loosening its traditional ISI policy, but with its mountainous debt it could not afford the imports of Northern technology needed to diversify their export mix and become competitive. The NICs had this technology free from US aid during their shift from ISI to EOI. Thus "the debt crisis arose inevitably from the export-oriented development strategies, which depended on heavy borrowing for infrastructure". However, South Korea borrowed heavily and has been relatively successful in paying back its commitments and is not caught in the debt crisis, but Brazil seems to have little hope of paying off its debt in the future, so it seems that the NIC model is closed to them. This closure is shown by the fact that Brazil's debt price on the secondary debt market has depreciated from 80 per cent of face value in March 1980 to about 30 per cent in March 1991. Other countries have even lower debt prices in March 1991 as

Peru's debt price was 3.5 per cent and Cote dIvoire's price was around 4.5 per cent. These countries likely see little future hope in being able to adopt an EOI strategy.

The "debt crisis" has led to a contraction of credit in the international loan market, and in the Eurocurrency sub-market. The fear of many large debtor LDCs reneging on their loans has made borrowers wary. Thus, interest spreads have been raised and credit has become more costly. This contraction is starkly shown as, "in 1983 international bank lending to developing countries, excluding offshore bank centres, totalled $ 35 billion. By 1985, a mere $ 3 billion trickled in". The World Bank does offer "structural adjustment loans" (SALs) to countries willing to accept the condition of pursuing the NIC path to help them out of the credit squeeze. However, these loans were usually in the sum of millions of dollars not in the sums of billions present in the debt crisis (only nine countries by 1982 had received loans larger than $ 50 million). Also, as the World Bank only targetted a certain group of countries (30 altogether) for these SALs, the Bank seems to be implicitly stating that the NIC's EOI (or, autonomous) model is not a worthy pursuit as a general development strategy for the whole South.

The second crucial external factor that the NIC experience needed in order to be successful was the presence of Northern demand for the light manufactured goods they turned out in the early stages of EOI. Effective demand can be defined as a customer who is willing and able to pay for goods or exports. The present situation in the world economy and especially Northern markets, seems to fail on both these counts. The example of the textile industry will illuminate a lack of willing demand. The lack of will seems to stem from the large employment benefits of the textile and clothing sector in Northern economies: the textile and clothing industries remain an important source of employment in the developed economies. In particular they employ many of the more 'sensitive' segments of the labour force — females and immigrants—often in tightly localized communities.

The Northern markets do not have an apparent comparative advantage in these labour-intensive sectors, yet the unemployment problems of the North make these industries politically powerful lobby groups whose abandonment to a free market would have huge political costs for the Northern states. This fact has led these two industries to be probably the most trade-managed industries in the world economy. The codification of this management of textile trade is the Multi-Fibre Agreement (MFA) first signed in 1973 and still in effect today. This MFA controls about 85 per cent of the textile and clothing trade and it sets national quotas for textile and clothing exports. These quotas are present in order to slow down the penetration of LDC exports into this Northern market. The Agreement is a definite departure from the spirit of GATT and the "rule of comparative advantage".

This example of the MFA can be expanded to many other areas of traditional EOI exports, imports regulated by new protectionist measures are in particular the traditional trio of clothing, textiles and miscellaneous products as well as some resource-based products and electronics components. In these categories, LDCs were able to capture significant shares of the markets of industrialized countries which failed to adjust to the change in comparative advantage in due time.

Thus, it seems that the new protectionism of the Northern markets has put up a large barrier to the successful copy of the NIC path. The NIC successes in exports has led to the North shutting off this path from other travellers, at least for the time being. Therefore, the NIC strategy can be argued to be an *exclusive development strategy, not a general one*. The (third) criterion of effective demand is the ability to pay for the imports. It is argued that the export-led model of development contains a "fallacy of composition". This phrase means that: while the model may work well if pursued by a limited number of countries, it may break down if a large majority of the Southern countries seeks to pursue it at the same time, because the resulting outpouring of manufactured exports may be more than Western markets could absorb.

This generalization of the NIC model would, "multiply LDC manufactured exports seven-fold. This expansion would imply a rise in LDC share of the market for manufactured imports from 16.7 per cent to 60.6 per cent for all seven industrialized countries and 74.4 per cent in the United States". This extrapolation of the NIC's EOI model is so numerically powerful that, "even if only a limited number of NICs were to follow the G-4 model, there would be a substantial risk of protectionist response". If one matches this with one other factor in the ability to absorb imports, that of the level of domestic demand for that good (domestically produced or imported), the story becomes even darker as: this general flattening out of demand for clothing has been greatly exacerbated since the early 1970s by the effects of the world economic recession. Consumer demand for clothing has stagnated in real terms.[17]

Northern markets' ability to absorb more imports from the Southern Countries does not seem to be very high and the NIC path once again seems to be a dead end. The EOI development of the NICs itself casts some dark shadows on the applicability of their experience to many other countries. The most important shadow may be viewed as the fact that part of the NIC development was dependent on the diversion of exports to Northern markets from other LDCs to the NICs. In other words, the NICs developed partly at the cost of other Southern countries. This parasitical development is best seen through the NIC experience under the General System of Preferences (GSP). The GSP was a GATT mechanism to give certain LDCs preferred access to certain Northern markets in order to help them develop economically. South Korea and Taiwan both received extensive GSP access in the 1970s and less so in the 1980s. The GSP is delegated by specific Northern countries to specific Southern states. The Northern country can decide which state gets its GSP offerings and in what sectors these are given out in. The GSP programme seems to have helped certain LDCs cost at the cost of others as, export gains under the GSP accrued primarily to the countries of Asia and Occania. The effect on Africa was negligible. The trade gains were obtained primarily at the expense of non-beneficiary developing countries rather than the industrialized countries.[18]

The Asian NICs, especially South Korea and China were on the positive side of this Southern zero-sum game as, "the export gains from the GSP accrue overwhelmingly to Hong Kong, Korea and Taiwan". According to Cook, 2002,[19] the call for the NIC model to become the general path of development for the South has many structural flaws which seem to be so fundamental that the generalization itself fails an analytical test of its validity. This failure comes primarily from the fact that the advocates of this extrapolation are mostly modernization and neo-classical liberal theorists. These two schools of thought, which are not mutually exclusive, tend to focus their analysis on the internal aspects of development, at the loss of a full systemic element. This missing element seems to be the downfall in this case as two systemic, or external, factors refute this generalization. This does not mean that ISI should thus be translated into a general development for the South; instead it seems to point to the fact that no one strategy will be useful to all LDCs.

Thus the "export promotion" strategy may be chiefly applicable to East Asia. But *the practice of presenting the Newly Industrialized Countries of Asia as models for economic development is unjustified since the factors which facilitated their development are not present in the current global economy. Systemic constraints such as Northern Protectionism, the debt crisis, and lack of demand negate the possibility of duplicating the economic development process of countries such as Taiwan and South Korea.* It might be very difficult to transfer the final export-led phase to countries elsewhere in Asia, in Latin America, Africa that have not already experienced massive infusions of US aid, the displacement of traditional elites and land reforms, and dictatorships that can enforce low wages and floating prices. Equalization of income through redistribution before industrialization, a crucial element of the East Asian model, is hardly an option in countries that have already industrialized, such as Brazil, Mexico and India. Although an outward-oriented strategy may be advantageous, neither of these two models, the export-oriented Neo-liberal or the East Asian Autonomous Models that are now dominant may be particularly applicable to the rest of the South. Thus the question remains: *Can the NICs be Duplicated? An Analysis of Systemic Constraints*?

(Secondly), this shift has had important consequences for the spatial distribution of economic activities both globally and within countries. Indeed, most of the so-called "global companies" are still strongly attached to their home countries. For example, in 1991 only 2 per.cent of the board members of big US companies were foreigners, in Japan foreign directors are said to be as rare as British Sumo wrestlers. This bias is manifested in the activities of even the most "global" firms. When total sales decline, home markets tend to be protected at the expense of the foreign ones. When firms expand abroad they continue to rely heavily on suppliers from their home countries.

(Third), within the national economies of the "core", the NIDL has led to both a reorientation in employment away from manufacturing to services and massive restructuring of regional specialization (if not collapse) within national economies because of the loss of "traditional" manufacturing industries and the failure of the new ones to replace them. In Britain, the "inner cities" of London and other centres of the nineteenth century industrialization in northern England, South Wales and Central Scotland all succumbed to the NIDL and got concentrated by the poor people.

This tendency of disorganized capitalism increases in "interaction-at-a-distance in industrial organization has brought a tight *space-process-relationship* model in which firm control and operational activities are completely separated geographically. In Table 7.2 below, the operational close space-process-relationship would take the form A⇒B⇒C. Even though *D, E,* and *F* are possibilities, they do not attract attention. In fact, a major metropolis may continue to dominate *all* levels in the hierarchy of functions in *absolute* terms even with the appearance of foreign operations, as Sayer (1985)[20] has pointed out with respect to the electronics industry in the London-Southeast region of England. This example shows that it is important not to overstress the role of unskilled labour-intensive production across all industries and overstate the significance of standardized mass production (or, its opposite, "customized production" for that matter).

Table 7.2 : Close Space-Process-Relationship

Level of Corporate Hierarchy	*Type of Area*		
	Major Metropolis (e.g., New York)	*Regional Capitals (e.g., Brussels)*	*Periphery (South Korea)*
1. Long-term Strategic Planning	A		
2. Management of Divisions	D	B	
3. Production : A Routine work	E	F	C

The NIDL model is contingent and historically-specific and it should not be presented as universal relationship.

Geographical distances among nations are no longer as important as they used to be in defining national identities or national advantages. In another, quite different sense, the word "*globalization*" or "deepening of internationalization" has become the new ideology for mainstream development which leads to the idea that development is inherently universal and increasingly global economic process. This is qualitatively a new one hold by development of telecommunications and information technologies and a new organizational networking which has integrated the world's financial markets, allowing capital to flow unimpeded to where returns are highest, so that global and the local are enmeshed in a shrinking world giving a dimension of importance to rise to superterritoriality. The world has shrunk and no nation can remain in economic isolation from the rest of the world without hurting itself.

In another quite different sense, globalization is used as a process that guides exchanges among countries on the basis of a "global set of rules" enacted to "transactions" by structures: in today's lexicon, they can be termed as *institutions* and *organizations*. Examples of formal institutions are constitutions, legal systems, government regulations at national level and at international level they are the trade treaties, human rights etc. for which there is now increasing consensus about their universality. Cultural norms and traditions are examples of informal institutions. Examples of formal organizations are legislatures, courts etc. at national level while their international entities are the organizations like IMF, WTO, World Bank etc. Examples of informal organizations might include kinship groups, extended families etc. at national level.

If we have a two-by-two matrix, with *institutions on the horizontal* and *organizations at vertical axis*, we get *four* policy spaces. All of these *four* spaces are important. The interaction between (*i*) formal institutions and formal organizations are the policy space which is most favoured in globalization because this is the space in which "policy reforms are introduced both by institutions and by organizations for undertaking reforms". The interaction between (*ii*) formal institutions and informal organizations—or between (*iii*) informal institutions and formal organizations—are the two spaces (or boxes) where many of the problems of "poor governance" occur. The fourth space (of informal institutions and informal organizations) defines the "informal traditional economy with self-regulating, system" where the formal apparatus is largely absent than national rules—Some of these rules such as those governing trade in terms of *tariff* etc. are frameworks for trade among nations—this involves policy reforms which has actually helped to encourage the opening of the economies of the countries of Global North to trade, capital flows, and technology transfer. All these represent a shift in domestic policies for the Countries which, in some ill-defined way, makes the national interest subsidiary to interests of other countries or multinational corporations (*the huge conglomerate enterprises, MNEs which straddle more than one nation*). Thus opening the country for international investment through this *international subcontracting* or such relocation of plants in low-wage economies for export to home Countries of Global North has been seen as the vehicle for a new variety of export-led growth. However, if one judges by present-day Mexico, Taiwan and Korea, the countries where international subcontracting has the greatest importance, the gross value for increasing the country's international export reach at most by 5-10 per cent.[21] Moreover, in the cases of both subcontracting to nominally independent concerns and production by subsidiaries or branches of MNEs for export purposes, the proportion of value-added in the host country to total price of products thus tends to be rather low. Very often, the subcontractors use only traditional skills and the lowest-paid labour. Thus Countries of Global North in order to overcome the crisis of worldwide inflation and stagnation has to search for low-wage locations which link the latter to the "world capitalist system" *through their Trojan horses of* MNEs. The powerful can then take advantage of the weak to siphon off profits and incomes from the latter. In a more political sense, greater participation in world markets is believed to represent economic dependence on dominant world powers, which in turn is expected to lead to political dependence or social and cultural absorption. It has been argued that an inevitable consequence of economic integration would be a clear loss of autonomy. This argument is part of a broader critique of market-based development, which is supposed to work only in favour of the middle and upper income group people in the receiving country. Such development, it has been argued, leads to marginalization of the weak and the poor. A reduction in controls and government intervention in the economy (and its ownership of assets) is likely to hurt the poor. It is also argued that the burden of economic stabilization policies is likely to fall more on the poor than the affluent sections of the society. The adverse impact of stabilization and liberalization policies on the poor can occur in a variety of ways—the poor may be hurt because of loss of employment in the public sector and/or closure of uncompetitive firms in the private sector; the poor may also suffer because of reductions in government subsidies and social expenditure during the period of stabilization. Finally, the poor may be hurt because of foreign competition and inflow of foreign goods if tariffs are lowered and controls on foreign investments are lifted. In the context of economic disparities by the globalizing changes, there have to be concerned efforts so that form of changes become less destructive for people who would otherwise be displaced. When an economic adjustment takes place, few tears are shed for the superseded methods of production and for the overtaken technology. "Thus in trying to correct the imbalances imposed on by the two hundred years of colonialism, the *resolution does not lie in yet another dose of forced orientation towards* international capitalist system: it lies in the development of economic structures, techniques and organizations that will foster the growth of these economies inward and will produce strong *internal* connections between different sectors of economy and different occupational groups. The logic of capitalist exploitation and domination has hindered such a development so far" (Bagchi, 1982).[22]

In whichever way the development is defined, so far as economic growth is concerned, it requires the production and distribution of a surplus and surplus is generated through production in excess of what is required to sustain society from day-to-day and generation to generation. This surplus must be consumed, redistributed and reinvested. The productive reinvestment which sustains the system is known as *accumulation*, and it is not easy to deny *the importance of capital accumulation or of industrialization.* Accumulation in turn does not take place in a social and political vacuum but in a mesh of social institutions like locality, class, ethnicity, and state governance which all interact in a complex way to sustain and sometimes to threaten this accumulation process. Not only does accumulation take a wide range of forms but also the accumulating class may not be homogeneous. Certainly, all accumulations requires the same logic of surplus appropriation and productive reinvestment, but there may be important divergences of interest within the capitalist class.

States have an important role to play in the nurture or suppression of divergent forms of accumulation—the legal and coercive framework cradling the market economy, the balance and directions of flows of revenue from taxes to the state, and the productive and distributive backwash from it may sustain one fraction of the accumulating classes over another, one type, one sector, one kind of site, one region over another.

Furthermore, what is really at issue is the crucial role of labour mobilization and use, and not whether the opportunity cost of labour is exactly zero. According to Prof. Amartya Sen (1984), "there is still much relevance in the broad policy themes which TRADITIONAL DEVELOPMENT ECONOMICS have emphasized. The strategies have to be adopted to the particular conditions and to national and international circumstances, BUT THE TIME TO BURY TRADITIONAL DEVELOPMENT ECONOMICS HAS NOT YET ARRIVED".[23]

Both institutions and organizations can bring a shift in domestic policies of the nation-state which, in some ill-defined way, makes the national interest subsidiary to interests of other countries or multinational corporations. It is feared that opening the country to international trade or international investment will "integrate" the economically weak developing countries to the group of economically powerful developed countries. The powerful can then take advantage of the weak to siphon off profits and incomes from the latter. In a more political sense, greater participation in world markets is believed to represent economic dependence on dominant world powers, which in turn is expected to lead to political dependence or social and cultural absorption. It has been argued that an inevitable consequence of economic integration would be a clear loss of autonomy. This argument is part of a broader critique of market-based development, which is supposed to work only in favour of the middle and upper income groups. Such development, it has been argued, leads to marginalization of the weak and the poor. A reduction in controls and government intervention in the economy (and its ownership of assets) is likely to hurt the poor. It is also argued that the burden of economic stabilization policies is likely to fall more on the poor than the affluent sections of the society. The adverse impact of stabilization and liberalization policies on the poor can occur in a variety of ways — the poor may be hurt because of loss of employment in the public sector and/or closure of uncompetitive firms in the private sector; the poor may also suffer because of reductions in government subsidies and social expenditure during the period of stabilization. Finally, the poor may be hurt because of foreign competition and inflow of foreign goods if tariffs are lowered and controls on foreign investments are lifted. In the context of economic disparities by the globalizing changes, there have to be concerned efforts so that form of changes become less destructive for people who would otherwise be displaced. When an economic adjustment takes place, few tears are shed for the superseded methods of production and for the overtaken technology.

States and Market-Access in World-System Economy

Territorial states are one type of concentration of social power that emerged in specific historical conditions in which state territoriality was practically useful in fulfilling the objectives of both dominant and subordinated social groups. More radically, the sovereign power of territorial states can be understood as resting on power "from below". In other words, the territorial state draws its power in capillary fashion from social groups and institutions rather than simply imposing itself (or coercing) upon them. From this point of view, the power of the state relies on the wide range of sources it can tap into. This is "*non-sovereign conception of power*" in contradiction from that view which sees power as flowing from a single (sovereign) source, such as the state. In this construction, power is best thought of as equivalent to the energy moving a circulatory system rather than as a mechanical opposition between a source of power, on one hand, and an obedient (or, "truculent") subject, on the other. There are multiple points at which consent and resistance come into play in expanding and restricting the interplay between states and subjects and, hence in defining the state's effective territoriality: how well it dominates its claimed block of space. The spatial monopoly of power exercised by a state is not and cannot be total when its power derives from that given up and potentially retaken by others.

Today, we see the emergence of local/regional governments and supraregional communities in the application of infrastructural power for such ends as economic development and political identity, usually without the coercive power traditionally associated with territorial states. Another example involves the contemporary decentering and deterritorialization (at a state level) of the means of production and communication. This reflects the opening up of the world economy for increased cross-border flows of trade and investment under American auspices since the eighties. Recent developments in financial markets and information technologies, however, have accelerated changes in the way people, places and states interact and how economic and political actors perceive these interactions. States (and others) must now manage these interactions for them to be concerned with attracting "foreign capital" and gaining access to global flows of information. Contemporary economic practices, therefore, point towards the incipient creation of a "*transnational territorial liberal hegemony*" in which territorial states are no longer the basic building blocks, they are being rapidly challenged by new spaces of networks and flows. *When states are situated in the context of a world of evolving social practices, they have inexorably led beyond state boundaries in pursuit of wealth from the deployment of "mobile property" that is, the capital.* This is like that a home territory also provides a base from which to launch attempts at acquiring property elsewhere. This requires that assets be reasonably liquid and transferable over space and across state boundaries when states effectively exchange control over economic flows emanating from their territories for increased access to flows coming from elsewhere. As a result, when increasing proportion of property are mobile beyond any one state's boundaries, individual states provide only a partial and tenuous protection for absolute property rights (*i.e., legal rights of ownership and use*) which originated the framework of the modern states in Europe in the seventeenth and eighteenth centuries and *without which global capitalism would not have been possible.* Other geographical levels of governance become attractive, as was the case with the Bretton Woods System[24] regulating international finance from 1944 to 1972, followed by annual G7 Summits between the leaders of the Big Seven Industrial Countries. But uncertainty as to future political actions and macroeconomic changes (tariffs, interest rates etc.) also gives an incentive to "property-holders" to spread assets around further than leave them pooled up in one state.

This process is not new. Its origin go back to the merchant capitalism (trading and collecting organizations) of the sixteenth century. What is new is the increased quantitative scale and the enlarged geographical scope of the mobile property (*i.e.*, the capital) now moving to and fro across the boundaries of the world's trading and investing states. In this context, the states and firms have changed their orientation from free trade to what has been called "market access". The underpinnings of the world trade regime that prevailed in the aftermath of World War II are being replaced by those of a regime in which a premium is placed on the "openness of borders". "Leakiness" in cross-border (*i.e.,* international) flows of goods and investment and in firm multinationality has become a torrent of capital, trade and corporate alliances. This consequence of corporate alliance contrasts the nature of the old mercantile regime with that of the new one by identifying the "six pillars of market-access regime" that they claim have underpinned each and the policy instruments associated with the new regime. The policy instruments reflect an abandonment of classical state sovereignty in return for guaranteed access to other states' territories. The world has moved away from the strict association of property rights with state territoriality. A range of non-territorial factors now determines the competitiveness of firms in many industries: access to technology, marketing strategies, responsiveness to consumers, flexible management techniques. All of these are now the internal assets of firms, not the state territories. Firms grow through deploying their internal assets as successfully as possible. States now compete with one another to attract these mobile assets (capital) to their territories.

Three features of the market access regime are particularly notable with respect to the changing geopolitics of power. (One) is the internationalization of a range of domestic policies to conform to global norms of performance. Thus, not only trade policy but also industrial, product liability and social welfare policies are subject to definition and oversight in terms of their impacts on market access between countries. (*Another*) is the increased trade in services, once produced and consumed largely within state boundaries. Partly this reflects the fact that many manufactured goods now contain a large share of service inputs—from R&D to marketing and advertising. But it is also because the revolution in telecommunications and internet means that many services, from banking to design to packaging, can now be provided to global markets. This represents a significant material challenge to the domestic/international distinction upon which the "realism" of the strictly territorial accounts of the spatiality of power rely. And, (finally), the spreading reach of transnational firms and the emergence of international corporate alliances have had profound influences on the nature of trade and investment flows, undermining the identity between territory and economy. Symptomatic of the integration of trade and investment are such frequently heard concerns as rules on international investment and unitary taxation, rules governing local content and place of origin to assess where value was added in the commodity chains of globalized production, and rules involving unfair competition and anti-trust or monopoly trading practices.

Transnational Liberalism and the New Geopolitics of Power: Market States

Outlining the above three main drawbacks of conventional accounts of the geopolitics of power brings into focus their central deficiency: taking the territoriality of states for granted as a fixed feature of the modern world rather than seeing it as the outcome of a number of historical contingencies. The contemporary "unbundling" of state territoriality provides the most direct evidence for reshaping of "hegemony away from the state-centred practices of the previous epoch". This does not mean to say that territorial states are (finally) "withering away", only that they must now operate in a global context in which their interactions with one another must take into account a changed military and economic environment. Indeed, in the absence of higher-level units for the enforcement of property rights and the delivery of public goods, states have a continuing and vital role to perform within the evolving world of networks and flows. For example, the deregulation of financial markets requires the deliberate action of governmental authorities—it does not simply "happen". During the "Cold War" between 1947 and 1990, USA in competing militarily and ideologically with the Soviet Union, "sponsored an unprecedented opening of the world economy", partly to spread its "political-economic message" and partly to take advantage of opportunities for its businesses. The net effort has been that markets have acquired powers heretofore vested in leading states. As this process has intensified and expanded, some localities and regions within states have been privileged within global networks of finance, manufacturing and cultural production to the disadvantage of others. The "market access regime" ties local areas directly into global markets. Successful ones are those which can enhance their position by increasing their attractiveness to multinational and global firms. A patchwork of places within a global node and network system therefore co-exists with but is slowly eroding the territorial spatiality with which we are so familiar. For example, the control and maintenance of a territorially uniform and exclusive currency is often regarded as one of the "main attributes of state sovereignty". It has been money as much as "literacy in a vernacular that has bonded state to nation in a process of mutual reinforcement". Yet since the late 1960s a number of trends challenge the notion that each state must have its own "territorial currency". The explosion of non-territorial state political identities associated with global and the increasingly decentralized world financial system, as illustrated by the emergence of projects like "deterritorialization" of currencies of 13 countries of Western Europe abolishing their individual territorial currencies in favour of supranational (or, world-regional) *EURO Currency* (or, European Currency Unit) which may intensify in years to come. Thus in contemporary France, often pointed to as a zone of changelessness by those sceptical of claims about an emerging world of flows challenging that of territories, a neo-Gaullist government support of this EURO currency and a withdrawal of that state from many areas of life. It was General de Gaulle who said that France only exists because of the state, the army and the franc. His disciples now stand watch over their dismantling.

Thus a new geopolitics of power based on global networks and flows is in the offing. All these are forerunner of a new constitutional order in its wake—in the twenty-first. century, "market states" could replace nation-states. Market states will have the same borders and political systems as the nation-states but will shift important responsibilities of the promise of ever-increasing well-being for all (which the government are unable to bear) from government to the private sector to find new ways to encourage these public goods: multinational enterprises MNEs will become "surrogate agents" of the government, filling roles that government can no longer play and blurring the boundaries between political and corporate leadership and a market state will shift reliance from the law- and regulation-based approach of the nation-state to the incentives of the marketplace.

Notes & References

1. *Cited from* Dickenson, John *et al.*, 1996, *"Geography of The Third World"*, Routledge, London.
2. NBER Studies in Business Cycles, OPEC.

3. World Bank, 1996, *World Development Report1996*, Oxford University Press.
4. John Maynard Keynes, 1936, *General Theory of Employment, Interest and Money.*
5. Roger Lee, 1995, *Social Relations and the Geography of Material Life.*
6. Foster, John Bellamy, 2002, *Monopoly Capital and the New Globalization*, Monthly Review.
7. "Thus the notion of global free market hegemony without the nation-state and without discernible centres of power (only highly visible instruments of the market) means a concept of capitalism that has become virtually synonymous with *globalization*. There is, it is proclaimed, no alternative because there is nothing outside the system, and no centre within the system".
8. In the last few years the idea of a "New Economy" has gained wide currency, almost rivalling "globalization" as a neologism that characterizes our era. This New Economy is seen, first and foremost, as consisting of those firms and economic sectors most closely associated with the revolution in digital technology and the growth of the Internet. The rapid convergence of information technologies—including computers, software, satellites, fibre optics, and the Internet—has, it is believed, fundamentally altered the economic landscape. Since the mid-1990s, these revolutionary technological developments have, it is argued, spilled over into the wider economy, generating higher productivity growth, a sustained acceleration of economic growth, lower unemployment, lower inflation, and an attenuation of the business cycle.
9. Polycentric multinational enterprises (MNEs) evolve from a number of epochs of the modern world economy. Using a normative model of firm and their locational behaviour, in the initial stage of *laissez-faire* capitalism, all firms has a small structure laid out over space; with rapid turnover, firms growing at a regional scale, can lead to regional national company (which sometimes) went multinational with control over foreign branches. In these two stages capital is nationally owned. But in the third stage, polycentric MNEs are found which operate with many divisions and allows subsidiaries in other countries considerable latitude.
10. The two NIC city-states of Singapore and Hong Kong before 1994 being such small and highly trade-dependent political units, their EOI experiences have not been as readily held up by certain Northern institutions, such as the IMF and World Bank as examples for other LDCs to follow. Both South Korea and Taiwan have been used more extensively as prototypes in the transition of the NIC experience into a general development theory.
11. Harris-White, 2002, "*Development and Intermediate Classes with special reference to India*", in , Vandana Desai and Robert, B. Porter, eds. The Companion to Development Studies, Arnold, London.
12. *Op. cit.,* Narula, Rajnesh, Page 416, footnote.
13. *Op. cit.,* 6 Chapter 4.
14. Richard, E. Barett & Soomi Chin, 1987, *The Political Economy of East Asian Industrialism*, Cornell University Press, Ithaca.
15. Peter, Dicken, 1986, *Global Shift in a Turbulent World*, Harper & Row, London.

16. In mid-1988, Brazil is again in a very delicate balance of payments situation. Total foreign debt was around $ 105 billion, and only token interest payments have been made on long-term commercial bank loans.
17. Garelik, Glenn, 1987, Newly Industrialized Countries, *Time Magazine*, October 19, 1987.
18. Drusilla, K. Brown, 1988, "Trade Preferences for Developing Countries : A Survey of Results", *Journal. of Development Studies*, April 1988.
19. Malcolm Cook, 2002, Can the NICs be Duplicated? An Analysis of Systemic Constraints.
20. *Cited from* Paul Knox and John Agnew, 1994, *The Geography of the World Economy*, Arnold, London.
21. In Pacific-rim countries of Taiwan and Korea and in present-day Mexico, where the international subcontracting has the greatest importance, export production reach at the most 5-15 per cent of their total value of exports which is much lower, for comparable products for similar subcontracting or export production by subsidiaries in the countries like Canada of the Global North", *cited from* Bagchi, A. K., 1982.
22. *Op. cit.*, 12, chapter 2, Bagchi, A. K., 1982.
23. *Op. cit.*, 15 Chapter 1, Sen, Amartya, 1984.
24. *Op. cit.*, 23, Chapter 4, Bretton Woods System.

8

THE CONCLUSION

This book has been about the "development" of underdevelopment and in that way it opens the way to the ideology which includes the "Countries of the South", known also as the less developed and developing countries, or the metaphor of Third World. The underdevelopment of these countries which got liberated in the aftermath of World War II from their previous colonial rulers of Europe, have been characterized by a low level equilibrium caused by a "trap" of low per capita income and high rates of population growth which result in a syndrome of capital-scarce and labour-surplus economy. And salvation of this underdeveloped South lies in their development to raise per capita income and to meet the basic needs of their people and place where *factors of production* interact by their proper symbiosis. The classic formulation about Countries of the South is that they do possess a "dual economy" reflecting *factor immobility*, the rigidity of which has been viewed as very different from the relatively smoothly functioning capitalist economy of the Countries of the North, and this metaphor continues to-date. It seemed that they are required to be cast also in an explicit context of interdisciplinary geographical-historiography approach by respecting the political economy's inclusive attitude rather than in mathematical terms of *monoeconomics'* claim. As a result the idea of development regarding its state of *process* rather than its *structure* as it has been unfolded in Countries of the South has been attempted in this book.

A few words are also in order about the meaning of "development". Although there are great differences in opinion on the matter, the majority of those who are concerned with economic patterns of Countries of the South, measure it in terms of increases in per capita GNP. However, few would consider this to be the sole indicator, since development is also taken to imply an improvement in income distribution and the reduction in poverty. Moreover, much more is implied in the standard of living than average product of income, since poverty can be defined as the failure to meet certain levels of consumption of such important commodities as food, clothing and education. Amartya Sen (1999) has linked *well-being* to doing certain things and to attain certain types of "being" (such as being educated, being free from avoidable morbidity, being well-nourished etc.) which he calls the "functionings" of a person. Development can then be taken to refer to an increase in the achievement of functionings of individuals, which depend not only on income and the public provision of goods, but also on the characteristics of goods which affect their ability to achieve particular functionings of individuals. Moreover, the list of functionings that are considered relevant can be extended to an alternative approach of collective goals like self-reliance, cultural independence, national sovereignty, sustainable and autonomous development. These do not exhaust all possibilities and the meaning of development is not only complicated by the existence of diverse measures but also by the fact that the process of development may itself entail changes in the yardstick which is used to measure improvement in the human condition. Thus development in today's world represents the uneven development of capitalism on a world scale. When entering into these and other complexities concerning the meaning and measurement of development, it is understood development has to mean increases in per capita production and income while also keeping in mind issues such as income distribution, poverty and expansion of capabilities. The first chapter of the book purposely presents extensive and inconclusive material about the meaning of development from which it is sidestepped to go into subsequent chapters as follows.

The contributions to development issue has been organized in a sequence. First, it is put chronologically to examine the substantive issues surrounding development, by placing it in the world context within a geographic historical overview of development or lack of it. Starting with the days of "mercantile trade" in the sixteenth century, when the world was immensely diverse in economic activity, the seafaring nations of Western Europe located new sources of wealth such as precious metals, desirable luxury goods and spices which they extracted from the new continents of Americas, Africa and Asia through plunder, extortion and trade. The Crown of these countries needed the merchants as allies and in exchange, the merchants required a powerful state to protect their international trade. This alliance between the Crown and the merchants brought *mercantilism,* or the "mercantile era" in Europe during which the favoured merchants were granted *charter* (like "East India Company") or the monopoly rights to control trading zones in the name of Crown and alongwith trade they combined acquisition of territories with force. Thus Europe which was so far *marginal in the affairs of the wider* world but soon European domination attested to some remarkable changes in the world history and economy which in terms of their first phase of extension of colonies, continued *till the mid-eighteenth century.* Feudalism during this time gave way to the series of innovations and inventions (which marked the "industrial revolution"), and the latter spurred by capital investment established the dominance of capitalist industrialization. By 1850, England was regarded as the "Workshop of the World" and the rapid growth of industrial capitalism which started in these countries brought the key phases in the making of the *international capitalist economy,* so that the old protectionism of the mercantile era called for a policy of "free trade" and for the first time in European history, commerce was to become subservient to industry. Thus in the sixteenth century when wealth was finite and typically appeared to be in the form of land ownership or in trade so that the amount of wealth (or lack of it) was known thing but today there simply are no boundaries to the creation of wealth. Yet, we live in a world characterized by deep inequalities between and within the industrialized nations of the North and the rest of the world (that is, the South) so that literally the countries of the North and of the South seem to be worlds apart!

In this considerable differentiation between the North and the South the identification of industrialization as necessary for development of the South has been the conventional wisdom in the post-Second World War period. Development covers both "theory and practice", *i.e.,* both ideas about how development should or might occur in *theory*, and in *practice* how the Countries of the South put efforts to various aspects of growth into practice for their development. In context of a theory it is a set of ostensibly general propositions, which aim to explain by virtue of exclusion or inclusion of different kinds of information and data regarding how theories as *normative models* have commanded support and influenced development policy in practice at different times since the 1950s and in different Countries of the South as well as at the international level. Development has occurred in the past, and/or how it should occur in the present context, remain the issue. Hence there is the need for a brief survey of development theories for understanding the influence of their theoretical innovations which has been done from Chapter Three onwards.

The post-war period till the rise in oil prices in 1973 was characterized by about 25 years of sustained growth in the world economy when economic growth, low inflation and full employment became the norm in countries of the North. This long boom period which created immense demand for trade and goods, witnessed the rapid dismantling of the great European colonial empires so that there emerged the politically independent countries of the South. During this period USA having the hegemony of the world economic and political power, eschewed from its pre-war position of isolationism and the practitioner of protectionism in trade, laid a framework for stimulating free trade and regulating international finance through establishing two world bodies, the World Bank and IMF.

A consensus evolved in the 1950s and 1960s around a philosophy of development in which the economic growth was assumed to be a function of capital formation and the USA economic historian W.W. Rostow in 1960 advocated a "big-push" conjecture in his *5-stage modernization model* for the South. Derived essentially from the Marshall Aid Plan for Europe's post-war reconstruction, in this schema "modernity" was assumed to be achieved through a rapid and sustained phase of industrial economic growth by availing large-scale capital and technological aid from the West. Rostow's scheme implied a continued reliance on the West (on USA) for capital and technology transfer and it ignored the wider role played by world market and the centre-periphery international division of labour. These were the issues developed by the Argentinian Raul Prebisch who as the Director of ECLA (and later he became the first General Secretary of UNCTAD), put forward an argument in 1950 that the international division of labour worked against the interests of the developing countries in

Latin America in a system of "unequal exchange"; the latter was defined as the tendency for the prices of primary commodity exports from countries of the South to decline, relative to the price of manufactured imports from the North (as war-time rationing made the most ingenious use of primary product substitution which gave the North the scope to do so). One consequence of this movement in prices was declining terms of trade, which meant that for primary commodity specialization (which had been associated with colonialism) and backwardness, the countries of the South had to export ever greater volumes of raw materials in order to import the same quantity of manufactured goods and machinery so that trade deficit became the structural characteristic of Latin American countries' trade with USA as their major trading partner. ECLA began to campaign for trade reforms and a fairer deal for primary commodity producers. It also began to promote a new and influential industrialization strategy, that is, ISI which drew on the experiences of countries like Brazil and Mexico in the pre-war years; this structural analysis of Prebisch influenced many newly independent countries in adopting this state-led ISI strategies throughout the 1950s and 1960s and they made use of central planning for their development. Prebisch used for the first time the terms "centre" (USA as "cyclical" centre), periphery (Latin America as "periphery" for USA) and "unequal exchange" which got incorporated in the Dependency theory of Frank and World-System theory of Immanuel Wallerstein.

Dependence is defined as a *continuing* situation in which the condition is created so that the development and expansion in economies of one group of countries are conditioned by the development and expansion of economy of another country to which the former is subjected. The relation of interdependence between two or more such economies, and between these and the world trade, becomes a dependent relationship when some countries (the dominant ones like USA) can expand through self-impulsion while others, being in a dependent position, can only do this or expand as a reflection of that expansion of the dominant countries, which can have either a positive or negative effect on their immediate development. Dependence thus defined leaves room for further interpretations (in dependency terms) of global interdependence, the arguments of the latter are complex set of associations in varying degrees of the internal ones and indeed, internal variables may very well reinforce the pattern of external linkages.

In this term of dependence, most dominant countries are very dependent on imports of raw materials, energy or minerals: speaking more generally, one can even argue that because of their greater integration in the world economy, the Countries of the North are more dependent among themselves and are very often affected by world market crises compared to those of the South, but since there is no political dominance their economic inter-dependence on one another help them to expand and to sustain their individual growth.

But dependency as a theory is a *peripheral insertion in the world-system* through which former colonies that is, the Countries of the South are exploited economically, and their backwardness is maintained over time. No doubt this economic exploitation requires and involves not only economic domination but to create or maintain dependency, political domination is required so that the whole question of power and, therefore, the political dimension is intrinsically linked with the notion of dependency. While political domination is required to create or maintain dependency, it is the degree of *economic exploitation*, and the extent to which this can be maintained over time, which determines the level of dependency. For example, in relations between countries *e.g.,* the United States and England, there is no asymmetrical political domination which does not lead to economic exploitation and it is difficult to say about their inter-dependency.

Thus economic exploitation remains the crucial element in any dependency debate. The most useful concept of dependency focuses precisely on the creation and maintenance of that sort of exploitation, that is, on the existence of a set of structural mechanisms which obstruct the growth of the economy and prevent its filtering down to the masses in "dependent" economies, while creating exactly the opposite conditions in "dominant" economies. The most important characteristic of dependency theory is its attempt to analyze the peripheral capitalism from the point of view of the interplay between internal and external structures and this interplay is analyzed in three different ways : the (first) is that of those following Andre, G. Frank (1967), who do not accept the possibility of capitalist development in the periphery (the Southern Countries), but only of the "development of underdevelopment". The (second) is the dependency as a formulation of ECLA for Latin American development through a *structuralist* approach of Rahul Prebisch (1950).

For understanding the dynamics of development, it requires theoretical conceptualization and the kind of theory needed is a theory of the totality of the process of development *i.e.*, that of world-system perspective which requires studying the development of world-(economic)-system in all its forms and to continue to theorize and reconceptualize. The history of the modern world is the history of capitalist accumulation and that of capitalist development (which started with Western European countries since the mercantile days of the sixteenth century) is the development of a single system, the modern world-system of Immanuel Wallerstein (1976), the dynamics of which we can grasp into three sets of processes : those (1) affecting and maintaining the structural features of the system, (2) representing the cycles of the system, and (3) that represent the secular trends of the world system. The understanding that history of the world-system theory distinguishes it from *modernization paradigm* but it does not distinguish from Marxist approach though it is differentiated by its own theoretical centrality of the other two basic structures like the core-periphery international-scale division of labour and the state system.

International trade has been an integral facet of the total growth phenomena but fluctuation in terms of trade or a deterioration in it is destabilizing and harmful for a country's economy. A major impact of colonial rule was restructuring of the economies in ways that involved the expansion of cash crop and mining of metals for export that is, these countries are known as *primary commodity exporter* but since "income elasticity of demand" is relatively low (when compared to manufactured goods), for a country to be a primary commodity exporter implies considerable disadvantage in the world-system of trade : that is, any increase in national income leads to a smaller demand in the quantity of primary products. Thus the trade-and-growth nexus doesn't become good for Countries of the South. As the most commodity dependent continent among Countries of the South, Africa was hard hit by the declining commodity earnings (the terms of trade of which declined considerably during 1970-1990s); thus continuing dependence by most African countries on a small range of export commodities like copper for Zambia (accounting 98 per cent of export, similarly oil for Angola (96 per cent), coffee for Uganda (95 per cent), cocoa for Ghana (60 per cent) led to a short of "balance of payments" (trade deficit) and a resort of these Sub-Saharan countries to borrowing from agencies like World Bank and IMF on the condition of implementing market-friendly "structural adjustment policies", SAP for addressing the deficit both in trade (in import) and in reduction or downsizing of public sector expenditure through curtailment of government programmes in every aspect of life. While UNCTAD came out of demands by Countries of the South for fairer deal in the global economy, debt-forgiveness, an end to protectionism practiced by the North for their agriculture as well as a commitment by the North to make available the equivalent of 0.7 per cent of their GNP for development assistance, but the mighty forces of globalization remained impervious to the clamourings of the South, and all these were dealt a blow by the GATT trade agreement which results in a 30 per cent cut in effective protection and that the lower tariffs that would result in savings to the buyer. Since the most protected agriculture through subsidies is to be found in Countries of the North so while the losers would be the Southern Countries, the gainers would be the North. Thus, in effect, the Countries of the South (the periphery) would be financing the rapid growth of the North (the "core") with the latter reaping the best of both the worlds at the expense of the former. This is the essence of the Prebisch-Singer thesis (1950). This problem does provide an insight, that the countries of South need to industrialize and diversify their economies so that they could move from the production of income inelastic primary products to more income elastic industrial goods. But this option is not available at will. However, in the wake of the successes of export-oriented East Asian NICs (namely, South Korea, Taiwan, Singapore and Hong Kong), there has been an argument in favour of export-promotion policies in contrast to ISI.

The phenomenal rates of growth experienced by the East Asian NICs brought the view that during the 1980s these emergent group were driving global rates of growth in the face of stagnation of the world economy. Their emergence in the world-economy as industrial producers and exporters from the ranks of "underdevelopment" and to be identified in terms of World Bank's income ranking of as "developed", requires explanation. These four rapidly developing NICs, accompanied by their rise in the level of productive forces, and diversification of economic activities, can be put forward as example for a sort of *intervening model* challenging the dominant stream of Dependency theory and demanding their reformulation in important respects.

Issues relating to capital flow, labour, trade and technology transfer come together in the analysis of the effects of economic interaction between the North and the South. Do such interactions close the gap through any *convergence between them*? The early literature on relation between the North and the South by Gundar Myrdal in 1957 showed that because of *dual economy* in the South, capital flows and trade earnings enter the *enclave sector* of the country without the *spillover effects* of manufacturing. This may be true for Latin American countries where they are in a dependent position in their relations with the core country (which is USA) but whether with the transfer of technology alongwith flow of capital and movement of the MNCs to the South, there would be any closening of the gap between the North and the South, has no definite answer, neither any clue being available at present!

Because of the relativity of development to the nature of society of a country, there is no universal theory also that can be taken as "one-size, fits-all types of policy prescriptions". In the meantime for a Country of the South which practices *dual economy* but not having any constraints of size (by population and by space) as well as not lacking in natural resources so that the country possess the capacity to manoeuvre, one can certainly expect a structural transformation in an orderly manner from a low-income economies into middle- and high-income economies by following Arthur, W. Lewis' *two-sector growth* model (1954).

GLOSSARY

Agglomeration The concentration of productive activities in a particular region.

Appropriationism The replacement of agricultural inputs with industrial alternatives. It forms a key process in the industrialization of agro.

Autarkic A self-sufficient region not having any dependence with other region by link or by economy. Bhutan was an autarkic state in the world few years back before the link road with India was opened. The economy which is self-reliant is economic autarky.

Balance of Payments, (BOP) The difference in value between a country's inward and outward payment of goods, services and other transactions.

Bands Societies of hunter/fruit gatherers typically numbering up to 500 individuals.

Billion Thousand million.

Biomass Plant and animal residues.

Bottom-Up Development Economic and social changes brought about by activities of individuals and social groups in society rather than by the state or its agents.

Branch-plant Industrialization The growth of manufacturing employment in economically depressed industrial regions of core economies as a result of the availability of reserves of skilled and semi-skilled labour suitable for employment in branch factories of manufacturing enterprises based elsewhere.

Capital Money put into circulation or invested with the intent of generating more money.

Capitalism An historically specific economic system in which mode of production and social organization as well as the material conditions of existence are dominated by production for profit and the exchange of goods and services is largely determined by market prices rather than by principles of interpersonal reciprocity or redistribution by a central authority; the system is characterized by the separation of those who owns the means (or, modes) of production and those who work for them. In capitalism, land and labour are the *commodities* which can be sold or exchanged in *market*. Capitalism aims to reduce all beings (including persons) and all cultural differences to a *common commodified base*.

Capitalization The process whereby capital-intensive inputs such as technology are deployed by large firms and replace labour-intensive methods associated with smaller-scale production in agriculture and manufacturing industry.

Carrying Capacity Used in the context of food and agriculture, this term refers to the maximum population that can be supported within a given territory at a minimum daily diet, given the quality of local soils and local climatic conditions, and assuming the availability of appropriate forms of mechanization.

Centralization A trend toward corporate economic integration that involves the ownership of private enterprise by progressively fewer corporations. It is the result of corporate mergers and takeovers, and it is characterized by diversified, conglomerate, transnational corporations.

Chiefdom Social formation based on societies characterized by the internal and unequal differentiation of both power and wealth and organized around the principle of kinship.

City-state Urban regions having political jurisdiction and control over a specific territory. A form of social organization associated with some of the earliest states.

Civilization Refers to an advanced form of social development characterized by such things as urban life, commercial activity, writing systems and philosophical thought.

Climacteric Cycles Long-term changes in relative levels of national economic development based on the exploitation of cheaper *factor costs* (particularly land and labour) in peripheral or semi-peripheral countries and the concurrent failure of formerly dominant core economies to successfully adjust to changing market, technologies, and international conditions.

Cold War A period extending from the end of the 1950s to the late 1980s during which two ideologically opposed blocks, USA and the Soviet Union emerged as the core within the world-system, both with nuclear capabilities.

Command Economy An economic system characterized by state-led central planning of economic activity combined with simultaneous suppression of market-type relations.

Communism A political theory attributed to works of Karl Marx and Friedrich Engels. Communism is characterized by the common ownership of the means of production. The foundations for full communism was supposed to be laid along a transitional period, known as socialism. The Soviet Union was the first state to be ruled by a communist government, from 1917 to 1991.

Common Market A form of international economic integration that involves the elimination of *tariffs* and other trade barriers between member-states, the removal of internal restrictions on the movement of factors of production, and the creation of a common set of trade barriers and trade agreements with non-member countries.

Comparative Advantage The principle used to explain patterns of trade and specialization; if each region or country specializes in those economic activities which they perform *relatively* better than others, each is likely to gain (transport costs and terms of trade notwithstanding).

Competitive Advantage (or, "factor endowment theory") The advantage acquired in economic competition by some locations because of the benefits that accrue from an early start in production of a particular good and the continuing defence of that historical base through superior organization and adaptability.

Concentration A trend toward the reduced numbers of firms in any given industry or economic sector. It is partly the result of the elimination of smaller, weaker firms through competition, and partly the result of corporate mergers and takeovers.

Consumer Services Personal services, including retailing, medical care, personal grooming, leisure and recreation, culture and entertainment.

Core According to Wallerstein, the core refers to those regions of the capitalist world-economy characterized by the predominance of core processes associated with relatively high wages, advanced technology and diversified production.

Counterurbanization Population increases in rural areas beyond the commuting range of major urban areas.

Crude Birth Rate Number of births in one year per 1,000 person.

Crude Death Rate Number of deaths per 1,000 persons in one year.

Culture Systems of shared meanings embedded in community, ethnic groups or nation through language, religion, customs and tradition to make them more interpret more sense of the world. Culture thus embedded in the material and social world provide a sense of one's own identity and it is a process rather than a thing, not divorced from power relations.

Deindustrialization Refers to a relative decline in industrial employment, it may also refer to an absolute decline in industrial output as well as employment.

Demographic Transition Model Traces the shift from high birth rate and death rates to low ones.

Diaspora Literally, the scattering of a population; originally used to dispersal of Jews in other than Jarusalem.

Disorganized Capitalism A form of capitalism since the mid-1970s. It is characterized by a relative decline in the importance of extractive and manufacturing industries, together with a relative increase in the importance of services. There is also an increasing tendency for production process to be dominated by small-scale and flexible forms of organization.

Debt Trap The cycle of borrowing that results when the productivity gains from investments undertaken wth borrowed capital are insufficient to meet interest repayments. Further loans and debt rescheduling provide temporary relief, but in the long-term make it even more difficult to achieve increases in productivity sufficient for self-sustaining growth.

Debt versus *Equity Financing* The use of capital borrowed externally (from banks etc.) *versus* capital generated inside of a firm through retained profits and sales of assets.

Deindustrialization This term is used rather loosely by many authors. At the heart of the concept is a *relative* decline in industrial *employment* in a nation or region where industry has traditionally been a significant component of the economy. It may be the result of *climacteric changes* or of secular shifts in an economy that are related to technological change and/or the globalization of economy. In some instances, such trends may involve not just a relative decline but an *absolute* one; and may involve declining industrial *output* as well as employment.

Demographic Transition The evolution of *vital rates*—birth rates and death rates over time, from high to low levels. The demographic transition model posits improved diets, public health and scientific medicine as causing a steady decline in death rates with increasing levels of economic development over time. Birth rates decline later, and more slowly, as socio-cultural practices take time to adjust to these new circumstances. The result is a sharp increase in population growth, until birth rates fall to relatively low levels.

Deskilling A reduction in the range and level of skills within a local labour market that is the result of two trends: the increased mechanization and computerization of production processes (including management and management-support functions), and the geographic consolidation and localization of higher-skilled activities in world cities, major control centres, and centre of innovation (which leaves other markets with a preponderance of routine jobs in local offices and branch plants).

Diagonal Integration Corporate mergers involving firms that are engaged in separate and distinct enterprises, producing different goods or services for different markets. The result is a diversification of corporate interests.

Diffuse Industrialization The growth of manufacturing employment in peripheral rural regions of core economies—a result of the availability of reserves of relatively cheap and less militant unskilled and semi-skilled labour suitable for employment in branch factories of manufacturing enterprises based elsewhere.

Diminishing Returns, Law of: The tendency for productivity to decline, after a certain point, with the continued application of capital and/or labour to a given resource base. A simple example is provided by agricultural productivity: a large farm will yield progressively higher levels of output with the addition of more farm hands and more machinery, but there will be a point at which productivity decreases as some of the labour and machinery is underemployed, people get in one another's way, and coordination of activities becomes costly.

Disorganised Capitalism The label given to the most recent phase of the political economy of capitalism, chiefly in order to distinguish it from the preceding "organized" phase. Under disorganized capitalism, the relationships between capital, labour and government are more flexible, largely because a great deal of corporate activity has escaped the framework of nation-states and their institutions that still constrain organized labour and most governmental functions.

Distributive Services Services that are related to trading and the movement and storage of raw materials, parts and finished products. Examples include haulage services, package delivery services and warehousing services.

Division of Labour The specialization of different people and groups in different occupations and between different economic sectors (agriculture, various industrial sectors) and hence the separation of tasks in the production process and has allocation to different groups of workers. The division of labour is the result of the way production is organized into different types of job and the creative adaptation of people to this.

Dual Economy The existence of traditional self-sufficient economy with the market economy.

Economic Involution The sub-division of tasks and the proliferation of labour-absorptive activities within a local economy with a limited resource base, with the result that very little income or capital is generated by "exports" to other localities.

Economic Union A form of international economic integration that involves the removal of all internal barriers to trade and the movement of factors of production, the creation of a common set of trade barriers and trade agreements with non-member states, and the coordination of integrated economic policies within the union.

Economies of Scale The cost advantages resulting from large-scale flexible organization by the ability to rapidly switch from one product or production process to another; these grow as the average cost of production falls with increasing output. When average cost begins to rise *diminishing returns* set in. Increasing returns to scale come from increases in efficiency internal to production facilities or plants (larger plant produces more output relative to inputs than would smaller plants) and external economies that derive from the spatial clustering of plants in the same industry.

Elasticity of Demand The degree to which levels of demand for a product or service change in response to changes in price induces a significant change in demand, elasticity is high; where levels of demand remain fairly stable in spite of price changes, demand is said to be inelastic.

Empire An extended territorial political unit or state built up, often by force, under a supreme superiority. Empires usually involve rule over alien or subject people.

Energy Ratio The relationship between rate of change in energy consumption and rate of change in economic growth, GDP over a period. A value greater than one indicates that the amount of energy required to create an additional unit of GDP is increasing, a value less suggests the reverse.

Enlightenment A philosophy or a movement based on the idea of social progress through the application of reason.

Epidemiological Transition The predominant causes of death noted in the movement from a pre-transition to a post-transition society, primarily from infectious to degenerative diseases.

Entrepreneurship The activity of developing and exploiting new products or innovations.

Ethnic Groups Groups of people defined according to a common security and culture.

Eurodollars United States Currency that is held in banks located outside United States, mostly in Europe. It is a pool of currency for which there is a distinctive and independent market, for it represents a fairly stable, hard currency that has been established beyond the control of US government and its financial institutions.

Export Processing Zones Small, closely-definable areas within which especially favourable investment and trading conditions are created by governments in order to attract export-oriented industries, usually foreign-owned. These conditions include the absence of foreign exchange controls, the availability of factory space and warehousing at subsidized rents, low tax rates, and exemption from tariffs and export duties.

External Control This term refers to situations where employment opportunities and decisions about investment and production in a given plant or locality are controlled by corporate managers based in other cities, regions, or countries.

External Costs and Benefits An external cost is the negative effect on people or a place of the action of a person or firm. An external benefit (or economy) is the opposite. It is the favourable effect of the actions of others. Agglomeration economies deriving from the clustering of firms in the same industry (*localization economies*) and firms that provide services to them (*urbanization economies*) are important instances of external benefits (see *Industrial Districts*).

Factors of Production The fundamental components of any economic system are: land, labour and capital. *Land* includes not only space or territory but also the associated soils and natural resources. *Labour* includes not only the use of available workforce but also its skills, experience and discipline. *Capital* includes not only money capital but also everything deliberately created for the purpose of production, such as factories and machinery (*i.e., fixed* capital). A fourth factor, *enterprise,* is often recognized, though it may be missing from some forms of economic organization (*e.g., substance*); while it may legitimately be regarded as one aspect of labour.

Factor Endowment Theory See *Comparative Advantage.*

Feudal Systems Forms of economic organization based on a mode of production wherein the surplus product (*i.e.,* the outputs of productivity in excess of subsistence levels) is appropriated through a hierarchy of socio-political ranks by institutionalized coercion. In classical feudal system, land is allocated to vassals by lords, in return for military service and/or labour on the lord's estate.

Feudalism A hierarchial social and political system common in Europe during the medieval period. The majority of people were engaged in subsistence agriculture with simultaneously having an obligation to fulfil certain duties for the landholder, at the same time the landholder owed various obligations to his overlord.

Flexible Accumulation A mode of economic regulation in which *economies of scope* supplant *economies of scale.* Small-batch changes in fashion and the emergence of the financial services industry are the vital mediator of all other economic transactions.

Flexible Production Systems This term refers to various practices whereby manufacturing operations achieve flexibility in what they produce, when they produce it, how they produce it, and where they produce it. These practices include the exploitation of various kinds of enabling technologies (*e.g.*, robotics, computerized inventory control systems), greater use of sub-contracting, the exploitation of different labour markets, the exploitation of different market niches for products, and the development of new labour processes using flexible working hours, part-time workers, etc.

Flexible Specialization The association of "new" methods of production (such as small-batch production, multi-purpose machines, etc.) with the clustering of firms specialized in the same product lines in small areas or *industrial districts.*

Foreign Direct Investment (FDI) Capital invested directly by firms from one country in production facilities to another. It is usually distinguished from bank and portfolio investment, which is foreign capital placed in bank accounts or invested in stock, bond and foreign exchange (currency) markets. While international trade flows as a percentage of GDP has remained relatively constant in the recent period, FDI has been increasing more than GDP indicating that global economic activity is being facilitated more by FDI than by trade flows.

Fordism A regime of accumulation than centres on the mutual reinforcement of mass production and mass consumption. Named after Henry Ford because of his innovations and philosophy concerning automobile manufacture, it features a highly specialized and differentiated division of labour with assembly-line production geared to the availability of standardized, affordable goods for mass markets.

Form-Process Confusion A form of international economic integration that involves the elimination of some (but not necessarily all) trade barriers between member states, but where each member state continues to set its own tariffs and quotas as trade barriers to non-member states.

Free Trade Association The tendency to infer a specific process from a particular geographical pattern, as in reasoning backwards from the "discovery" of a hexagonal grid in the distribution of settlements to the specific logic of a central-place theory.

General Agreement on Tariffs and Trade (GATT) A United Nations Agency founded in 1948 with one of its main purposes being to facilitate free trade through the general reduction of tariff levels.

Gemeinschaft A form of community said to be common in traditional societies (as distinct from industrial societies) associated with notions of stability and informal personal contact.

Geographical Historiography The variegated terrain of uneven geographical development are constituted and constitutive of *socio-ecological and political-economic processes* through time.

Geographical Path Dependence The historical relationship between the present economic activities associated with a place and its past experience.

Geographies of Exclusion The spatial processes by which a powerful grouping consciously seeks to distance itself from other less powerful groupings.

Geopolitics The term which has been used to refer to many things, including a tradition of representing space, states and relations between them, emphasizing the strategic importance of particular places. This spatializing of political strategies is crucial for any form of emancipatory politics.

Gesellschaft A form of association common in urban-based industrial societies (as distinct from traditional societies) and associated with non-permanent and utilitarian social relationships.

Global Cities Those cities performing a dominant role in the world economy and characterized by specialized service-type functions such as financial markets.

Global Currency A currency used as a medium for international transactions. Since the World War II, the US dollar has been the main global currency, though now its use is challenged in some respects by Eurodollar and Japanese yen.

Global Economic System An envelope within which all the various other systems like economic, social and cultural values of the localities, regions and nations get combined to give rise to economic uniformity instead of its diversity. In global economy there has been a marked increase in the power of finance capital over production by its mobility.

Global Production System Refers to dominant form of production identifiable at a global scale. It has evolved over many centuries into its current form, referred to as post-Fordism.

Global Triad The term describes how tne world economy is essentially organized around a tri-polar, macroregional structure whose three pillars are North America, Europe, and East and South East Asia.

Global Sourcing The use of multiple sources in different countries for the components of a particular product that is assembled elsewhere.

Globalization A contested term relating to transformation of spatial relations that involves a change in the relationship between space, economy and society from a pervasive internationalization to a growing scattering over globe (*i.e.*, global economy). It is a contemporary phenomenon that integrates economy across national boundaries of markets, finance, technologies and nation-states to a level that has not witnessed before. This has enabled nation-states, MNEs and individuals to extend their reach around the globe faster, further, deeper and at lower cost than could ever have been imagined even a decade ago.

Glocalization The interaction between the particular character of places or regions and the more general processes of change represented by globalization.

Green Revolution Most commonly refers to the period during the 1950s and 1960s when new high-yield crop varieties especially corn (maize), rice and wheat and associated inputs (chemical fertilizers, pesticides) and technologies (irrigation, planting techniques) introduced into many less-developed countries in the late 1960s.

Gross Domestic Product (GDP) The value left after removing the profits from overseas investments and those profits from the economy that go to foreign investors.

Gross National Product (GNP) A broad measurer of a national economy's performance, it is the value of the final output of goods and services produced by the residents of an economy plus primary income from non-residential sources.

Heartland Identified by the British Geographer, Halford Mackinder (1904) as the zone in East-Central Europe, control of which would be a key to world domination. The term has since been appropriated by Latin American geopolitics.

Hegemony The domination over the world-economy exercised by one in a particular historical epoch through economic, military, financial and cultural means.

Horizontal Integration Corporate mergers involving firms formerly competing in the same market(s) with similar goods or services.

Human Asset Includes innate skills, talents, competencies and abilities of individuals as well as the effects of education and health.

Human-made-Assets Physical products, particularly those used in production like machineries, buildings and physical networks, as well as financial assets.

Hyper-Urbanization A situation in which the population of cities is growing more rapidly than the ability of the cities to absorb the population increase in terms of housing and employment.

Ideal-Type A theoretical construct or model through which actual situations can be understood by comparison. A potential danger of ideal-types lies in forgetting that they are rational abstractions rather than true accounts of particular situations.

International Monetary Fund (IMF) International financial institution that originated from the 1944 Bretton Woods Conference. Its main roles include international monetary exchange and controlling fluctuations in exchange rates in order to alleviate balance-of-payments problems.

Internationalization Refers to the spread of economic activity beyond national boundaries which is not a new phenomenon but the rate at which it has been accelerated since the 1980s that this term is getting merged with *globalization* so that changes occurring in economic life in one part of the world are rapidly diffused to others.

Import-Substituted-Industrialization (ISI) The policy of emphasizing export and minimizing import for industrialization.

Import-Penetration This describes the result of a significant share of domestic markets for a particular product or service being lost by domestic firms in the face of competition from foreign sources.

Imperialism A relationship of political, and/or economic, and/or cultural domination and subordination between geographical areas.

Import-Substitution The development of domestic firms capable of producing goods or services formerly provided by foreign firms.

Indigenous Peoples Those peoples native to a particular territory that was later colonized, particularly by Europeans.

Industrial Revolution A term that is often taken to the marked transformation of productive forces, initially within the British economic system between the mid-eighteenth and mid-nineteenth centuries. It resulted in the movement of Britain from a largely rural-based economy to one that was dominated by manufacturing and industrial production. Such a transformation resulted in substantial social, political as well as economic changes.

Infant Mortality Rate (IMR) Number of deaths of infants aged under 1 year per 1,000 live births.

International Division of Labour A term referring to the tendency for particular countries and regions of the globe to specialize in particular types of economic activity.

Increasing Returns Constant returns to scale assumes that doubling all inputs in production would exactly double total product until a point at which *diminishing returns* might set in. In dynamic economic development, however, the phenomenon of increasing returns is to be expected. If all inputs are increased in size the product grows *more* than proportionally. Each location cannot develop the size to achieve increasing returns. Hence, there is a geographical pattern to the distribution of economic activities characterized by increasing returns.

Independent Variable A tabulation of the occurrence of a phenomenon that is held to predict or "explain", at least in part, that of a dependent variable.

Industrial District Areas of limited geographical extent specializing in a particular industry because of the external economies (*e.g.*, labour pooling, service activities, information flows) associated with specialization.

Industrial Restructuring The process whereby in response to pressures from technological change, new forms of firm organization, and new patterns of demand, industries (such as steel, clothing, automobiles, electronics, etc.) undergo changes in the typical structure of firms, links between management and production activities, emphasis on *vertical integration* relative to the use of subcontractors, and reliance on different types of labour process (overall degree of capital or labour intensity, types of labour force —permanent / part-time, male / female, skilled / unskilled, etc.).

Inelasticity of Demand See *Elasticity of Demand.*

Infant Industries Industries at an early stage of development that are often protected from foreign competition by means of tariff and other barriers to trade until they can survive without such competition.

Informal Sector Economic activities that are undertaken without any formal systems of regulation (or, remuneration). In addition to domestic labour, these activities include strictly illegal activities such as drug-peddling and prostitution as well as a wide variety of legal activities such as casual labour in construction crews, on docks or on farms; domestic piece-work, street trading, scavenging, and providing personal services such as shoe-shining or letter-writing.

International Monetary Fund (IMF) Set up in 1947 by the Charter of Bretton Woods Conference of 1944 for a global method of economic co-operation with a fund facility to support world countries with balance of payment difficulties. IMF is having a dual purpose of economic systems on the ground of "good governance" by stabilizing most of the world under capitalism and by bringing the Countries of the South under it.

Isotropic Plain The essential element of most static models of economic landscapes; the featureless plain without barriers to movement and diffusion, having no local differences in soil fertility or other locational bias of any sort.

Keynesianism Specially, this refers to a doctrine of macroeconomic management that is closely associated with the British economist John Maynard Keynes, who advanced the use of fiscal policy (*e.g.* budget deficits) and the explanation of *economic multiplier effects* in order to achieve and maintain full employment. The term is also used to denote the mode of regulation associated with the Fordist regime of accumulation.

Knowledge Assets "Codified knowledge" which are transferable across space and time unlike "tacit knowledge" like as individual's learnedness, judgement and experience which are not easily transferred until codified.

Kondratiev Cycles Long waves of 50-55 years in duration that have characterized the cycles of boom and bust evident within the capitalist world-economy for the past 250 years. They take their name from the Soviet economist N. D. Kondratiev, who established their existence in the 1920s. Their origins and significance remains controversial, but in recent years they have been widely recognized to be closely tied into distinctive phases of political-economic development.

Kuznets Cycles Named after Ukrainian-born economist Simon Kuznets, who established their existence in the 1920s, indicating a cycle of regular changes in the rate of economic growth of a country as measured by per capita Gross National Income (GNP)

Laissez-faire The view that every person pursuing selfish ends is led, as if by an "invisible hand", to achieve the best good for all. This was one of the basic insights of the Scottish philosopher-political economist Adam Smith in his book, *The Wealth of Nations* (1776). This ideal lies behind economic models that either ignore the role of states or treat them as "separate" from the workings of an economy. Only *competitive capitalism* has come anywhere close to corresponding to the idea of *laissez-faire.*

Land Tenure A system of land-use rights and transfer mechanisms. The major types of land tenure include owner-occupation, cash tenancy, share-cropping (a form of land tenancy in which rent is paid in kind), use rights (where there is no codified legal owner and a person or group establishes a right to land by using it), and collectivism.

Legitimacy With regard to nation-states, a term meaning that the majority of people accept the rule of law of the governing political organization.

Leost Developed Countries, (LDCs) Countries at a disadvantage in today's global competitive environment because their comparative advantage in natural resource or cheap labour have become subordinated to knowledge-based factors. For example, knowledge is the key to innovations in production that ultimately make products more competitive. LDCs suffer from poor productive capacities and competitiveness related, for example, to weak linkages between manufacturing, services and infrastructural sectors such as telecommunication, insufficiently developed human resources, deficiencies in physical infrastructure; and an inability to generate adequate resources to invest in reducing economic and social problems in ways that would help to improve productive capacity. Examples are Nigeria, Ethiopia.

Liberalism A political philosophy distinguished by the importance it attaches to the civil and political rights of individuals. May also refer to belief in the importance of free market.

Life Expectancy (at birth) Average number of years of life expected on the basis of age-specific mortality schedules for the specified year.

Locality A place or region of sub-national spatial scale.

Logistic Cycles Long-term (150–300 years) cylces of growth and stagnation in overall levels of economic development.

Machinofacture A form of organization of industrial production that is capital-intensive, with labour tending machines rather than operating them directly. It was the basis for the regime of accumulation that preceded Fordism.

Maquladoras Literally, "mills" in Spanish. These are the export-processing zones located along the Mexican-side of the US-Mexico border.

Market-based States Market States where the market is the dominant means by which land, labour, capital and goods are exchanged and has a major influence over social and political organization.

Marxism A form of socialism derived from the teachings of Karl Marx (1818-83). Marxism regards capitalism as an inherently unjust system with the capitalists (those who owns the means of production) exploiting the proletariat (those who must sell their labour in order to live). It aims to replace capitalism with a fairer system, socialism maturing into communism.

Merchant Capitalism or *Mereantilism* Refers to an early phase of capitalist industrial development dominant in the larger urban regions of Europe from the late fifteenth century. Merchants were the principal actors engaged in both the provision of capital and movement and trade of goods (predominantly bulky staples such as grain, lumber and manufactured goods).

Minisystems The form that can be used to describe local societies with a simple division of labour within a single cultural framework, such as hunter-gatherer and some agricultural activities. These societies are integrated through the principle of reciprocity in interpersonal transactions rather than through redistribution by a central authority or market-exchange relationships.

Modernism A term typically associated with the twentieth century reaction against realism and romancism within arts. More generally, it is often used to refer to a twentieth-century belief in the virtues of science, technology and the planned management of social change.

Models Idealized or simplified representations of a specific phenomenon (*e.g.*, agricultural land use) used to produce hypotheses about how the phenomenon came about.

Modernity Refers to a period extending from the late sixteenth and early eighteenth centuries (in the case of Europe) to the mid to late twentieth century characterized by the growth and strengthening of a specific set of social practices and ways of doing things. It is often associated with capitalism and notions such as progress.

More Developed Countries (MDCs) Countries with significant competitive advantages in today's globalizing economy. They have well-developed, increasingly knowledge-based and strongly inter-connected manufacturing and service sectors that provide a significant proportion of employment and contribute to significant national and individual wealth. Indices such as literacy levels, incomes and quality of life are high and these countries exercise considerable political influence at the global scale. Examples are UK, France, Germany.

Multiculturalism Refers to a belief or policy which endorses the principle of cultural diversity and supports the right of different cultural and ethnic groups to retain distinctive cultural identities. It has often been criticized for being too symbolic and not politically radical enough in challenging racism.

Mode of Production A fundamental form of economic organization with distinctive relationships between the main factors of production (land, labour and capital). The concept is derived from Marxian economies but now has much wider use. The identification of different modes of production can become problematic, but at the most general level it is possible to recognize a subsistence mode (or, "primitive communism"), a slavery mode, a feudal mode, a capitalist mode and a socialist mode.

Monetarism A doctrine of macroeconomic management that disavows demand management and regards the money supply as the most important determinant of economic stability. Important in the US and UK in the late 1970s and early 1980s, it reasserted the relevance of price theory and the importance of free markets.

Monopolistic Competition A form of imperfect competition in which certain producers have advantages in setting prices and reaping profits because of control over knowledge, markets, and resources or some combined thereof. This is the prevailing mode of competition in the world economy.

Monopoly Colonization Along the eastern seaboard of North America colonization by different European states did not initially produce competition between the various gateway ports that were established. Rather, each port monopolized its hinterland. This method of settlement must be distinguished from the more competitive process of port establishment and hinterland capture that often prevailed elsewhere, at least in the early years (*e.g.* West Africa).

Multilateralism This most commonly refers to increased openness of international trade, freedom of capital flows and full convertibility of currencies within the world economy. More specifically, it includes the institutions that have been major instruments of these trends: the General Agreement on Tariffs and Trade (GATT), the IMF, and the World Bank. It is to be distinguished from the bilateral (two party) links that characterize many inter-state relationships and the regional multilateralism implicit in such trading blocks as the European Community.

Multiplier Effects The extra industries, firms, incomes and employment generated by a new economic activity can be said to result from that activity's multiplier effects. Such effects may be localized and give rise to a growth pole or occur elsewhere than in the vicinity of the original stimulus.

Nationalism The ideology and sentiment belonging to a "nation" and the claim that the "nation" should be expressed in a "state".

Nation-State (*and State*) A symbolic system of institutions claiming sovereignty over a bounded territory.

Natural Increase Surplus of births over deaths in a population.

Natural Resources (or, *assets*) Renewable and non-renewable resources which are having "source" functions that has inputs into production and utility as well as "sink functions" to accommodate the unusable outputs of production due to consumption by way of pollutions.

Neo-classical Economy Neo-classical world of perfect competition, in which producers and consumers freely enter into price-fixing markets as sovereign individuals where their exchanges are regulated by *frictionless intersection of supply and demand schedules,* so as to sustain a partial or general equilibrium.

Neo-Fordism Sometimes referred to as "post-Fordism" or "flexible accumulation", this term identifies the regime of accumulation that has succeeded Fordism within parts of the world's core economies. Rather than being predicated on the mutual reinforcement of mass production and mass consumption, it depends on flexible production systems to explicit specific market segments and/or niches.

Neo-liberalism An economic doctrine promoting market-led growth, deregulation and the privatization of state-owned enterprises.

New Industrial Division of Labour (NIDL) refers to the emerging world-wide division of labour associated with the globalization of production. The most noted manifestations of this are the spread of industrialization to a number of rapidly-growing *newly-industrializing countries* (NICs) and the increase of *global sourcing.* NIDL occurs when the process of production is no longer constructed primarily around national economies.

New International Economic Order (NIEO) Raised by the Countries of Global South in UN General Assembly of 1974 for redefinition of *terms of trade* so that the prices of their primary commodities are tied to those of manufactured goods they have to import. But this "wish list" of international reform eventually did not become a reality.

New World Order The notion of a post-Cold War structure of global power, led by USA which was proclaimed by President Bush, "the end of the Cold War has loosened the paralyzing grip of ideology on the world's resources and imagination. There is an emerging awareness that now the real problems confronting humanity can and must be addressed."

Newly Agriculturing Countries (NACs) Refer to the handful of low-middle income countries which have come to specialize in the export of high-value foods. These include Brazil, Mexico, China, Argentina and Kenya.

Newly Industrialized Countries (NICs) Countries where there has been a recent and significant shift away from primary activities towards manufacturing production. In some cases the proportion of manufacturing products is similar to that of UK or the US. Examples are South Korea and Taiwan.

Niche Markets Refer to the existence of consumer groups with identifiable tastes and life-styles.

Non-tariff Barriers Policy instruments (other than import taxes) designed to protect domestic industry from foreign competition. Examples include special standards and regulations, quotas and licensing requirements.

Normative A term that refers to what "ought to be". Many static models in economic geography rest on specific assumptions about values (especially that the best eonomy is one with unrestricted competition) and how an economic landscape should work. They address themselves to the question of what is the *best location* for a particular activity rather than to the more mundane task of discovering or positing the *actual determinants* of location. The danger with normative models lies in trying to empirically validate them when their predictions derive from assumptions about economic transactions that are far from actual practice. This is an example of the normative policy, confusing what one wants with what is.

Oligopolistic This refers to the domination of an industry by relatively few sellers. In the US the aluminium, steel, auto, breakfast cereal, and many other industries, are concentrated in the hands of four or fewer companies.

Organization of Economic Cooperation and Development (OECD), A cartel of 29 countries of the North, which include the 15-member countries European Union, in addition to USA and UK, is having a similar factor endowments of developed economy. These countries are having most of their trade among themselves.

Organisation of Petroleum Exporting Countries (OPEC) The most successful commodity-producer cartel (price-fixing organization) in the history of world economy. Established initially at the initiative of the governments of Venezuela and Iran in the 1960s, OPEC became in the 1970s an effective instrument for dramatic increases in the world price of petroleum through controlling the rate of oil production. OPEC has lost much of its influence, at least temporarily, because of political conflicts between members (*e.g.* Iraq *versus* Iran, then *versus* Kuwait and Saudi Arabia), differences over strategy (highly-indebted members favoured higher rates of pumping), and the expansion of oil production by non-members (such as Mexico, Britain and Norway).

Organized Capitalism The phase of industrial capitalism, from about 1890-1970, that was characterized by comparatively highly-structured relationships between labour, government and corporate enterprise, all scaled to the framework of nation-states. These relationships were mediated through legal and legislative instruments, formal agreements and public institutions.

Overaccumulation A distinctive phase in the long-term dynamics of capitalist economies, characterized by unused or underutilized capital and labour. It is an inevitable outcome of the difficulty of matching supply to demand under changing conditions, and it represents a critical moment for the political economy of capitalism. It can be recognized by the appearance of idle productive capacity, excess inventories, gluts of commodities, surplus money capital and high levels of unemployment.

Pastoral Nomadism A form of social organization which is based on livestock husbandry for largely subsistence purposes. Pastoral nomads are characterized by a high level of mobility which allows them continually to search for new pastures in order to maintain their herds of animals.

Peasant The term has a number of different interpretations over the years and in different parts of the world. In general it usually refers to those whose livelihood is largely dependent on the land and on real subsistence-type of activities. The term peasant is usually reserved for those living in organized states, thus distinguishing them from band and tribals etc.

Periphery According to Wallerstein, refers to those regions of the world capitalist economy characterized by low wages, simple technology and limited production.

Political Economy An approach to study of national economic policies, initiated in the seventeenth century in England which called the polity (the civil order of the society) through the emphasis on the political/social construction and the resulting consequences of economic activity. In spite of the sophisticated analytical techniques, and often highly esoteric specialisms which characterize the present-day economic science, political economy still retains its perennial political dimension rather than to conjure up any ideological overtones.

Population Explosion Population which is generally charcterised by growth as the difference in birth and mortality rates in an areas or country may reach an explosive stage when the increase in number of persons surpass their means of production of resources.

Positional Goods Consumer goods that are acquired (in part, at least) in order to denote affluence, social status and/or style.

Post-Fordism Production System Comprises a mix of different ways of organizing production at a number of spatial scales. Common to all these types of production is some form of flexible production.

Post-modernism A philosophy which holds that the traits associated with twentieth century modernism, such as belief in the possibility of managing social change according to sets of agreed principles are now in retreat in the face of increasing individualism, pluralism and eclecticism.

Post-productivist Transition A term used to describe the movement away from productivist agricultural systems. This new phase of agricultural production is characterized by extensive and diversified patterns of farming and the growing importance of non-agricultural activities in the countryside.

Power Geometry The ways in which different social groups/ individuals are placed in relation to the forces of globalization, enabling some to benefit and others to be disadvantaged.

Prehistoric Societies Societies which have left no written records.

Primacy This refers to the situation in urban systems in which the largest city is more than twice the size of the next-largest city.

Producer Services Services that enhance the productivity or efficiency of other firms' activities or that enable them to maintain their specialized roles. Examples include advertising, personnel training, recruitment, finance, insurance and marketing.

Productivist Agriculture Strongly associated with North America, it is characterized by the development of agro-industrial complexes based on grain-fed livestock. It was at its most influential during the period from the 1920s to 1970s.

Proto-Industrialization Refers to the early phase of capitalist industrial development in Europe. Characteristics of this period include a significant level of rural-based industrial activity and a low level of technological application in the production process.

Proven Reserves Those deposits of a resource that have already been discovered and are known to be economically extractable under current demand, price and technological conditions.

Purchasing Power Parity (or, PPP) The purchasing power of different currencies of countries reflect the number of units of a country's currency required to buy the same quantity of comparable goods and services, like food, health and housing in the local market as one US dollar would buy in that country. This comparison brings out the fact that how a particular currency of a country can be taken regarding how overvalued or undervalued it is. From a PPP perspective, the share of activity by the Countries of Global South is larger than is reflected in a market-based exchange of their currencies to a common currency, such as US dollar. For example, in 1992, according to World Development Report 1992 prepared by World Bank, while China's GNP per capita was $ 370, *i.e.,* around one US dollar a day per person, its purchasing power per capita was $ 2,120.

Race Refers to the division of human beings into supposedly recognizable groups based predominantly on physical characteristics.

Ranked Society Society in which there is an unequal division of status and power between its members, where such divisions are based primarily on such factors as family and inherited social position.

Reaganomics The application of supply-side economics to the management of the US economy in the 1980s. Supply-side economics is similar to monetarism in its disavowal of demand management, rather the key to economic stability and well-being is seen to be the enhancement of aggregate supply. Reaganomics consisted of a set of objectives that included tax reduction, deregulation of business, increased government spending on social welfare.

Redistribution In pre-capitalist societies a system whereby accumulated wealth and goods were taken and divided among members of society by a central ruler or organization, or kept by the centre for its own purposes.

Regionalism Regional economic planning at the national level requires the designation of regional units and organization of agencies through which regional policy can be put into operation. This regionalism has been of considerable importance in many countries. Recently it has been eclipsed somewhat by the emergence of competitive strategies of economic development pursued by lower-tiered governments.

Regulated States Social formation organized on the basis of territory. The regulated, or pre-modern, state was not dominated by the market system with political, social and religious considerations usually being of greater importance.

Renewable Energy Energy sources such as winds, waves and tides that are naturally replenished and cannot be used up.

Renewable (flow) Resources Those resources that are naturally renewed within a sufficiently short time span to be of use to human society. The continued availability of such resources is increasingly dependent upon effective management.

Rent This is the return or surplus generated by the comparative productivity or profit-yield of a piece of land. The highest rent is produced by the "highest and best use". To the nineteenth century English economist David Ricardo this was seen in an agricultural context as a function of differences in soil fertility. To the Prussian landowner J. H. Von Thunen locational attributes were more important. The return to different land uses (or, *differential rent*), which goes to the owner, is supplemented by the returns that can come from adjacency to a highway, etc.) and monopoly ownership of the parcels that are candidates for particular uses.

Replacement Rate Fertility Number of babies that an average woman should have to replace her generation, given prevailing levels of mortality of 2.1 per woman in post-transition societies.

Reserves/Production Ratio, R/P If the reserves remaining at the end of any year are divided by the production in that year, the result is the length of time those remaining reserves would last if production were to continue at that level.

Resource A substance in the physical phenomena that has value or usefulness to human beings and economically feasible and socially acceptable to use.

Rurality Functionalist, critical political economy and social representation approach for distinguishing between rural and urban economy and society.

Rust Belt A region of the north-eastern USA which suffered substantial industrial decline, especially after the World War II.

Socialism Socialism is a vision (alternative to *capitalism*) regarding how a society has to work, how socio-ecological relations will unfold, how human potentialities can be realized albeit within a geography of difference.

Space-Process Relationship The idea that different types of industrial activity (management, assembly, etc.) are carried on in totally different locations in conformity with a functional separation that takes the form of hierarchy of places, from world cities or centres to various peripheries.

Second Industrial Revolution A term sometimes used to account for the profound technological and accompanying social changes which affected industrial capitalism from the late nineteenth century. During this period, Britain's industrial position was challenged by USA and Germany.

Semi-periphery Accoring to Wallerstein, refers to those regions of the world capitalist economy which,while exploiting the periphery, are themselves exploited by the core countries. Furthermore, they are characterized by the importance of both core and peripheral processes.

Social Asset Known also as relational asset, it includes impersonal trust and networks plus understanding and shared values. This alongwith *human asset* can enhance a person's "freedom to be and to do".

Social Exclusion The exercise of systematic economic and political power in a society so as to exclude individuals from life styles open to the majority.

Spatial Division of Labour The division of labour is manifested geographically in different ways depending upon particular conditions of production. Consequently, the spatial division of labour can change over time. When Fordist conditions were more dominant, for example, the spatial division of labour tended to take the form of regional industrial specialization. Earlier an industry's activities were regionally concentrated. Today, separation of function by location and regional dispersal have become relatively more important. A danger lies in seeing spatial divisions of labour as discrete and singular "waves" that totally displacc older forms through successful displacement.

Spatial Interaction A term used to indicate interdependence between geographical areas. Covers the movement of people, goods, information and money between places.

Stagflation Episodes of economic recession accompanied by comparatively high rates of price inflation.

State A political unit having recognizable control (claiming supreme power) over a given territory. Unlike earlier social formations like bands and tribes, states always base their power on their ability to control a specified territory and its inhabitants.

Stochastic A process generating a statistical series (such as city-size distributions) can be thought as random or as containing such a large number of causal influences that it is effectively random in nature. In either case it can be referred to as stochastic.

Stratified Society Society in which there is an unequal division of material wealth between its members.

Strategic Trade Policy Under conventional trade theory comparative advantage as revealed by trade statistics is seen as determining patterns of international specialization. However, if comparative advantage is created rather than revealed then there must be some historic basis to it. Advocates of strategic trade theory and policy argue that the relative power and intervention of states have played important roles in establishing patterns of trade. In particular, governments can help create conditions under which industries and firms operating within their territories can gain competitive advantages over their competitors elsewhere. The difficult policy issues concern precisely how governments can do this.

Structural Adjustment Loans designed to foster structural adjustment in LDCs by promoting market-led growth and a reduction to government intervention in the economy.

Substitutionalism Refers to the increased use of non-agricultural raw materials characterized by replacement of food and fibre with industrial alternatives. It forms a key process in the industrialization of the agro-food system.

Suburbanization Refers to the movement of middle- and skilled working-class people into residential areas located some distance away from their paid employment.

Sun Belt Refers to the major growth areas of southern and western parts of USA during recent years in contrast to the contracting and declining industrial base of the north-east. Can also be used in other parts of the developed world to describe fast-growing regions, *e.g.*, the M4 corridor in England.

Supranational Political Union A form of international economic integration that involves the removal of all internal barriers to trade and the movement of factors of production, the creation of a common set of trade barriers and trade agreements with non-member states, the coordination of integrated economic policies within the union, and a unified fiscal and monetary system controlled by a supranational authority with executive, judicial and legislative powers.

Sustainable Development A pattern of resource-use and economic development that does not jeopardize non-renewable resources, damage existing ecosystems, or harm individual species.

Tariffs Schedules of duties or taxes imposed by a government on imported or exported goods and materials.

Taylorism The name given (after analyst F. W. Taylor) to forms of organization in manufacturing industries wherein the planning and control of work is given over entirely to management, leaving production workers to be allocated specialized tasks in a finely-tuned division of labour that is subject to careful analysis—"scientific management" using techniques such as time-and-motion studies.

Technology Systems Distinctive "packages" of technologies, energy sources, and political-economic structures that represent the most efficient means for the organization of production at any given phase of economic development. Based on key sets of interdependent technologies, they represent the underpinnings of successive regimes of accumulation and modes of regulation.

Technopolies Growth Poles or industrial districts based on high-tech industries. In France the term refers to high-tech complexes established initially under national government sponsorship.

Terms of Trade The ratio of the prices at which exports and imports are exchanged. When the price of exports rises relative to the price of imports, the terms of trade reflect an improvement for the exporting country.

Third World A rather vague term used to describe those regions of the world in which levels of development, as understood by such measures as GDP, are significantly below of the economically more advanced regions. The term is increasingly seen as an adequate description of the prevailing world situation since it disguises a significant amount of internal differentiation.

Top-down Development Economic and social changes brought about by activities of the state and its agencies.

Total Factor Productivity Associating labour with productivity in economy by recognizing labour to different levels of education and skills.

Total Fertility Rate (TFR) In general terms equal to the average number of live births per woman within a population.

Trade Block A group of nation-states that act together in order to further their influence over world trade, *e.g.*, the European Union (EU), or the North American Free Trade Area (NAFTA).

Trade Creation Effects The positive effects of international economic integration, resulting from the free movement of factors of production and free trade, which allows each region to specialize according to its comparative advantage, thus leading to a greater overall productivity and internal trade.

Trade Diversion The displacement of pre-existing trade flows as a result of transnational economic integration.

Trade Preference Associations Loose forms of transnational economic integration that involve reduced trade barriers between member states.

Transnational This refers to activities and organizations that span international boundaries and that are not closely assciated with states in general or any one state in particular.

Tribe A type of social formation usually said to be stimulated by the development of agriculture. Tribes tend to have a higher population density than bands and are also characterized by an ideology of common descent or ancestry.

Tributary Mode of Production Introduced by the spread of European mercantilism, in this mode of production the society is reorganized into two classes, (1) a ruling elite of surplus takers, and (2) an underclass of surplus producers. Unlike *capitalism* in this system of production, people do not produce for wages, but rather, because they are coerced by military might or because the elite controls an essential part of production process, such as the land (or irrigation). This mode encompasses a wide range of variations, from the tight control of "monarchs and colonial viceroys" to the loose domination of local strongmen. In some cases, the *kin-ordered mode* could subsist, in modified form with tributary mode supplanted the kin mode, transforming many aspects of society. The fur trade and the introduction of horse from Europe changed a formerly sedentary people with a subsistence economy into widely ranging buffalo hunters heavily involved in commerce. In the tributary mode, the underclasses may still control their own land and resources, whereas under the "capitalist mode", it is the elite that owns the means of production. Also, takers of tribute simply demand a surplus, whereas under capitalism, workers must sell their labour for wages, otherwise lacking access to the means of production, they cannot survive. In addition, capitalist competition requires that in order to stay in business, the owner of production must constantly increase output and cut costs. This puts the worker in the precarious position of being threatened with losing his job to new technology (and having no fallback for survival) or having his pay cut. Unlike the kin-ordered mode, both the tributary and capitalist modes of production require an apparatus of coercion; this is the state, which provides the laws, courts, the police, and the armies to maintain the system.

Two-Way Gap Scope for investment through internal savings as well as import-purchase capacity by earning foreign exchange from exports.

Urban Village A residential sector of a city believed to possess social characteristics typical of a village-type community.

Urbanization Refers to the increasing importance of the urban relative to the total population. It is initially stimulated by movement of people from rural to urban areas.

Vertical Disintegration The tendency for specialized firms to be created as sub-contractors and suppliers within industries formerly dominated by large functionally-integrated firms.

Vertically-Integrated Firms Corporate mergers involving firms that are engaged in different aspects of the same industry or enterprise.

Vital Trends Changes in fertility and mortality.

World Bank International financial institution established in 1944. Its main role is to provide development funds to LDCs in the form of loans and assistance.

Wage Labour When people work in exchange for monetary payment rather than for bartered goods, military protection or as a result of enslavement, they can be said to engage in wage labour.

World Economy The modern world (capitalist) economy coming into existence in Europe from the sixteenth century onwards and progressively incorporating the world into its ambit. As it has expanded geographically it has also changed institutionally and shifted in terms of the economic-locational principles upon which it operates.

World-Empire This is a *world-system* with a single division of labour and multiple cultural-systems but with a common political system.

World-System Any spatially-extensive economic system (as distinct from a *minisystem*) that has a single division of labour but multiple cultural system.

World-Systems Theory Immanuel Wallerstein's conceptualization of the changing nature of the world socio-economic system into three distinct historical categories.

BIBLIOGRAPHY

Abler, Ronald Abler, John S. Adams and Peter Gould, 1971, *"Spatial Organization : The Geographer's View of the World"*, Prentice-Hall, N.J.

Adam Smith, 1776, The *Wealth of Nations : An Inquiry into the Nature and Causes.* [17]*. Book IV, Chapter VII,* republished, Oxford Clarendon Press, 1976.

Apter, David, E. (1965), "*The Politics of Modernization*", University of Chicago Press, Chicago.

Arrighi, Giovanni, 1994, "*The Long Twientieth Century : Money, Power and the Origins of Our Times*", London.

Allen, Tim and Thomas, Alan, eds. Oxford University Press. "*Poverty and Development into the* 21st Century",

Appadurai, Arjun, 1997, *Modernity At Large : Cultural Dimensions of Globalization,* Oxford University Press, Delhi.

Bagchi, A. K. 1982, "*The Political Economy of Underdevelopment*", Cambridge University Press, Cambridge.

Barett, E. Richard, & Soomi Chin, 1987, "*The Political Economy of East Asian Industralism*", Cornell University Press, Ithaca.

Bernstein, (1971), Modernization and the Structures of Society, Princeton University Press, Princeton.

Brandt Commission, 1983, "*Our Common Crisis : North-South Cooperation for World Recovery* ", Pion, London.

Brandt, Willy, (1980), "*North – South : A Programme for Survival,*" Cambridge, Mass. MIT Press.

Worsley, Peter, 1984, "*The Three Worlds : Culture and World Development* ", University of Chicago Press, Chicago.

Colman, David and Fredrick, Nisson, 1986, 2nd Edi., "*Economics of Change in Less Developed Countries*", Philip Allan Publishers Limited, Oxford.

Cook, Malcolm, 2002 Can the NICs be Duplicated? An Analysis of Systemic Constraints David Harvey (1996) in his book, "*Justice, Nature and the Geography*", Blackwell, Oxford.

David Ricardo Alongwith Adam Smith, David Ricardo, the doyen of political economy in the 18th century, in his "*Law of Comparative Advantage,*" (1817), showed the way to economic prosperity through trade.

Desai, Vandana and Robert, B. Potter (eds.2002), "*The Companion to Development Studies*", Arnold, London.

Dicken, Peter, 1986, "*Global Shift in A Turbulent World*", Harper & Row, London.

Dickenson et al, 1996, "*A Geography of the Third World*", Routledge, London.

Drusilla, K. Brown, 1988, "*Trade Preferrences for Developing Countries : A Survey of Results*", Journ. of Development Studies, April 1988.

Durkheim, Emile, 1893/1894, "*The Division of Labor in Society*", *translated* by W.D.Halls, London.

Economic Commmunity of Latin American and Carribean, 2000

Eisenstadt, S. N. 1973, Tradition, Change and Modernity, Wiley, New York.

Fei, John, C.H. and Gustav, Ranis, 1964, Development of the Labor Surplus Economy : Theory and Policy, Yale University Press, Yale.

Forbes, D. K. 1984, The Geography of Underdevelopment : A Critical Survey, Johns Hopkins University Press, Baltimore.

Foster, John Bellamy, 2002, "*Monopoly Capital and the New Globalization*", Monthly Review.

Frank, A. G., 1971, "*Capitalism and Underdevelopment in Latin America*", Penguin

Frank, Andre Gundar, (1967), The Sociology of Development and the Underdevelopment of Sociology".

Frank, A. G. 2002, "*Immanuel and Me With-out-Hyphen*",

Frank, A. J., S. Amin, G. Arrighi and I. Wallerstein, 1982, Dynamics in Global Crisis, Monthly Review Press, New York.

Friedman, John, 1963, Regional Development Policy : A Case Study of Venezuela, MIT Press, Cambridge, Ma.

Friedman, John, 1972, Urbanization, Planning and National Development, Beverley Hills, California.

Galbraith, J. K., 1977, "*The Affluent Society* ", Houghton Mifflin, Boston.

Garelik, Glenn, 1987, Newly Industrialized Countries, Time Magazine, Oct. 19, 1987.

Goulet, D. 1971, "*The Cruel Choice : A New Concept on the Theory of Development*", Atheneum, New York.

Habib Irfan, 1969, "Problems of Market Historical Analysis in India", Reprinted in S. A. Shah, 1973 (ed.) Towards National Liberation : Essays on the Political Economy of India, Montreal.

Harris-White, 2002, "*Development and Intermediate Classes with special reference to India*", in , Vandana Desai and Robert, B. Porter, eds. The Companion to Development Studies, Arnold, London.

Hayami, Yujiro, 1997, "*Development Economics : From the Poverty to the Wealth of Nations* ", 2nd. Edition, Oxford University Press, Oxford.

Hicks, N. and Streeten, P.,1981, "Indicators of Development : The Search For A Basic Needs Yardstick", *in,* Streeten, P. and Jolly (eds.), *Recent Issues in World Development*, Pergamon.

Hirschman, Albert, O. 1981, "Eassays in Tresspassing : Economics in Politics and Beyond", Yale University Press, New Haven.

Hirschman, Albert, O., 1958, "*The Strategy of Economic Development*", Yale Univ. Press, New Haven.

Hobson, John, A., 1948, "*Imperialism*", Allen and Unwin, London.

International Monetary Fund, 2000, "*World Economic Outlook*", Oxford University Press, Washington D.C.

International Monetary Fund, 2001, World Economic Outlook, 2000, Oxford University Press, New York.

Jalan, Bimal, 1999, "*India's Economic Policy* : *Preparing for the Twenty-First Century*", Viking, Delhi.

Jones, Charles, I., (2002), Introduction to Economic Growth, Norton & Co., New York.

R.J. Johnston, 1993, "The Challenge For Geography – A Changing World : A Changing Discipline", Blackwell, pp250, London.

Johnston, R. J., 1984, The World is Our Oyster, *Transactions, Institute of British Geographers,* Vol.9, in Rapley, John, 1996, "*Understanding Development : Theory and Practice in the Third World* ", Lynne Rienner Publisher, Boulder.

Jones, Charles, I. (2002), Introduction to Economic Growth, Norton & Co., New York.

Karl Marx, (1867), *Das Capital* The legendary work in 3 volumes. [10]. Marx's *Labor Theory of Value*[11]. *op. cit* 5 in Chapter One.

Keynes, John Maynard, 1936, *General Theory of Employment, Interest and Money,.*

Knox,Paul and John Agnew, 1994, "*The Geography of the World Economy* ", Arnold, London.

Krueger, Anne, O. : *Cited from Dickenson et al,* 1996, "*A Geography of the Third World*", Routledge, London.

Krugman, Paul, 1991, Theorization of Uneven Regional Development, MIT Press, Cambridge, Ma.

Martin, Ronald, Peter Sunley, 1998, "Paul Krugman's Geographical Economics and Its Implication for Regional Development Theory : A Critical Assessment", Economic Geography, Vol. 74.

Lee, Roger ,1995, Social Relations and the Geography of Material Life.

Lewellen, Ted, C., 1995, "*Dependency and Development : An Introduction to The Third World* ", Bergin & Garvey, Westport. The problem in Morris' PQLI is that life expectancy and infant mortality are significantly correlated.

Lewis, W. A., 1954, "Economic Development with Unlimited Supply of Labor", Manchester School of Social and Economic Studies", Vol. 22, No.2.

Liang, Hong, 1997, "*Rationales of ISI – A Strategy*" , University of Denver.

Marglin, Stephen, A. and Juliet, D. Schor, 1990, eds, *Golden Age of Capitalism* : Reinterpreting the Postwar Experience, Oxford University Press, London.

Marsh, William, M. and Grossa, John M. Jr., 2002, "*Environmental Geography : Science, Land Use and Earth Systems*", Wiley, New York.

Massey, Doreen, (1995 2nd edition), *Spatial Divisions of Labor: Social Structures and the Geography of Production,* Routledge, New York.

Mihaly Simai, 1999, "*The Democratic Process and the Market: Challenges of the Transition*" , UN University Press, Tokyo.

Morris, M. D., 1979, "*Measuring the Condition of the World's Poor : The Physical Quality of Life Index*", Pergamon Press, New York.

Myrdal, Gunar, 1957, "*Economic Theory and Underdeveloped Regions*" , Duckworth, London.

Myrdal, Gunar, 1957, Asian Drama : Rich Lands and Poor, Harper and Row, New York.

Narula, Rajneesh, 2002, "*Switching from Import Substitution to the "New Economic Model" in Latin America : A Case of Not Learning from Asia*", Research Memoranda, Maastricht.

Naqvi, Syed Nawab Haider (2002), "*Development Economics – Nature and Significance*", Sage Publications, New Delhi.

OECD, 1983, "*Industry in Transition*", OECD, Paris.

Pollin, Robert, 1997, "*Contemporary Economic Stagnation*" , Cited from Surin, 1998, "*Dependency Theory Reanimation in the Era of Financial Capital* ", Cultural Logic, Vol 1, No.2.

Potter, R. B. and Llyod-Evans, S. 1999, "*Geographies of Development*", Prentice-Hall, Harlow.

Prebisch, Raul, 1950, "*The Economic Development of North America and Its Principal Problem*" , United Nations, New York.

Public Sector Enterprises Survey, 1998-99, Dept of Public Enterprises, Ministry of Heavy Industries and Public Enterprises, Govt. of India, New Delhi.

Rees, Judith, 1989, "*Natural Resources, Economy and Society*" in, Gregory, Derek and Rex Walford, eds. *Horizons in Human Geography*, Macmilan, London.

Rapley, John, 1996, "*Understanding Development : Theory and Practice in the Third World*", Lynne Rienne Publishers, Boulder, Colorado.

Rimmer, P. J. and Forbes, D. K. 1982, "*Underdevelopment Theory : A Geographical Review*", Australian Geographer, 15, 4, pp197-211.

Rose, Gerry, 1979, "*Deciphering Sociological Research*", Macmillan.

Rostow, W. W., 1960, "*The Stages of Economic growth : A Non-Communist manifesto*"", Cambridge University Press, MIT.

Rudolph, Lloyd, I. and Susanne Hoeber Rudolph, 1967, *The Materials of Tradition : Political Development in India,* Chicago University Press, Chicago.

Samir, Amin, 2001, "*Imperialism and Globalization*", Monthly Review Press, New York.

Samir Amin, 1992, "*Delinking towards A Polycentric World*", London.

Sauvy, Alfred, ( ), *Tiers Etat* (or *Monde*) or the "Third World" in an article in *Tiers Etat* (or *Monde*) or the "Third World".

Schor, Juliet, B. 1990, eds. "*Golden Age of Capitalism : Reinterpreting the Postwar Experience*", Oxford University Press, London.

Sen, Amartya, (1999,2002, 2nd Edition), "*Development as Freedom : Human Capability and Deprivation* ", Clarendon Press, Oxford.

Sen, Amartya, (1984), "Development Which Way Now?" *In,* Sen, Amartya, (1984), *Resources, Values and Development*, Chapter 19, pp485-508, HUP, Cambridge, Ma.

Slater, David, 1992, Internal Atomization, Monthly Review Press, New York.

Spraos, J., 1983, "*Inequalizing Trade? A Study of Traditional North / South Specializing in the Context of Terms of Trade Concepts*" , Oxford University Press, Cambridge, Ma.

Stiglitz, Joseph, 2002, "*Globalization and Its Discontents*", Penguin Books, London.

Streeten, P., *1980,* "Development Ideas in Historical Perspective" in, *A. D. Hirschman* et al. *Toward A New Strategy for Development, Pergamon, Oxford, pp* 21-52.

Taylor, Peter, J. (1989), "The Error of Developmentalism in Human Geography" in, Gregory, Derek and Rex Walford, (eds. 1989), Horizons in Human Geography, Macmilan.New York.

Taylor, Lance, 1991, "*Economic Openness : Problems to the Century's End*" in, Tariq Banuri, ed. Economic Liberalization : No Panacea (The Experiences of Latin America and Asia), Oxford University Press, London.

United Nations Conference on Trade and Development, UNCTAD 2001.

UNCTAD Secretariate, 2002, "*Least Developed Countries At A Glance*", Bangkok.

UNDP, (1998), Human Development Report, 1998

Wallerstein, Immanuel, 2000, The Essential Wallerstein, The New Press, New York.

Subsequent two volumes are : *The Modern World-System II,* subtitled "Mercantilism and the Consolidation of the European World-Economy, 1600-1750," was published in 1980, and Volume III, bringing the story into the 1840s, appeared in 1989.

Walter, L. Goldfrank, 2000, "paradigm Regained? The Rules of Wallerstein's World-System Method", Journal of World System Research, Vol. VI, No.2, New York.

Weber, Max 1922, 1967, "*The Protestant Ethic and the Spirit of Capitalism*", originally in German, Unwin, London.

Wolf, Eric, R., 1982, "*Europe and the People without History*" , University of California Press, 1982.

World Bank, "*World Development Report*, 1996", Oxford University Press, New York.

World Bank, 1992, "*World Development Report*" , Oxford University Press.

World Bank, "*World Development Report*, 1991", Oxford University Press, New York.

World Trade Organizations, WTO, Trade Statistics, 1997).

INDEX